VIKING

THE
VIKING
ATLAS
OF
EVOLUTION

ROGER OSBORNE
AND
MICHAEL BENTON

CONSULTANT EDITOR
STEPHEN JAY GOULD

VIKING

VIKING

Published by the Penguin Group
Penguin Books Ltd, 27 Wrights Lane, London W8 5TZ, England
Penguin Books USA Inc., 375 Hudson Street, New York, NY 10014, USA
Penguin Books Australia Ltd, Ringwood, Victoria, Australia
Penguin Books Canada Ltd, 10 Alcorn Avenue, Toronto, Ontario, Canada M4V 3B2
Penguin Books (NZ) Ltd, 182–190 Wairau Road, Auckland 10, New Zealand
Penguin Books Ltd, Registered Offices: Harmondsworth, Middlesex, England

First published 1996
1 3 5 7 9 10 8 6 4 2

Filmset in Nottingham, England

Printed in Great Britain by Bath Press Colourbooks, Glasgow, Scotland

A CIP catalogue record for this book is available from the British Library

ISBN 0–670–85827–7

Preface

Enough books have been published on evolution to fill a substantial library. Many of them, including Charles Darwin's *On the Origin of Species*, have been written for both a general and a specialist audience. Similarly, there are plenty of books, many beautifully illustrated, about the stupendous diversity of life on Earth and its spectacular variation from place to place.

One of our aims in *The Atlas Of Evolution* is to build a bridge between these two vast bodies of literature, for general readers and specialists alike. We want to show how the action of evolutionary mechanisms at particular times, and in particular places, has led to the patterns of life that cover the surface of our planet. To do this, we have used maps of distributions, of organisms, of migrations, of territories, of biogeographic regions, of dispersal patterns and of voyages and journeys. This preponderance of maps emphasizes our concentration on the geography of evolution. Although this is a rich area of research, it is rarely presented to the general reader in any detail.

Evolution is most often explained with reference to time. The history of life on Earth began around 3,500 million years ago with microscopic protozoa. Since then, millions of different life forms have emerged through the operation of evolutionary processes. Evolution starts with microbes and ends up with mammals. The emergence of human consciousness comes right at the end of the timescale. Along the way there are insects, fish, amphibians, dinosaurs, flowering plants and birds. All have their allotted place on the evolutionary time spectrum. But we know that life is not really like that. When we look at the world around us, we see plants and animals of different shapes, sizes and apparent complexities, all coexisting. And when we look closer, we see that each life form is connected to others in intricate ways. Furthermore, when we investigate the development of these organisms we find that they have evolved in parallel with the other life forms around them. Their evolutionary development is a kind of intimate dance in which one element takes a step and a host of others move to keep pace.

These dynamic webs of life are spread all over the Earth and have arisen as evolution has proceeded both in time and space. History and geography have worked hand-in-hand. Geologists have been untangling the history of the development of life since well before Darwin's time. Biologists and geographers have been plotting the present-day patterns of vegetation and animal distributions. More recently, the science of biogeography has emerged in an attempt to knit together knowledge from these and other disciplines. In particular, the acceptance of continental drift (mapped out in *The Historical Atlas of the Earth*, the sister volume to this atlas) and its effects on the present pattern of life on Earth has given biogeography an enormous boost, bringing scientists from many schools scurrying to its doors. This atlas therefore represents our interpretation of the current state of knowledge in what is a rapidly developing subject.

The Atlas of Evolution does not begin with microbes and end with humans. Nor is it a comprehensive review of the evolution of every 'important' life form on the planet. Instead, we have tried to select case studies that show the evolutionary process at work in particular geographical settings, and then to explore some of the resulting patterns of life. We are aware of the dangers of this approach. Natural history is not a laboratory science, and using real life forms in real settings to demonstrate simple principles is to invite the exception that can disprove any rule. We therefore ask that the examples we provide be seen as a way of illuminating some of the profound complexities of the natural world, rather than shoe-horning nature into a predetermined framework. Our aim is simply to show some aspects of the fascinating interplay between geographical space and evolutionary time.

Our researches for this book have led us into many of the by-ways of biology, geology, biogeography, anthropology and medicine. In all of these fields we have relied on the primary research of hundreds of scientists, most of whom appear in the Bibliography at the end of this book. In addition to thanking them, we would like to extend our thanks to the staff of the library of the Linnean Society in London for their invaluable help.

Roger Osborne, Scarborough
Michael Benton, Bristol
1996

Contents

Introduction

There have been many theories of evolution, but it was Charles Darwin who discovered the correct solution – evolution by means of natural selection – when his studies of geological and geographic distributions demonstrated how new species could originate.

Charles Darwin's explanation of evolution by natural selection is rightly marked as a key event in the history of science and perhaps *the* most important discovery in terms of how humans view themselves and the world. Darwin was not the first to propose a theory of evolution, but he was the first to piece all the evidence together, and the power of his ideas has been demonstrated time and again in the past 150 years by observations that he could never have predicted, using techniques that he could not have dreamed about.

Evolution literally means 'unfolding', and for centuries the word was used in a different sense from that understood today. Until 1830 or 1840, it was used to refer to a popular theory of embryology, the changes in embryonic form during the formation of an individual animal. Anatomists had long been fascinated by the way in which an animal developed from the egg, and reproduction was seen as an almost mystical process. There was a debate between preformationists and developmentalists about what was contained within the egg. Preformationists thought that there was a little animal, a miniature version of the adult, which simply became larger by taking nourishment from its mother's blood or from yolk in an egg. Developmentalists rightly said that this idea was not supported by the evidence. When they dissected eggs, they found simply dividing cells inside.

The debate was over by 1800, but German anatomists then began to compare different stages of embryonic development. The eggs of all animals look very similar. In the 1820s, Karl Ernst von Baer found that it was also impossible to tell apart the very early embryos of fish, frogs, birds, cows and humans. After a few days, some distinguishing characteristics – small fins in the fish and limb buds in the others – began to appear. After a little longer he could make out hooflets in the embryo calf, tiny fingers in the human and feather traces in the bird. General characteristics seemed to appear early in development, and special characteristics appear later.

Theories of evolution

In the late 18th and early 19th century, another kind of evolution had been noted – the apparent unfolding of life through time. Living organisms seemed to be linked in a great chain of being or *scala naturae,* from single-celled amoebae to (of course) humans. Some French naturalists, Jean Baptiste de Lamarck in particular, suggested that the *scala naturae* represented a real pattern of linear change through time. Other naturalists, however, roundly opposed such evolutionary ideas.

Swedish naturalist Carolus Linnaeus drew up detailed schemes for the classification of plants and animals that form the basis of our modern international schemes, and showed how nature is arranged in a hierarchical way. Species fit in genera, genera in families, and so on. To our eyes, and to those of Darwin, this hierarchical pattern tells us that nature has evolved by the continuous branching and splitting of species, but Linnaeus and others saw the pattern as a manifestation of God's will. They saw classification as perhaps the best way to read God's mind – hence the success of natural history as a proper pursuit for country parsons in many countries at the time.

By 1830, geologists had also gathered fossil evidence to suggest that simple organisms seem to be found in the oldest rocks and that there is an apparent progression to ever more complex fossils as the rocks get more recent. These ideas might be seen to oppose the biblical story of creation, since they provided increasing evidence for the vast age of the Earth, and hinted at former worlds populated by strange and exotic life.

This was the mix of ideas that the young Darwin encountered when he began to think seriously about the patterns of nature. He knew about the German embryologists, Linnaeus and the unorthodox evolutionary views of the French naturalists, yet he set sail on the *Beagle* as an Anglican and creationist. His subsequent insight, however, was to break through old certainties and change our view of the world for ever. Darwin proved that species are not immutable; that life has diversified by a process of speciation (the splitting of species); that life today is just part of an enormously long history of origins, extinctions and diversification; that evolution will continue; that organisms are reasonably well adapted to their environments but that adaptation seldom reaches perfection because environments vary continuously; and that the whole process is guided by natural selection, the weeding out of organisms in order that those that are best adapted to the transient environment survive to breed.

The story of Darwin's insights, and of the critical role played by geography, is told in the first section of this book. The geographical aspects of evolution are then demonstrated in sections on natural selection in action; the origin of species; the larger-scale picture of the history of life; and the way in which the distributions of modern organisms reflect their long-term history and, in some cases, the influence of the more recent ice ages. Aspects of modern biogeography are then explored: the variations of diversity with geography, the distribution of life zones on Earth, the special role of islands, and the beneficial and catastrophic effects of human interventions. Recent advances in biology and paleontology have provided fascinating insights into the geography of evolution, confirming again and again the strength of Darwin's 1859 book *On the Origin of Species by Means of Natural Selection or the Preservation of Favoured Races in the Struggle for Life*.

Part One
Change and Movement,
The Elements of Evolution

Diversity and Adaptation

Natural Selection and Variation

The Origin of Species

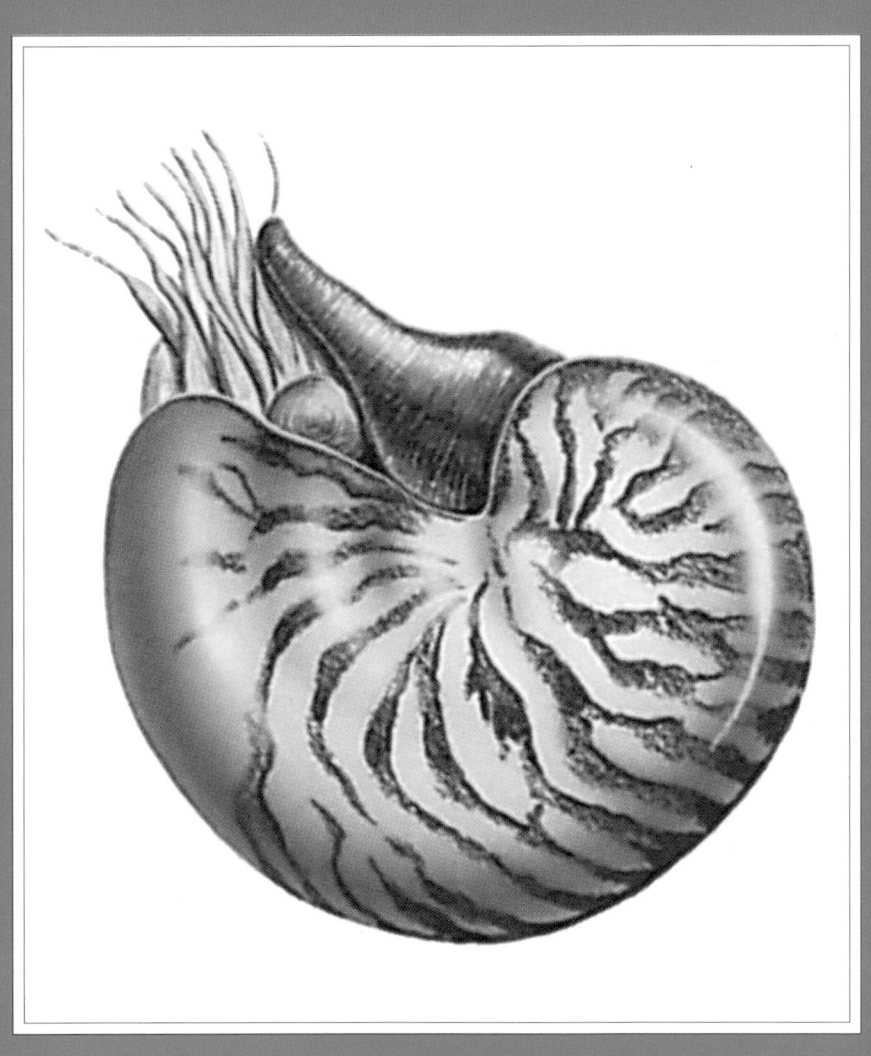

Section I: Diversity and Adaptation

Two key concepts in the theory of evolution are diversity – the number and range of living species – and adaptation – the ways in which organisms are suited to their roles. Darwin showed how diversity and adaptation result from natural selection.

Naturalists in the 1830s were well aware of the vast diversity of life, even if no one had been able to count the millions of plant and animal species on Earth. Thousands of species had been formally named, but there were clearly many more species yet to be named, and the total was estimated at a million or more living species. Today over 1 million species have been named and described, and the number keeps rising. Ten years ago, naturalists estimated that there might be 5 million species on Earth. New studies suggest that the total may be ten times that. This is based on the rate of discovery of new tropical beetles, new deep-sea organisms and new species of creatures that live among sand grains on the seabed. If there are 30–50 million species on Earth today, diversity is clearly a key feature of life, and explaining diversity is as important as early naturalists thought.

Adaptation is both a state and a process, according to modern evolutionary interpretations. In the 1830s, naturalists viewed adaptation as a state in which there was a perfect matching of organisms to their role. William Paley, the great promoter of natural theology, argued in his *Natural Theology* that God was the ultimate watchmaker. Every plant and animal was so ingeniously designed to do the biological task assigned to it that there must be a supreme being who had created life. Paley quoted numerous examples of this: the exact relationship between flowering plants and their insect pollinators, the detailed arrangements within ant and bee colonies, the high quality of bird skeletons and their marvellous ability to fly. These adaptations exceeded the skills of the best early Victorian engineers. Plants and animals had clearly been designed by a creator of infinite wisdom, whose achievements were far in excess of anything that human beings could dream of.

As was normal at the time, Darwin studied and was examined on his knowledge of the Bible and Paley's *Natural Theology*. He accepted Paley's views, and when he set sail on the *Beagle*, he shared the typical views of his class and country. He was deeply influenced by the new work on geology by Charles Lyell, and was aware that the Earth was ancient, but he was in no sense an evolutionist at that time.

The voyage of the Beagle

The 5-year voyage of the *Beagle*, and Darwin's reading and thinking immediately on his return, were to change everything. During the voyage, Darwin saw and collected numerous mammal fossils from South America. These confirmed that animals could become extinct, since the bones belonged to species that no longer existed. More important-ly, the fossil forms only resembled mammals that still lived in South America. If life had been created, why should there be evidence of a relationship between extinct and living animals in one part of the world? When he visited the Galapagos Islands, Darwin was again shaken by two geographical observations. First, plants and animals on the islands were like those on the South American mainland. Why should that be, if life had been created in one act? Second, the tortoises and finches on each of the islands looked very similar to one another, although there were clear differences from island to island. Darwin thought that the finches belonged to several unrelated groups, but when he showed some skins to an ornithologist on his return, he was assured that they were all finches, but of different species. Darwin had to admit the impossible – that species are not immutable.

The carefully erected edifice of natural theology crumbled. If species were not permanent, then they could evolve and split into separate species – God had not created all life forms in one act. If species could evolve to the extent that he had seen on the Galapagos Islands, then Darwin had to accept that the reason the South American fossil mammals resembled modern forms was that they were related by descent. Ultimately, all life may have evolved from a single common ancestor. The vastness of geological time was already accepted, forming a time-scale in which this process could have taken place.

The final brick was fitted into Darwin's shocking new edifice when he read Thomas Malthus's *Essay on the Principle of Population*. This showed that human populations breed more quickly than the rate of increase of the food supply, and that certain processes must come into play in order to maintain correct population at a level to match available resources. Darwin borrowed this idea to formulate his theory of evolution by natural selection. Animals and plants produce too many young to survive, and in general only the strongest survive. The features that enable them to survive (bigger teeth, stronger legs, brighter feathers) must be inherited in some way and passed on in turn to their offspring. In time, the make-up of the whole population may change, or evolve, in the direction of features that most promote survival at the time.

Diversity and adaptation are still of great interest. It is especially important to understand diversity, or biodiversity as it is often called today, because of human threats to life on Earth. The true number of species alive today is still poorly known. Biologists still have only the vaguest ideas about the range of genomes and morphologies. Paleontologists can only point to the most general aspects of the history of life. No one can yet say how well any plant or animal is adapted, nor how fast it can change its adaptations. Darwin pointed the way, and we are only now beginning to understand the implications of the new way in which he taught us to view the world.

Galapagos Finches
Variation and the origin of species

Darwin and evolution always go together. The story of Darwin's long voyage around the world aboard the sailing ship, the *Beagle*, is also well known, and it is often said that his visit to the Galapagos Islands set him on the track to a full modern explanation of how evolution works.

Darwin was a young man, only 22 years of age, when the *Beagle* set sail. He had studied medicine for two years at Edinburgh University, and had transferred to Cambridge and received his degree when the chance of a voyage around the world was presented. During the five years of the journey, Darwin collected fossils and specimens of plants and animals, and made extensive notes. These formed the basis of several books when he returned to England.

Darwin visited the Galapagos Islands in 1835. He mistakenly thought that the different kinds of finches living on the various islands belonged to several different families. On bringing the skins home, however, he submitted them for study to John Gould, the eminent ornithologist, and learnt, to his amazement, that they were all finches, and that these Galapagos finches were most like birds on the South American mainland.

The Galapagos group is home to four genera of finches, divided into 13 species. These have specialized in a broad range of diets (seeds, cactus, insects and fruit) in comparison with the normal diet of seeds. Darwin realised that the finches had reached the islands long ago (the islands had a geologically recent origin as volcanoes), perhaps in a small breeding population of one species, which then spread over the whole archipelago.

Above: **Charles Darwin** (1809–1882) is rightly regarded as one of the leading scientists of all time. While still in only his late twenties, he resolved the problem of how evolution had occurred.

The finches were crucial to Darwin's formulation of the principle of evolution by the splitting of species (speciation). When Darwin set sail on the HMS *Beagle*, he accepted most of the standard beliefs of the time and thought that species were fixed and unchangeable. Small variations could occur, but only within species; they did not lead to the origin of new species. The Galapagos finches, so similar and so close to each other geographically, were separate species. This demonstrated that species were not immutable, and that the splitting of species was a result of geographical isolation.

Below: The *Beagle* was used by the British Admiralty to survey the coastlines of the southern continents.

Finches that feed primarily on insects, plus some plants

Warbler finch, small insects

Darwin (Culpepper) `1` `4` `13`

Wolf (Wenman) `1` `4` `13`

Tree finch, large insects

Tree finch, medium insects

Tree finch, small insects

Tool-using finch

Mangrove finch

Finches that feed primarily on plants, plus some insects

Large ground finch

Medium ground finch

Small ground finch

Sharp-beaked ground finch

Cactus ground finch

Large cactus ground finch

Vegetarian tree finch

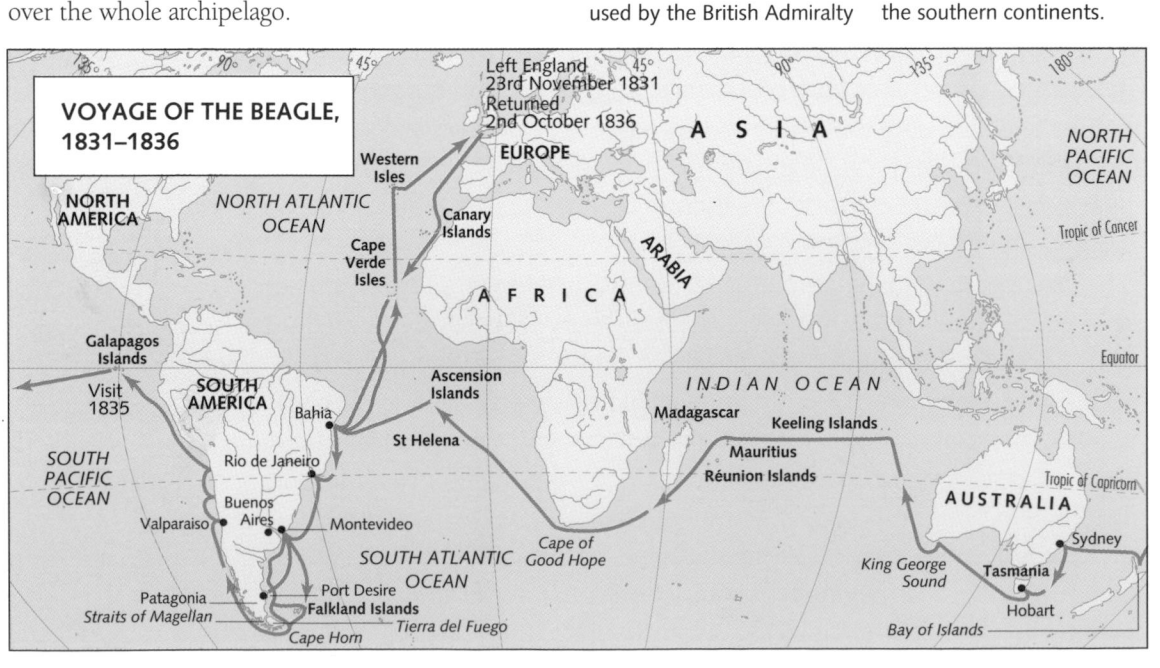

VOYAGE OF THE BEAGLE, 1831–1836

Left England 23rd November 1831
Returned 2nd October 1836

EUROPE

ASIA

NORTH AMERICA

NORTH ATLANTIC OCEAN

Western Isles

Canary Islands

Cape Verde Isles

AFRICA

ARABIA

NORTH PACIFIC OCEAN

Tropic of Cancer

Galapagos Islands

Visit 1835

SOUTH AMERICA

Bahia

Ascension Islands

St Helena

Madagascar

INDIAN OCEAN

Keeling Islands

Mauritius

Réunion Islands

Equator

SOUTH PACIFIC OCEAN

Rio de Janeiro

Buenos Aires

Valparaiso

Montevideo

Cape of Good Hope

King George Sound

AUSTRALIA

Tropic of Capricorn

Sydney

Patagonia

Straits of Magellan

Port Desire

Falkland Islands

Cape Horn

Tierra del Fuego

SOUTH ATLANTIC OCEAN

Tasmania

Hobart

Bay of Islands

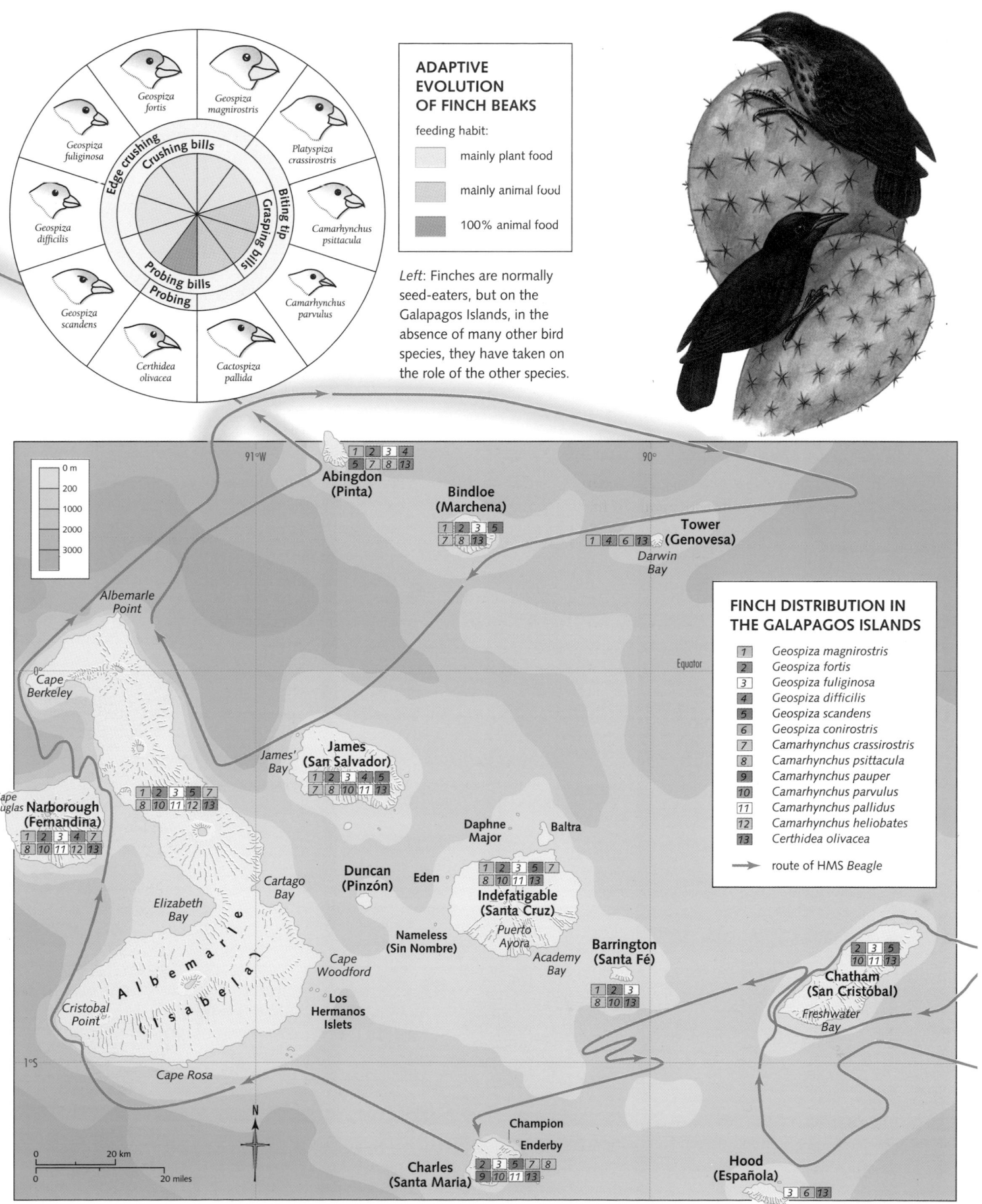

ADAPTIVE EVOLUTION OF FINCH BEAKS

feeding habit:

mainly plant food

mainly animal food

100% animal food

Left: Finches are normally seed-eaters, but on the Galapagos Islands, in the absence of many other bird species, they have taken on the role of the other species.

Finch beak wheel labels:
- Geospiza fortis
- Geospiza magnirostris
- Platyspiza crassirostris
- Camarhynchus psittacula
- Camarhynchus parvulus
- Cactospiza pallida
- Certhidea olivacea
- Geospiza scandens
- Geospiza difficilis
- Geospiza fuliginosa

Crushing bills — Edge crushing
Biting tip — Grasping bills
Probing bills — Probing

FINCH DISTRIBUTION IN THE GALAPAGOS ISLANDS

1 *Geospiza magnirostris*
2 *Geospiza fortis*
3 *Geospiza fuliginosa*
4 *Geospiza difficilis*
5 *Geospiza scandens*
6 *Geospiza conirostris*
7 *Camarhynchus crassirostris*
8 *Camarhynchus psittacula*
9 *Camarhynchus pauper*
10 *Camarhynchus parvulus*
11 *Camarhynchus pallidus*
12 *Camarhynchus heliobates*
13 *Certhidea olivacea*

→ route of HMS *Beagle*

Map labels:

Abingdon (Pinta) — 1 2 3 4 / 5 7 8 13

Bindloe (Marchena) — 1 2 3 5 / 7 8 13

Tower (Genovesa) — 1 4 6 13 / Darwin Bay

Albemarle Point

Cape Berkeley

James' Bay / James (San Salvador) — 1 2 3 4 5 / 7 8 10 11 13

Cape Douglas / Narborough (Fernandina) — 1 2 3 4 7 / 8 10 11 12 13

Cape Douglas — 1 2 3 5 7 / 8 10 11 12 13

Daphne Major / Baltra

Duncan (Pinzón)

Eden

Indefatigable (Santa Cruz) — 1 2 3 5 7 / 8 10 11 13

Cartago Bay

Elizabeth Bay

Nameless (Sin Nombre)

Puerto Ayora / Academy Bay

Barrington (Santa Fé) — 1 2 3 / 8 10 13

Cape Woodford

Los Hermanos Islets

Chatham (San Cristóbal) — 2 3 5 / 10 11 13

Freshwater Bay

Cristobal Point

Cape Rosa

Champion / Enderby

Charles (Santa Maria) — 2 3 5 7 8 / 9 10 11 13

Hood (Española) — 3 6 13

Equator

0° Cape Berkeley

1°S

91°W   90°

Elevation scale:
0 m
200
1000
2000
3000

0   20 km
0   20 miles

N

Linnaeus
The father of classification

The first step in understanding evolution is to make some sense of the huge variety of plants and animals around us. The earliest naturalists, such as Aristotle, understood that nature does not consist of an uncontrolled mass of varieties. They could see that there are dogs, cats, oak trees, daffodils and other specific forms and that in each of these obvious groups, or genera, there are particular forms, or species. Thus, foxes, wolves and domestic dogs are separate species of a genus that includes all dogs. The key to understanding nature was therefore to have a system of classification.

The first efforts at classification were made over 2,000 years ago. The earliest naturalists made the important observation that species do not follow a linear pattern, like a series of numbers in order. The classification was clearly an inclusive hierarchy – small units (species) fitted within larger units (genera) and these in turn fitted into ever larger units, (families, orders, classes and phyla.)

The father of systematics (the science of classification) and all related aspects of natural history was Swedish naturalist Carolus Linnaeus, whose seminal books *Species Plantarum*, *Genera Plantarum* and *Systema Naturae* (1753, 1754 and 1758) are accepted internationally as the starting points for the naming of plants and animals respectively. He established the 'binomial system' in which every species is given a generic and a specific name, such as *Quercus ruber* (pedunculate oak) or *Homo sapiens* (humans). The generic name comes first, and is given a capital letter; the specific name is second, with a lower-case first letter. One genus may contain many species.

Linnaeus also clarified the inclusive hierarchical nature of classifications in his books. He developed these ideas in parallel with schemes to allow naturalists to identify specimens. These, known as dichotomous or branching keys, are still used today, especially in botany. The botanist answers a series of questions about the unknown plant, each of which may have one of two answers. Linnaeus did not develop evolutionary ideas to explain this branching pattern, but to modern eyes, and to those of Darwin, this represents splitting in evolutionary time.

Linnaeus undertook numerous expeditions in Scandinavia and noted how floras were controlled by environmental conditions. These were fundamental surveys, carried out for the first time, in which Linnaeus was able to show the diversity of modern life in his native lands, and to show how that diversity was controlled by ecology, geography and climate.

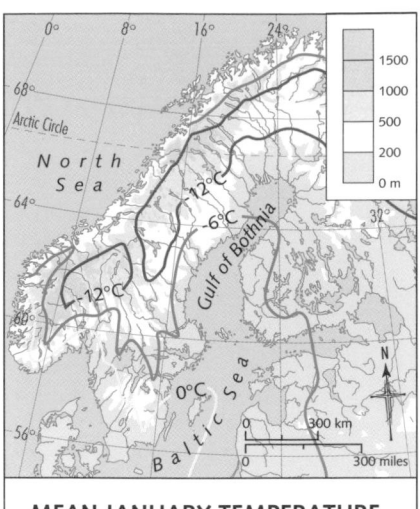

MEAN JANUARY TEMPERATURE

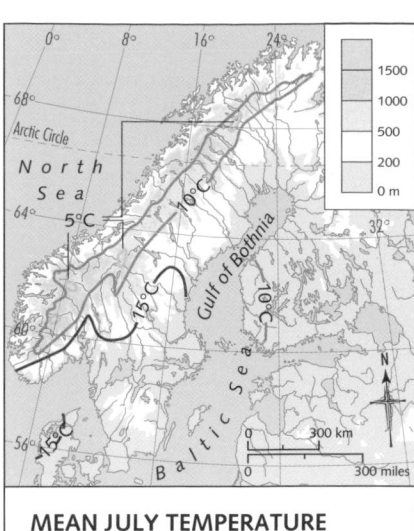

MEAN JULY TEMPERATURE

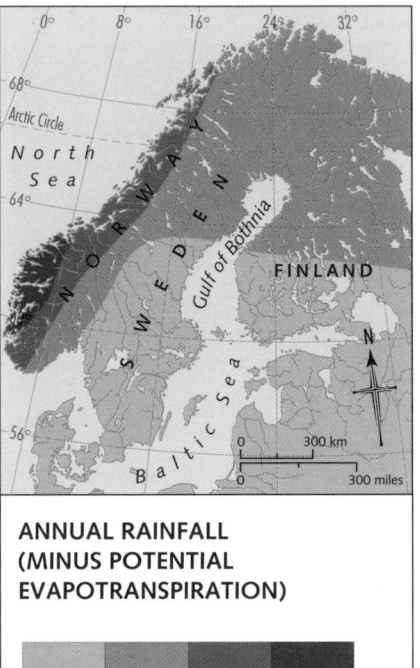

ANNUAL RAINFALL (MINUS POTENTIAL EVAPOTRANSPIRATION)

0   250   500   1000 mm

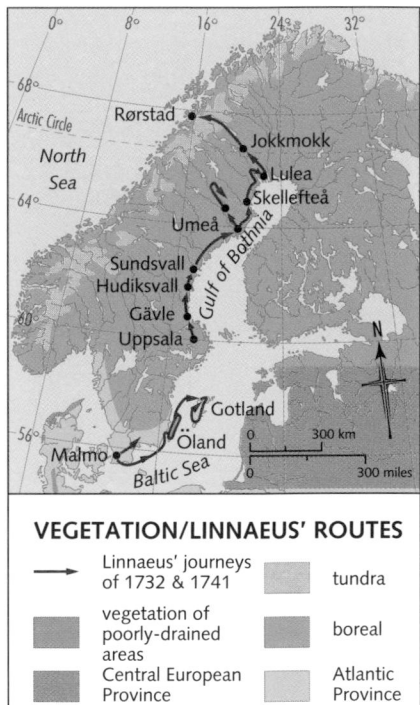

VEGETATION/LINNAEUS' ROUTES

→ Linnaeus' journeys of 1732 & 1741

  vegetation of poorly-drained areas

  Central European Province

  tundra

  boreal

  Atlantic Province

Above: **Linnaeus** (1707–1778) in Lapland's national costume, obtained during one of his many expeditions to northern Scandinavia.

The flora of Scandinavia
Carolus Linnaeus (or Carl von Linné) undertook several plant-collecting expeditions to northern parts of Scandinavia and to the major islands in the Baltic Sea. He noted great variations in the plants, which was a result of both variation in climate (temperature, rainfall and winter freezing) and variation in topography (which ranges from lowland coastal zones to high mountains). Typical are these plants of a sphagnum moss bog (*right*): pink bog rosemary (*top*), dwarf birch behind hare's tail (both *top right*) and cloudberry (white flower, *centre*) in a mass of sphagnum moss.

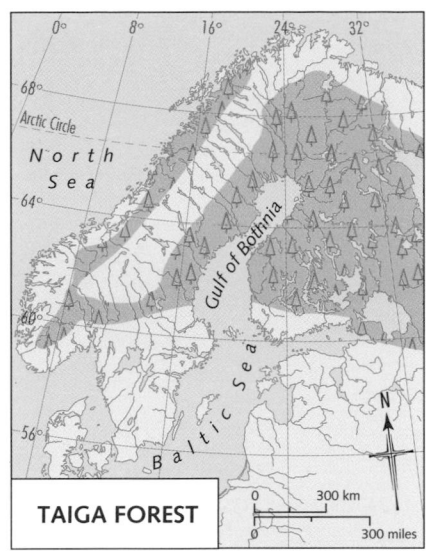

TAIGA FOREST

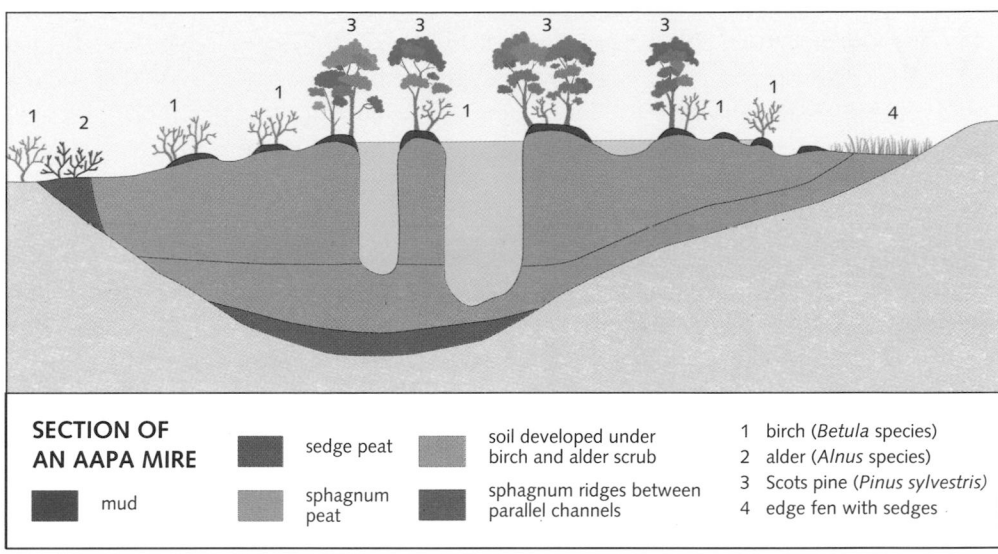

SECTION OF AN AAPA MIRE

- mud
- sedge peat
- sphagnum peat
- soil developed under birch and alder scrub
- sphagnum ridges between parallel channels

1 birch (*Betula* species)
2 alder (*Alnus* species)
3 Scots pine (*Pinus sylvestris*)
4 edge fen with sedges

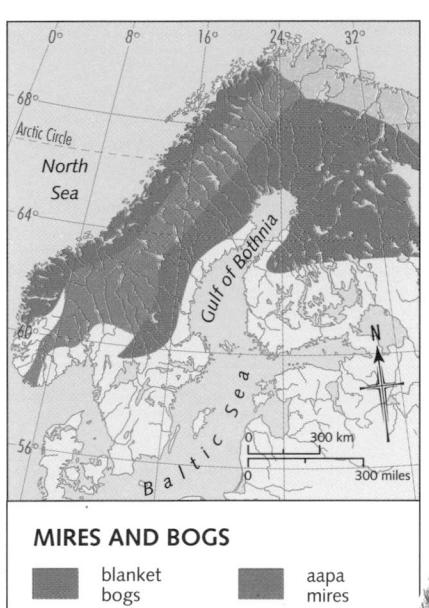

MIRES AND BOGS

- blanket bogs
- mountain mires
- aapa mires
- palsa mires

Typical sphagnum moss bog plants

Above: Much of Linnaeus' plant collecting concentrated on aapa, or string, mires. These are distinguished by a series of ridges separated by deep grooves, produced partly by ice movements. The commonest ground cover is sphagnum moss. Small birch and alder trees grow on more mature parts.

Linnaeus focused on the reproductive structures of plants as a guide to the major divisions (*right*). Describing pollination as a sexual act, he devised a 'sexual system' for the classification of plants. This system was extremely useful but led some detractors to accuse him of 'botanical pornography' and outraged the clergy of the time.

LINNAEUS' CLASSIFICATION OF FLOWERING PLANTS

marriage of plants (flowers)

- public marriage (flowers visible)
  - husband and wife in one bed (flowers hermaphrodite)
    - husbands not related (stamens free)
      - all males equal (stamens level)
        - 1 male
        - 2 males
        - 3 males
        - up to 20 males
        - many males
      - males of different ranks (stamens of different height)
        - 1 group
        - 2 groups
        - many groups
    - husbands related (stamens joined)
  - husband and wife in separate beds (male and female flowers)
    - 1 house (monoecious plant)
    - 2 houses (dioecious plant)
    - many houses
- clandestine marriage (flowers small or absent)

Cuvier and Extinction

The life of past ages

While Linnaeus was attempting to put the modern world of plants and animals into order, naturalists were debating the life of the past. Fossil shells and fish had been found in many places, and there had been lengthy discussions in medieval times about their origin. Were these petrifactions related to modern shells and fish, or were they simply odd pebbles that happened to look like the remains of plants and animals? By 1700, most naturalists accepted that fossils were indeed the remains of ancient organisms that had somehow been buried and turned to stone. One big question remained: were these the remnants of extinct plants and animals, of forms that had been created by God and died out because of some inadequacy?

Until 1750, most naturalists accepted that fossils were the remains of species that were either still living or would soon be found in some hitherto unexplored part of the world. Isolated fossil bones of vertebrates had been found in Ireland, Britain, France and elsewhere. In the 1750s, explorers in North America began to dig up the remains of elephants (mastodons and mammoths) and sent some of the bones to Paris and London, where distinguished anatomists and naturalists such as William Hunter,

Georges Louis Leclerc, Comte de Buffon and others debated the specimens. Were these the remains of extinct animals, or were such elephants yet to be found in the unexplored Far West of North America? As more specimens were found, and as more of the Americas were explored, it became clear that they were in fact the remains of forms that had recently become extinct.

Baron Georges Cuvier of the Muséum d'Histoire Naturelle in Paris was instrumental in clarifying this question in a series of books and papers from 1796 onwards. He made the point convincingly by the use of the science of comparative anatomy, a discipline in which he established new standards of precision. He demonstrated that the fossil animals were different from modern elephants, rhinos and other large mammals by a painstaking comparison of every bone, in the course of which he noted similarities and differences between equivalent elements in the skeletons of a variety of extinct and living forms. By the 1820s, Cuvier had honed his skills in comparative anatomy to such a pitch of perfection that it was fancifully said that he could identify any animal from a single bone and could reconstruct any unknown fossil form from a single bone.

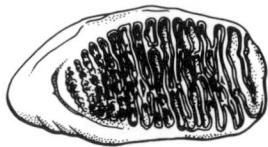

molar of *Mammuthus primigenius*

Mastodon

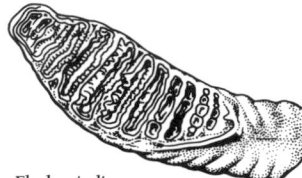

Elephas indicus
(modern Indian elephant)

Elephant teeth (*above*)
These three side views of elephant teeth show similarities between the extinct mastodon and mammoth and the modern African elephant.

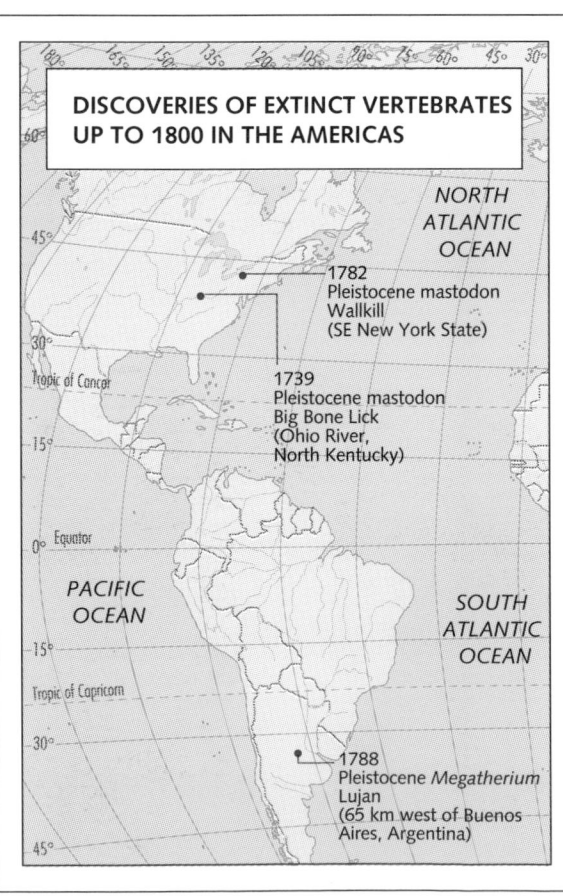

DISCOVERIES OF EXTINCT VERTEBRATES UP TO 1800 IN THE AMERICAS

NORTH ATLANTIC OCEAN

1782
Pleistocene mastodon
Wallkill
(SE New York State)

1739
Pleistocene mastodon
Big Bone Lick
(Ohio River,
North Kentucky)

PACIFIC OCEAN

SOUTH ATLANTIC OCEAN

1788
Pleistocene *Megatherium*
Lujan
(65 km west of Buenos Aires, Argentina)

Left: Early finds of giant bones in North and South America settled the 18th-century debate about extinction. Mammoth and mastodon bones showed that elephants had once roamed North America. The complete skeleton of a giant ground sloth from Argentina caused a sensation when it was exhibited, as it was much larger than any living mammal in South America, but was related to modern tree sloths.

Right: Extinct mammal remains were collected in China from the 12th century but were not widely known to scientists. Thought to be dragon's bones, they were used in medicines. From the 16th century, mammoth bones and tusks were found in Siberia in increasing numbers and were traded throughout Asia and Europe.

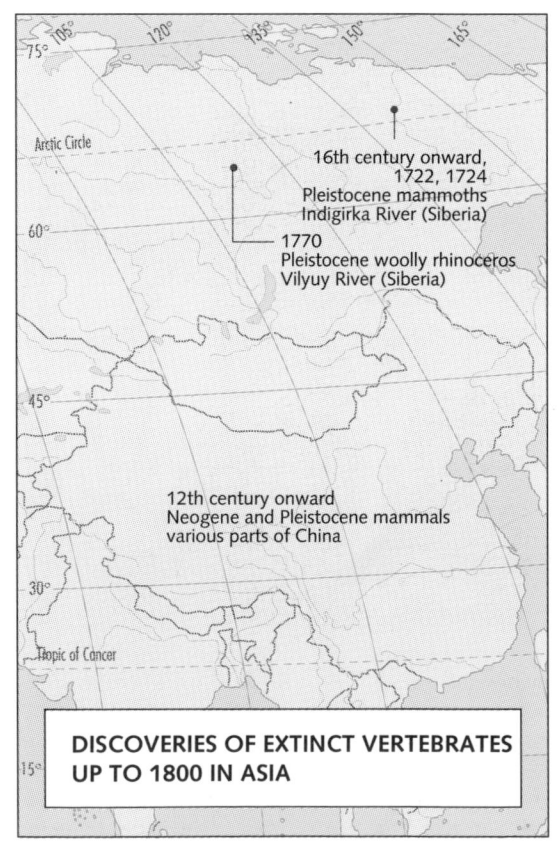

16th century onward,
1722, 1724
Pleistocene mammoths
Indigirka River (Siberia)

1770
Pleistocene woolly rhinoceros
Vilyuy River (Siberia)

12th century onward
Neogene and Pleistocene mammals
various parts of China

DISCOVERIES OF EXTINCT VERTEBRATES UP TO 1800 IN ASIA

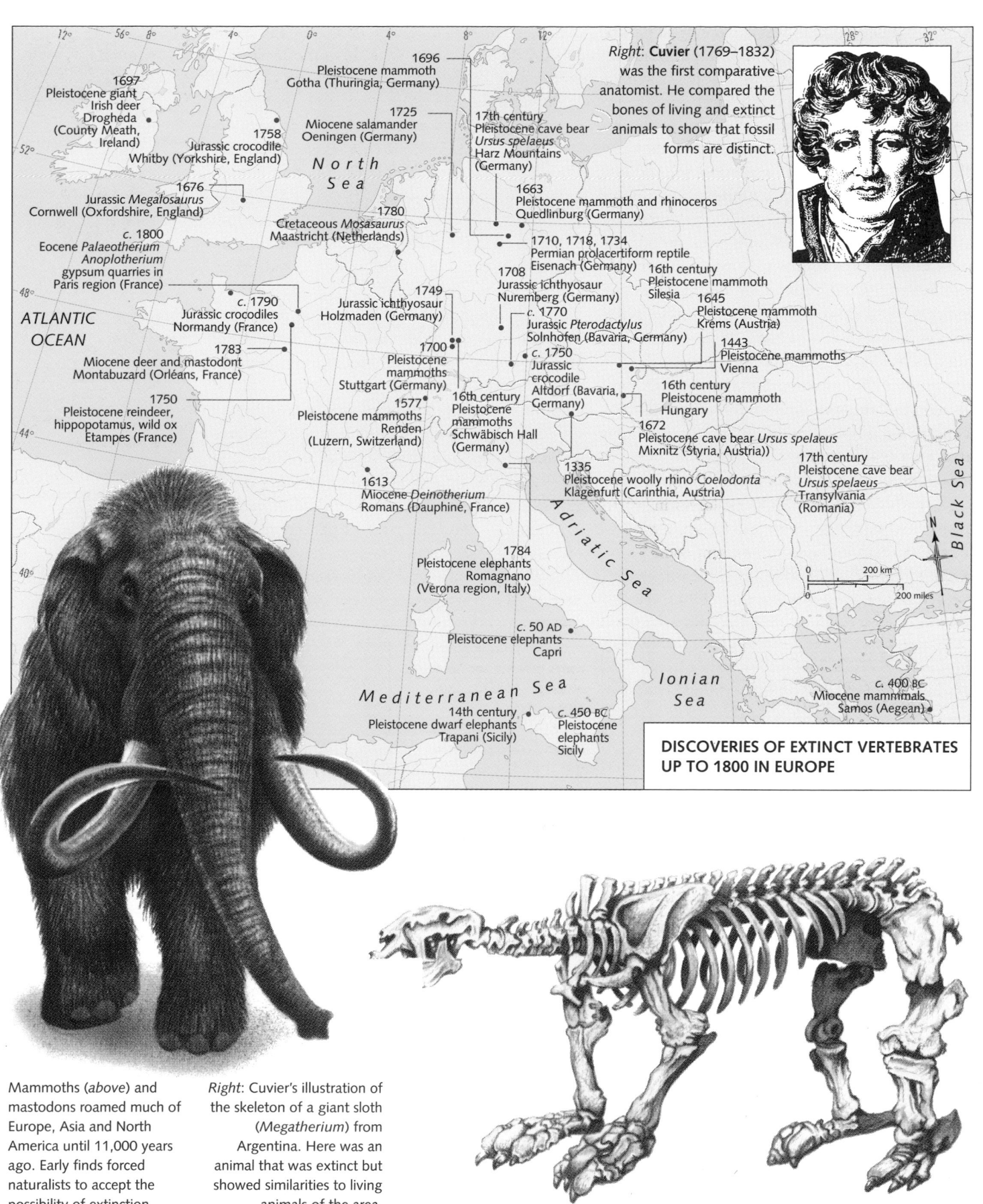

1697
Pleistocene giant
Irish deer
Drogheda
(County Meath,
Ireland)

1758
Jurassic crocodile
Whitby (Yorkshire, England)

1696
Pleistocene mammoth
Gotha (Thuringia, Germany)

1725
Miocene salamander
Oeningen (Germany)

Right: **Cuvier** (1769–1832) was the first comparative anatomist. He compared the bones of living and extinct animals to show that fossil forms are distinct.

17th century
Pleistocene cave bear
Ursus spelaeus
Harz Mountains
(Germany)

1663
Pleistocene mammoth and rhinoceros
Quedlinburg (Germany)

1676
Jurassic *Megalosaurus*
Cornwell (Oxfordshire, England)

1780
Cretaceous *Mosasaurus*
Maastricht (Netherlands)

c. 1800
Eocene *Palaeotherium*
Anoplotherium
gypsum quarries in
Paris region (France)

ATLANTIC
OCEAN

c. 1790
Jurassic crocodiles
Normandy (France)

1783
Miocene deer and mastodont
Montabuzard (Orléans, France)

1750
Pleistocene reindeer,
hippopotamus, wild ox
Etampes (France)

1749
Jurassic ichthyosaur
Holzmaden (Germany)

1708
Jurassic ichthyosaur
Nuremberg (Germany)

1710, 1718, 1734
Permian prolacertiform reptile
Eisenach (Germany)

16th century
Pleistocene mammoth
Silesia

1645
Pleistocene mammoth
Krems (Austria)

c. 1770
Jurassic *Pterodactylus*
Solnhofen (Bavaria, Germany)

c. 1750
Jurassic
crocodile
Altdorf (Bavaria,
Germany)

1443
Pleistocene mammoths
Vienna

16th century
Pleistocene mammoth
Hungary

1672
Pleistocene cave bear *Ursus spelaeus*
Mixnitz (Styria, Austria))

1700
Pleistocene
mammoths
Stuttgart (Germany)

1577
Pleistocene mammoths
Renden
(Luzern, Switzerland)

16th century
Pleistocene
mammoths
Schwäbisch Hall
(Germany)

1335
Pleistocene woolly rhino *Coelodonta*
Klagenfurt (Carinthia, Austria)

17th century
Pleistocene cave bear
Ursus spelaeus
Transylvania
(Romania)

Black Sea

1613
Miocene *Deinotherium*
Romans (Dauphiné, France)

Adriatic Sea

N

0   200 km
0   200 miles

1784
Pleistocene elephants
Romagnano
(Verona region, Italy)

c. 50 AD
Pleistocene elephants
Capri

Ionian
Sea

c. 400 BC
Miocene mammals
Samos (Aegean)

Mediterranean Sea

14th century
Pleistocene dwarf elephants
Trapani (Sicily)

c. 450 BC
Pleistocene
elephants
Sicily

**DISCOVERIES OF EXTINCT VERTEBRATES
UP TO 1800 IN EUROPE**

Mammoths (*above*) and mastodons roamed much of Europe, Asia and North America until 11,000 years ago. Early finds forced naturalists to accept the possibility of extinction.

Right: Cuvier's illustration of the skeleton of a giant sloth (*Megatherium*) from Argentina. Here was an animal that was extinct but showed similarities to living animals of the area.

Dinosaurs and Deep Time
The immensity of geological time

The mammoth, mastodon and giant ground sloth fossils found in the late 18th century were clearly not very ancient, and geologists were happy to accept that some of these animals had perhaps died out only a few thousand years ago. Other fossils seemed much more ancient, however. The bone or shell material seemed to be filled with crystalline minerals that must have taken time to enter the pores and solidify, and in many cases the fossils seemed to come from plants and animals with no obvious living relatives. Geologists debated the true antiquity of the Earth during this period, but the outlines of what had happened in the past became clear only from 1800 onwards.

Geologist James Hutton argued in his *Theory of the Earth* (1788) that the Earth was enormously ancient. He based his arguments on observations of modern-day processes which were then applied to ancient examples according to his principle of uniformitarianism, or 'the present is the key to the past'. Hutton argued that if sediments accumulate at rates of only a few centimetres per year, then the vast piles of ancient sedimentary rocks, hundreds or thousands of metres thick, must represent millions of years.

These vast spans of time could be subdivided according to recognizable events. At the turn of the 19th century, canal engineer William Smith, mapping out prospective routes in various regions of England, observed that certain rock types and fossil assemblages always occur in a predictable order. In his map of the geology of England and Wales (1815), Smith established a practical system of stratigraphy, the foundation of the geological time-scale. He named various divisions of the Mesozoic rocks of England in particular, and showed that they could be identified by their characteristic fossils. His scheme has been extended worldwide, and fine divisions of geological time are routinely identified, and correlated from place to place, by means of their fossils.

The first dinosaur discoveries (1824–50) bore witness to an astonishing fauna of giant reptiles in the Mesozoic. This was the final part of the jigsaw that laid the foundations of a modern understanding of the history of life. The dinosaurs showed early naturalists that plants and animals quite unlike anything now living had existed in the past. Since 1825, collecting has provided evidence of many other extinct organisms extending back well into the Precambrian. These ancient organisms were seen by Darwin to be linked by a pattern of evolutionary splitting, forming a single great evolutionary tree of life.

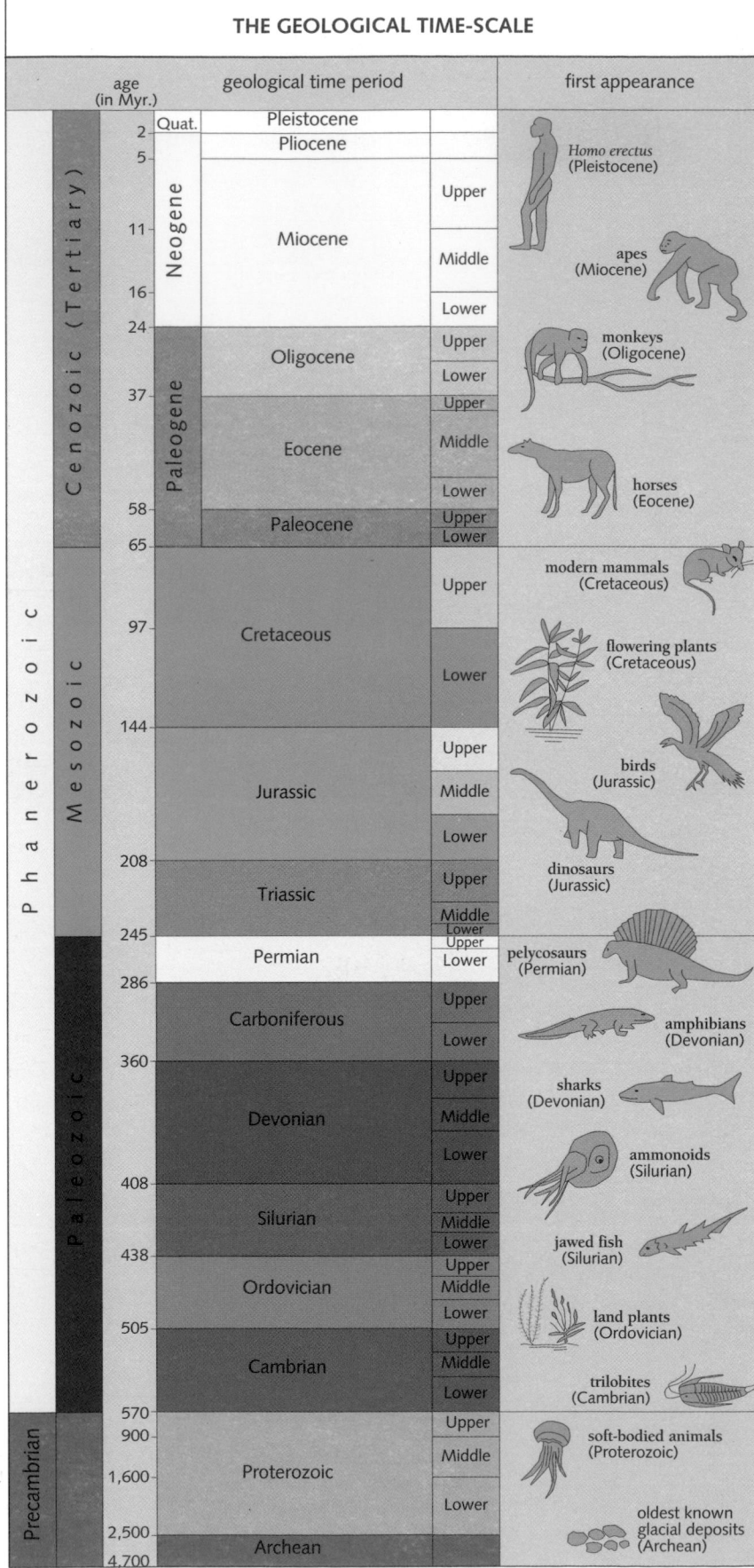

THE GEOLOGICAL TIME-SCALE

age (in Myr.)		geological time period		first appearance
		Quat.	Pleistocene	
2			Pliocene	
5			Upper	*Homo erectus* (Pleistocene)
11	Neogene	Miocene	Middle	apes (Miocene)
16			Lower	
24		Oligocene	Upper	monkeys (Oligocene)
			Lower	
37	Paleogene		Upper	
		Eocene	Middle	horses (Eocene)
58			Lower	
65		Paleocene	Upper / Lower	
			Upper	modern mammals (Cretaceous)
97		Cretaceous		flowering plants (Cretaceous)
			Lower	
144			Upper	birds (Jurassic)
		Jurassic	Middle	
			Lower	dinosaurs (Jurassic)
208			Upper	
		Triassic	Middle	
245			Lower / Upper	pelycosaurs (Permian)
286		Permian	Lower	
		Carboniferous	Upper	amphibians (Devonian)
360			Lower	
		Devonian	Upper	sharks (Devonian)
			Middle	
408			Lower	ammonoids (Silurian)
		Silurian	Upper	
			Middle	
438			Lower	jawed fish (Silurian)
		Ordovician	Upper	
505			Middle	
			Lower	land plants (Ordovician)
		Cambrian	Upper	
			Middle	
570			Lower	trilobites (Cambrian)
900			Upper	soft-bodied animals (Proterozoic)
1,600		Proterozoic	Middle	
2,500			Lower	oldest known glacial deposits (Archean)
4,700		Archean		

(Left-hand spanning labels: Cenozoic (Tertiary), Mesozoic, Paleozoic under Phanerozoic; Precambrian)

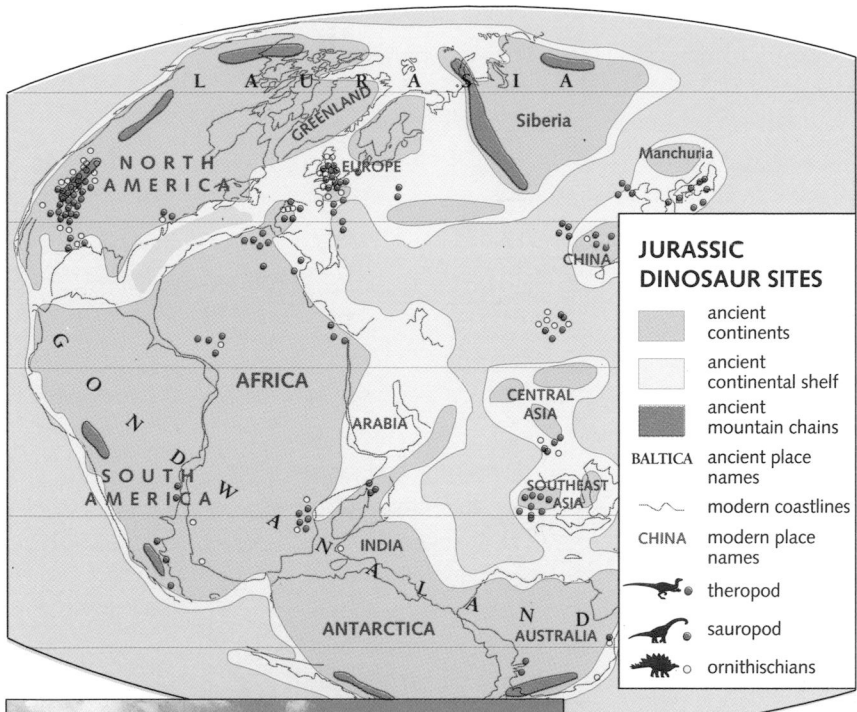

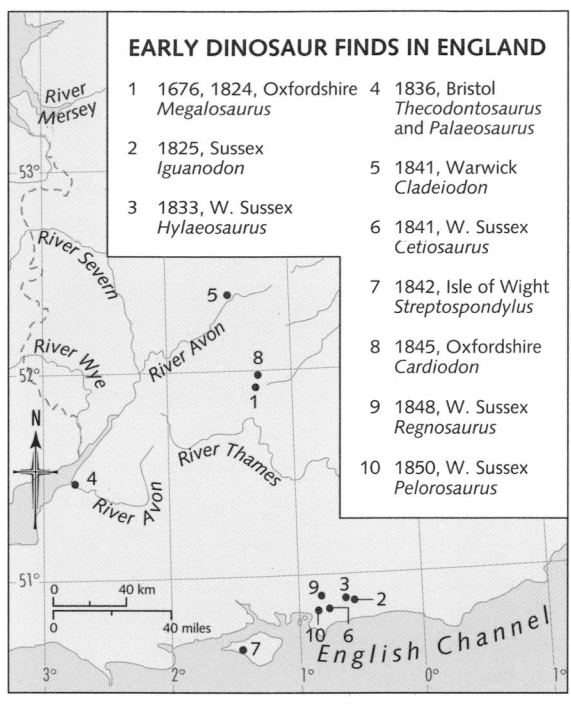

EARLY DINOSAUR FINDS IN ENGLAND

1 1676, 1824, Oxfordshire
   Megalosaurus

2 1825, Sussex
   Iguanodon

3 1833, W. Sussex
   Hylaeosaurus

4 1836, Bristol
   Thecodontosaurus
   and *Palaeosaurus*

5 1841, Warwick
   Cladeiodon

6 1841, W. Sussex
   Cetiosaurus

7 1842, Isle of Wight
   Streptospondylus

8 1845, Oxfordshire
   Cardiodon

9 1848, W. Sussex
   Regnosaurus

10 1850, W. Sussex
   Pelorosaurus

Above: Dinosaurs were abundant worldwide in the Jurassic period. The first dinosaur to be named, *Megalosaurus*, in 1824, was a typical Jurassic meat-eater. Similar dinosaurs lived worldwide because they could move between the land masses.

Early discoveries (*above*) The first dinosaur specimens were found in England between 1824 and 1850. These included examples from Triassic, Jurassic and Cretaceous rocks. In the same period, dinosaur finds in Germany and France were also reported.

Below: Cretaceous dinosaur sites extend from the North Pole to the South Pole, and it seems likely that the polar regions were not glaciated then. The most abundant finds come from northern continents, but new expeditions are extending knowledge in the south.

Left: The geological time-scale is an international standard scheme. The order of geological time units was established from the fossils that each contained. Most of the names were given between 1820 and 1870, and were based on the geographic areas where the rocks had first been studied. Cambrian, Ordovician and Silurian, for instance, are named after ancient Welsh tribes, and Devonian after Devon, England. The exact ages (given in millions of years or Myr) are based on studies of radioactive decay.

Above: The long-necked plant-eater *Apatosaurus* is one of the best-known dinosaurs, and has also been called *Brontosaurus*. It probably used its long neck to forage widely for food – a lower-energy strategy than moving all its bulk around. Dinosaurs are one of the most spectacular groups of extinct animals. They were first collected in England in the 1820s and were initially thought to be giant lizards. By 1842, however, they had been correctly recognised as a unique group of reptiles, unlike anything living.

CRETACEOUS DINOSAUR SITES

see JURASSIC DINOSAUR SITES above for key

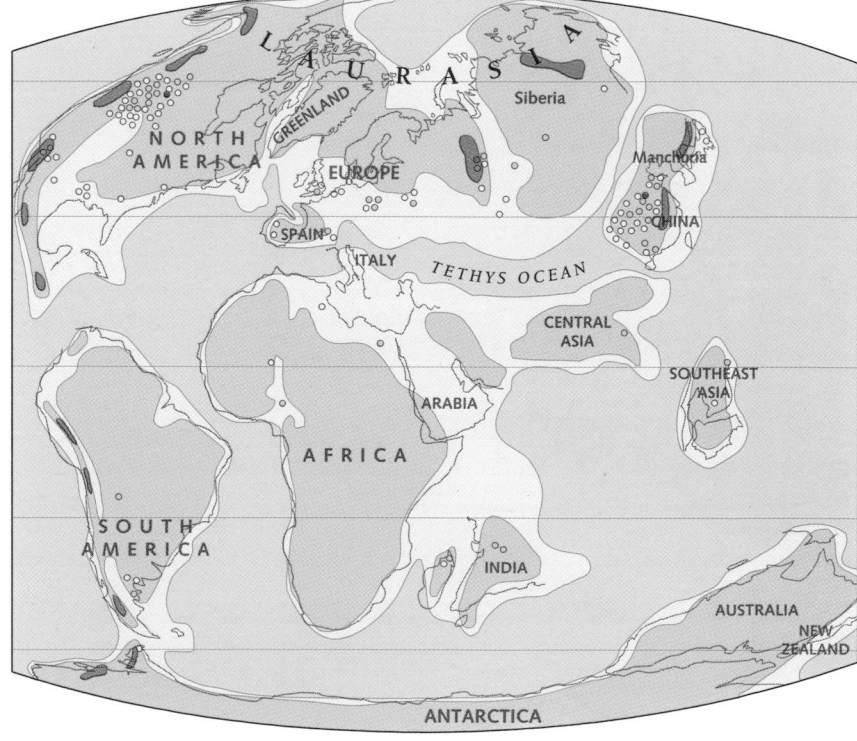

The Tree of Life
Finding the shape of evolution

By 1830, paleontologists had noted that the sequence of fossils in rocks shows a progression from simple to complex. The oldest rocks seem to contain only microscopic creatures, sponges, corals, and the like. In younger rocks, fish become abundant, then amphibians and reptiles. By the Mesozoic, the Earth was dominated by dinosaurs and the seas by ichthyosaurs and plesiosaurs. Some small mammals are known from the Mesozoic era, but modern-style mammals only appear in rocks dating from the Tertiary era.

Paleontologists proposed the idea of a progression of life, a kind of evolution, but no one in the 1830s could explain how it had happened. Perhaps, as French biologist Lamarck had proposed in the 1790s, life had followed a great chain of being, a kind of moving escalator along which animals and plants became more and more complex. Lamarck argued that there was a preordained line of change, and that humans had evolved from chimps. Chimps had evolved from monkeys, monkeys from shrew-like mammals, and so on back to the primeval slime.

Progress would continue: present-day chimps would evolve into humans in the future and present-day humans would eventually become angels.

Darwin could not square such a simple idea of linear change with what he saw in nature, and focused on the branching and splitting of species – diversification – in developing his ideas of evolution (*see pages 18–19*). He was the first to see evolution as a branching tree, and the only illustration in the *Origin* (1859) is a diagram showing evolution as a series of branches.

The branching tree of life, or phylogeny, can be discovered in various ways. Classically, evidence of general resemblance was used to group animals and plants into species, genera, families and higher groups. The key principle was homologies – structures that had originated once and then evolved along different lines. A classic example is mammals' arms. The arm is homologous in all mammals (and, indeed, in amphibians, reptiles and birds), however different it might look. A bat's wing, a whale's paddle and a mole's digging paw share the same basic

Cladistics
Cladistics is a technique for discovering monophyletic groups (groups that have a single common ancestor and include all the descendants of that ancestor), or clades. Clades may be characterized by one or more unique features, such as the feathers of birds or the hair of mammals. In constructing a cladogram of dinosaurs (*below*), it is found that all share a particular pattern of hip girdle, but that a subclade, the Ornithischia, has a derived, or more evolved, structure that is absent in the others. The other dinosaur subclade, the Saurischia, is characterized by a particular kind of hand. Features of the skull, the teeth and the feet allow for further subdivision.

Successful clades generally split into many species (*below*). In many cases, this diversification continues at a high rate and the group becomes very diverse. This is true of birds, flowering plants and insects. In other cases, a dramatic extinction event may wipe out a huge swathe of species, leaving a few survivors, at random. These survivors may in turn diversify further.

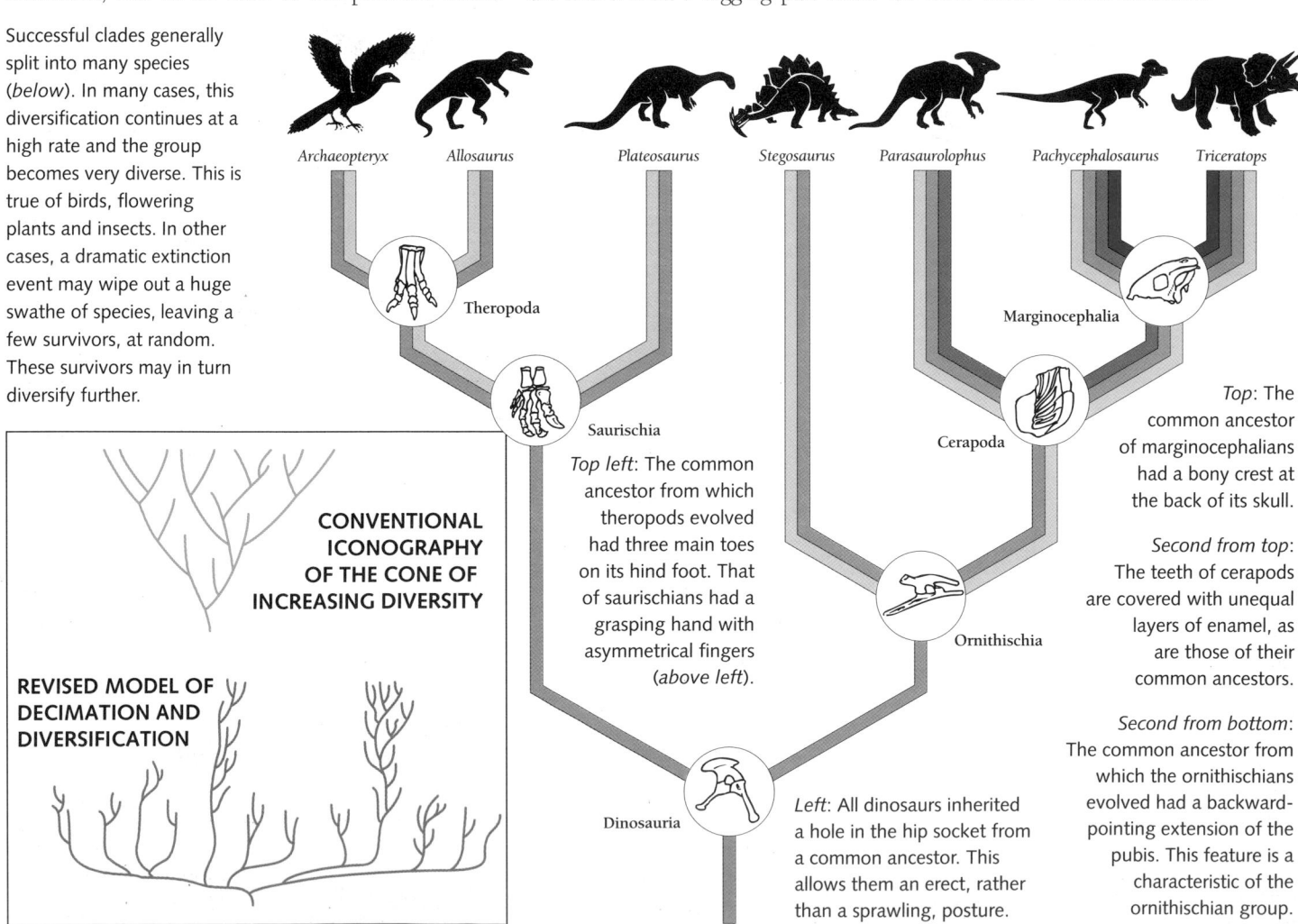

CONVENTIONAL ICONOGRAPHY OF THE CONE OF INCREASING DIVERSITY

REVISED MODEL OF DECIMATION AND DIVERSIFICATION

Archaeopteryx   *Allosaurus*   *Plateosaurus*   *Stegosaurus*   *Parasaurolophus*   *Pachycephalosaurus*   *Triceratops*

Theropoda

Saurischia

Marginocephalia

Cerapoda

Ornithischia

Dinosauria

Top left: The common ancestor from which theropods evolved had three main toes on its hind foot. That of saurischians had a grasping hand with asymmetrical fingers (*above left*).

Left: All dinosaurs inherited a hole in the hip socket from a common ancestor. This allows them an erect, rather than a sprawling, posture.

Top: The common ancestor of marginocephalians had a bony crest at the back of its skull.

Second from top: The teeth of cerapods are covered with unequal layers of enamel, as are those of their common ancestors.

Second from bottom: The common ancestor from which the ornithischians evolved had a backward-pointing extension of the pubis. This feature is a characteristic of the ornithischian group.

pattern of bones, so all mammals can be proved to belong to one group that originated at a single point. Other homologies are hair, the production of maternal milk, large brains and extended parental care.

Two methods are now used as independent approaches to the reconstruction of phylogeny: cladistics (the assessment of morphological characters for the most parsimonious [shortest] tree linking the species); and molecular phylogeny reconstruction (using proteins or RNA/DNA). Since 1970, there has been increasing clarity in some parts of the tree of life, and the two methods can produce results that can be tested against each other. If they agree, then the phylogeny is probably correct.

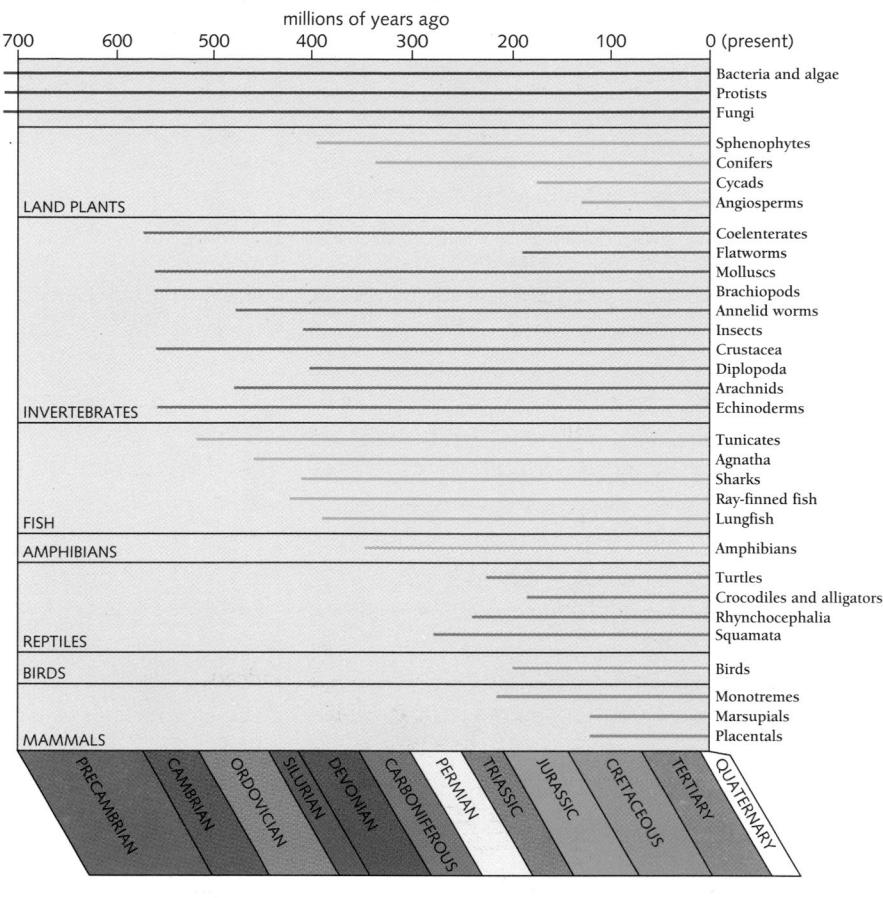

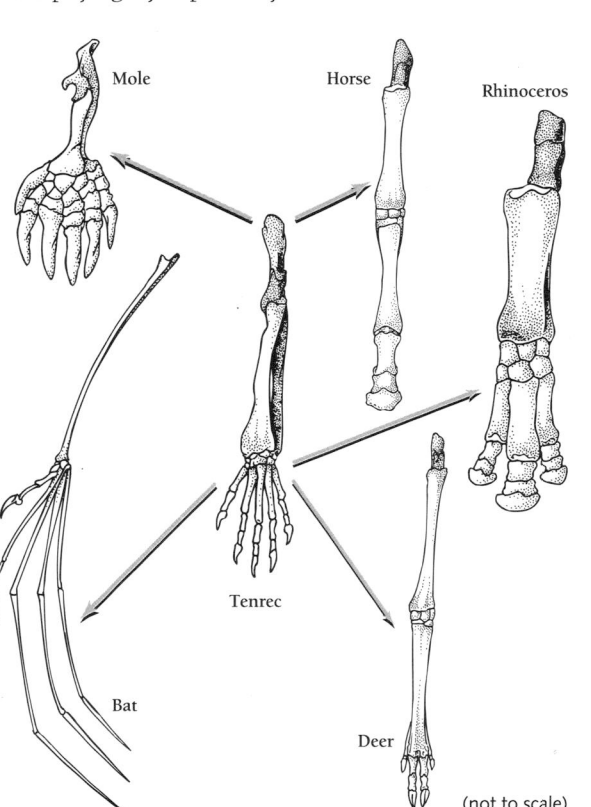

A range of kinds of evidence is put together to make a phylogeny. First, detailed information on relationships, based on homologies, is required. Homologies, like the mammal arms above, may look superficially different, but their fundamental structure is identical and has a single point of origin. Information on the known geological range of each group (*top right*) is added to produce a phylogeny (*right*).

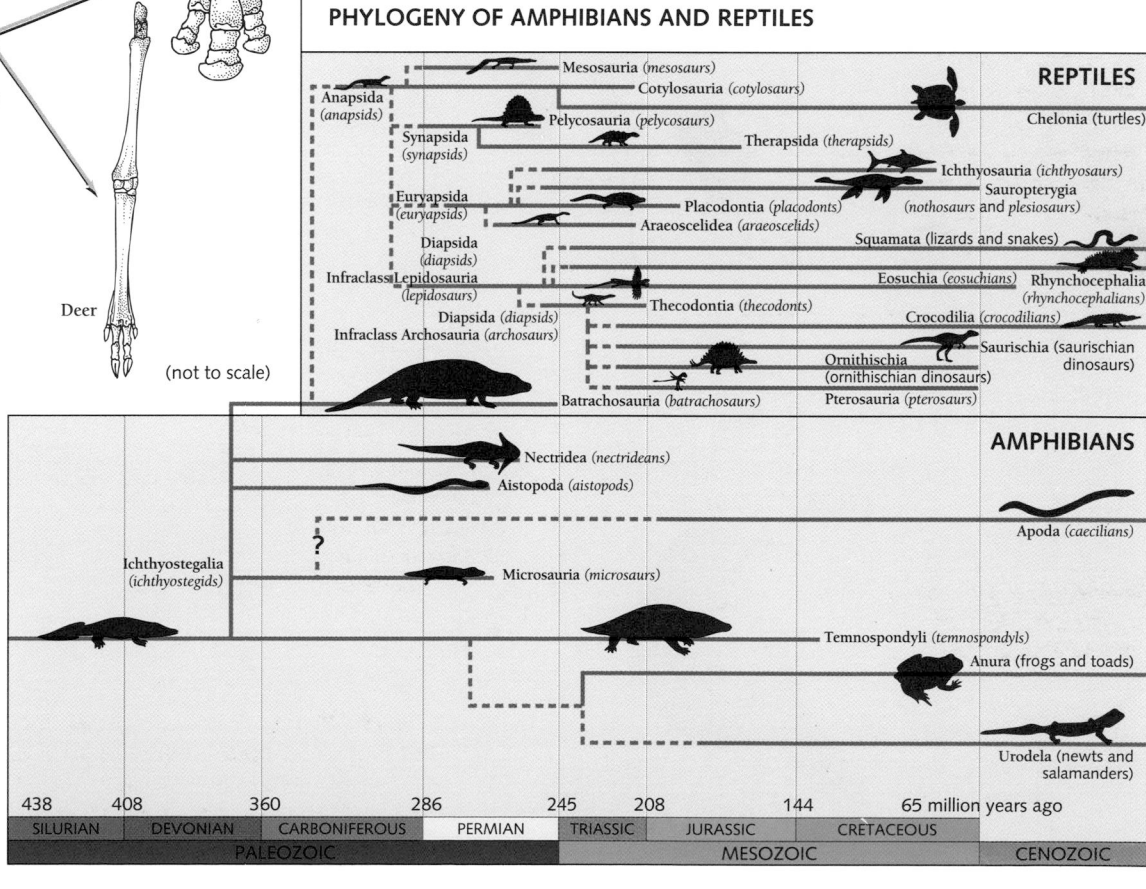

PHYLOGENY OF AMPHIBIANS AND REPTILES

Diversity
The huge but unknown variety of life

The total diversity of life is difficult to estimate. Almost 1.5 million species have been given formal Latin names (*see page 20*), and until recently biologists scaled up from this figure to arrive at an estimate of 3–5 million species alive today. However, several new studies have suggested that the true total might be ten times that: in other words, between 30 and 50 million species on Earth.

The higher estimates are based on the astonishing diversity of unknown species found in tropical rainforests. Scientists shook trees and counted the species of beetles that fell out. Each species of tree seemed to harbour 100 or more unique beetles, and every tree that was shaken produced more new forms. Every day, the biologists found hundreds of species that had not yet been named or described, and the work of updating the guidebooks would clearly take the efforts of hundreds of entomologists working day and night for decades. Scaling up from these observations gave a global species diversity of 30 million.

This estimate has been confirmed by studies of species found on the deep seafloor and in the spaces between sand grains. Marine biologists collect seabed life by trawling for specimens. They too found that the rate at which they turned up new species was so great that the true diversity of life has to be measured in tens of millions. The same conclusion was reached by marine biologists studying meiofauna, the small animals that live between sand grains on the seafloor. Again, this relatively unstudied section of life turned out to be so diverse that the global estimates had to be revised dramatically upwards.

Modern diversity is not uniformly distributed, and it seems that the tropics have more species than temperate and polar regions. The first maps to show the patterns of diversity distribution are now being compiled, and these demonstrate that beetles, seed plants, reptiles and mammals are most diverse in tropical latitudes, particularly the rainforests of Southeast Asia and South America. These maps highlight 'hot spots' of

Generalists

In regions of low diversity, each species has a broader niche, or range of activities. Thrushes (*right*), typical of temperate areas, are the ecological equivalents of ten tropical bird species.

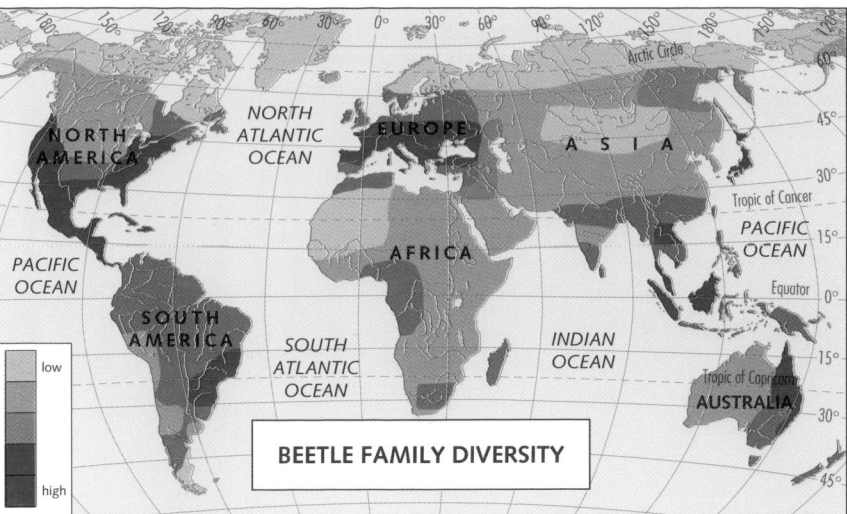

BEETLE FAMILY DIVERSITY

low

high

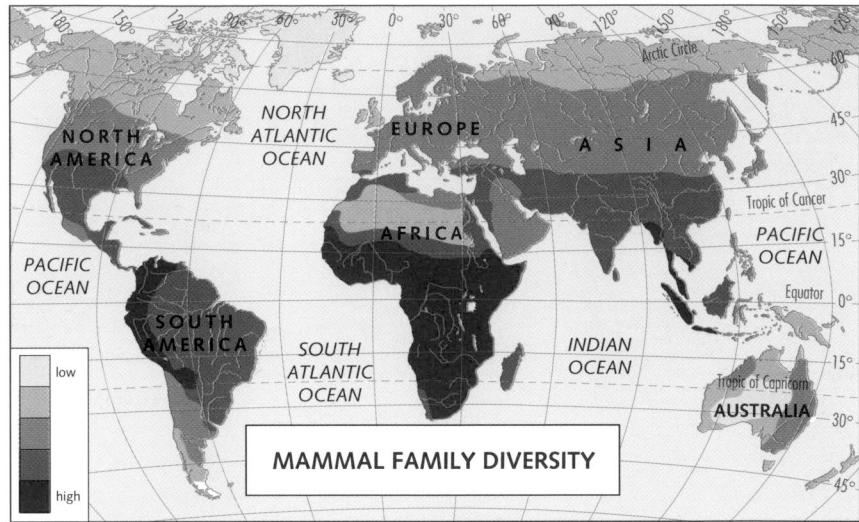

MAMMAL FAMILY DIVERSITY

low

high

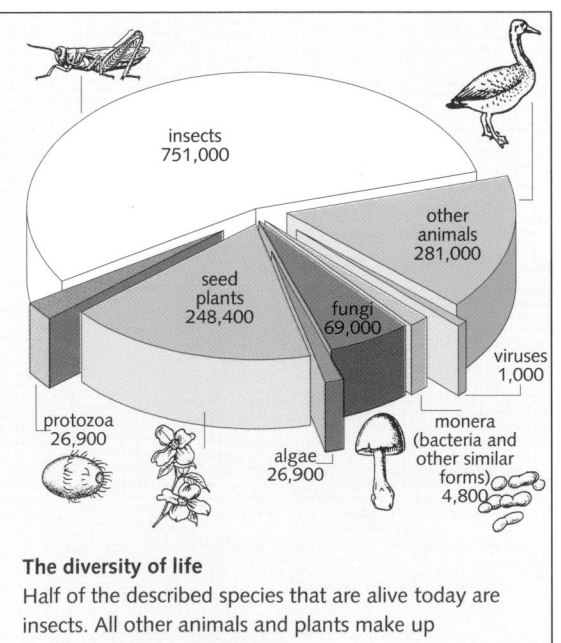

insects
751,000

other animals
281,000

seed plants
248,400

fungi
69,000

viruses
1,000

protozoa
26,900

algae
26,900

monera (bacteria and other similar forms)
4,800

The diversity of life
Half of the described species that are alive today are insects. All other animals and plants make up approximately a quarter each.

Diversity

Different groups of plants and animals differ enormously in their modern diversities (*left*). By far the most diverse group are the insects, represented by over three quarters of a million named species. Next come the higher plants (flowering plants, conifers, mosses, and so on), with a quarter of a million. Other groups, such as bacteria, viruses, protozoans, algae and fungi represent very long-lived groups whose species diversities are probably grossly underestimated. All remaining animals, including the backboned animals, and ourselves, form only a small wedge in the cake of modern diversity.

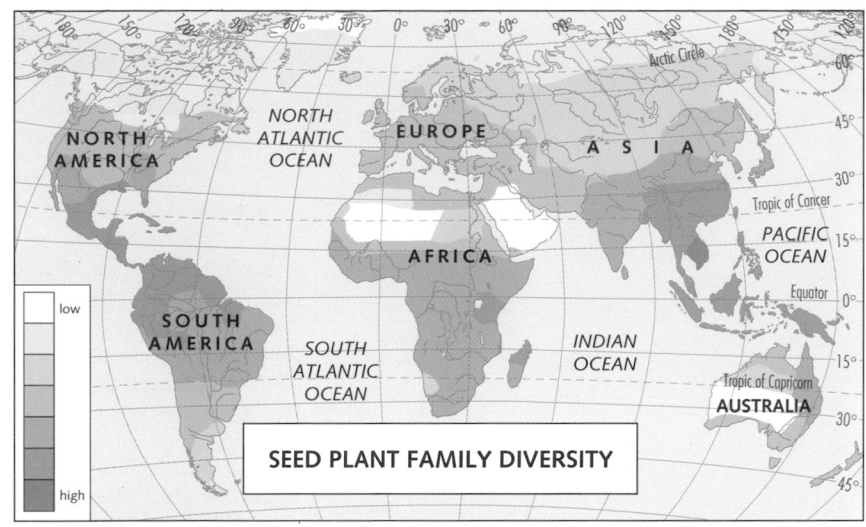

SEED PLANT FAMILY DIVERSITY

low
high

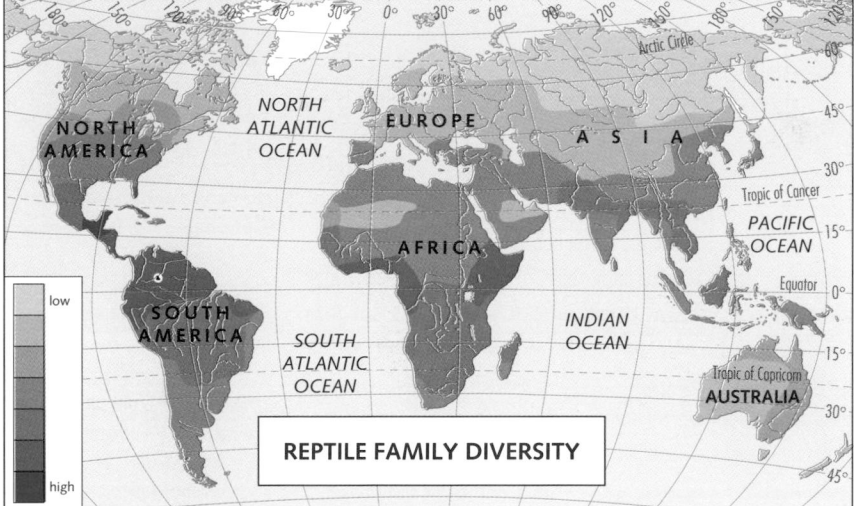

REPTILE FAMILY DIVERSITY

low
high

diversity, where species numbers are highest, and this technique provides critical information for conservationists fighting for the preservation of certain habitats. The hot spots clearly merit top priority.

If the present diversity of life cannot be estimated, it is even more difficult to estimate the diversity of life in the past, or to say how many species have ever lived. All living and extinct species evolved from a single common ancestor, perhaps 3,500 million years ago. Life has diversified in some way from a single species to the present total of 30–50 million. Did it diversify along a straight line (regular step-by-step increases), an exponential curve (regular doubling of diversity) or a concave curve in which a high level was reached early on? Present evidence, based on studies of the history of marine animals, vascular land plants and non-marine tetrapods, suggests that different groups diversified in fits and starts, with rapid rises, hiatuses, and fallbacks that were caused by mass extinctions.

Hot spots

The diversity of life is not evenly spread over the globe (*above*). Biologists have always assumed that species diversities are highest in tropical regions, both in tropical rainforests on land and in coral reefs in the sea, and that levels of diversity tail off in temperate and polar regions. New diversity maps for beetles, seed plants, reptiles and mammals confirm this assumption. It is important to note that these maps, which are of widely different kinds of organisms, show broadly similar patterns. The highest diversities are observed in the tropical rainforests of southeastern Asia and South America.

Plant and animal diversity patterns are comparable. Marine animals increased their diversity rapidly in the first 100 million years of the Paleozoic era, then apparently remained at an equilibrium level of around 250 families until the dramatic end-Permian mass extinction event (*pages 72–73*). Diversification has followed a straight-line pattern of increase for the past 250 million years. The diversification of vascular (land) plants shows similar bursts of diversification, interspersed by a couple of plateaux. Non-marine tetrapods (amphibians, reptiles, birds and mammals) seem to show an exponential pattern of increase.

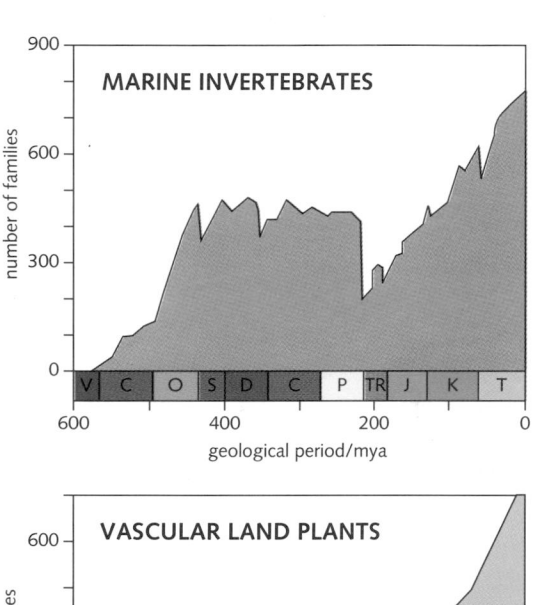

MARINE INVERTEBRATES

number of families

geological period/mya

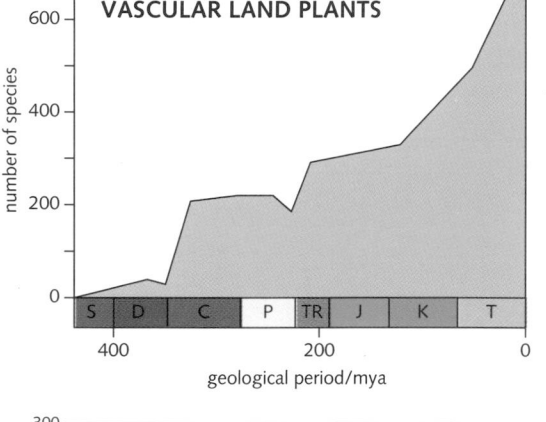

VASCULAR LAND PLANTS

number of species

geological period/mya

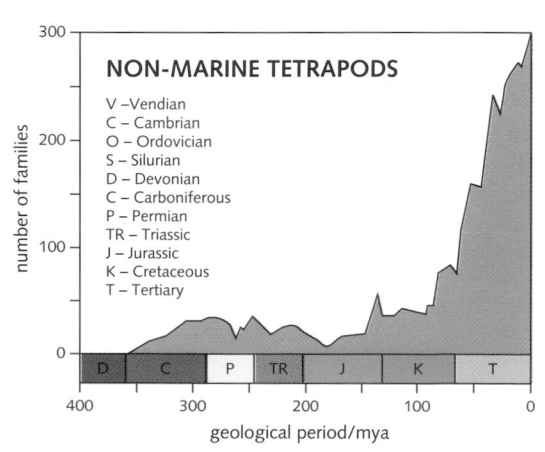

NON-MARINE TETRAPODS

V – Vendian
C – Cambrian
O – Ordovician
S – Silurian
D – Devonian
C – Carboniferous
P – Permian
TR – Triassic
J – Jurassic
K – Cretaceous
T – Tertiary

number of families

geological period/mya

Human Evolution
How new studies have changed our views

Human evolution is of intense interest. Since 1850, hundreds of paleontologists, anthropologists, explorers, anatomists, geneticists and molecular biologists have devoted their lives to unravelling our origins. However, new evidence from work undertaken in the 1980s and 1990s is changing our perceptions of human evolution enormously.

Our current knowledge is based on evidence from fossils, molecular and genetic techniques, geographic distributions and artefacts. This extraordinary combination of information sources is increasingly important in modern studies of plant and animal evolution. Anatomists now have a clearer understanding of how to distinguish variation between individuals of a single species from variation between species, and this has led to radical new ideas about the numbers of human species in the past. Refined dating techniques are also helping paleontologists to calibrate the evolutionary tree correctly.

The study of artefacts and of the evolution of languages is helping to unravel the last stages of human evolution, the spread of *Homo sapiens*. But the most exciting new developments have come from studies of genetics and molecular biology. Each human being has a different genetic constitution from every other (except in the case of identical twins). Close relatives are most similar genetically, and members of a single racial group are generally more similar to one other genetically than they are to members of other races. Further, it has been shown

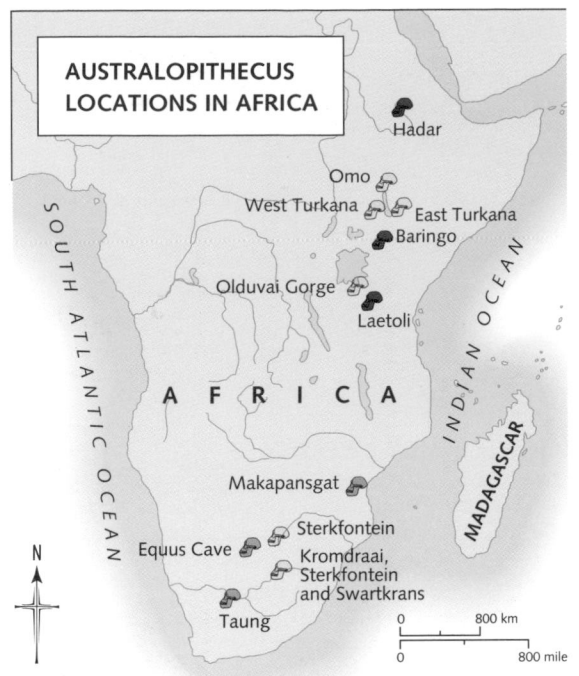

AUSTRALOPITHECUS LOCATIONS IN AFRICA

that members of different species have different genetic constitutions. Indeed, the compositions of proteins, and of the ribonucleic acids in particular, differ in proportion to the amount of geological time that separates any pair of species from their latest common ancestor. The proteins of humans and chimpanzees are virtually identical, while those of humans and cows differ much more. The proteins of humans and oak trees are even more dissimilar.

Early species of *Australopithecus*, *A. ramidus* and *A. afarensis* were small hominids, about 1 metre tall, walking upright but with ape-sized brains.

The most widespread australopithecine, *Australopithecus africanus*, was first reported from South Africa in 1924. At that time, few anthropologists accepted that humans could have arisen in Africa.

The australopithecines *Paranthropus robustus* and *P. boisei* were found in South Africa in the 1940s. They were larger and more heavily built than *Australopithecus*, but they apparently formed an evolutionary dead end.

Human evolution (*below*) This provisional evolutionary tree is based on the knowledge that we have in the mid 1990s. Dramatic new discoveries still occur. *A. afarensis* was found in 1974, and *A. ramidus* was found in 1994. Together they have added 2 million years to human history.

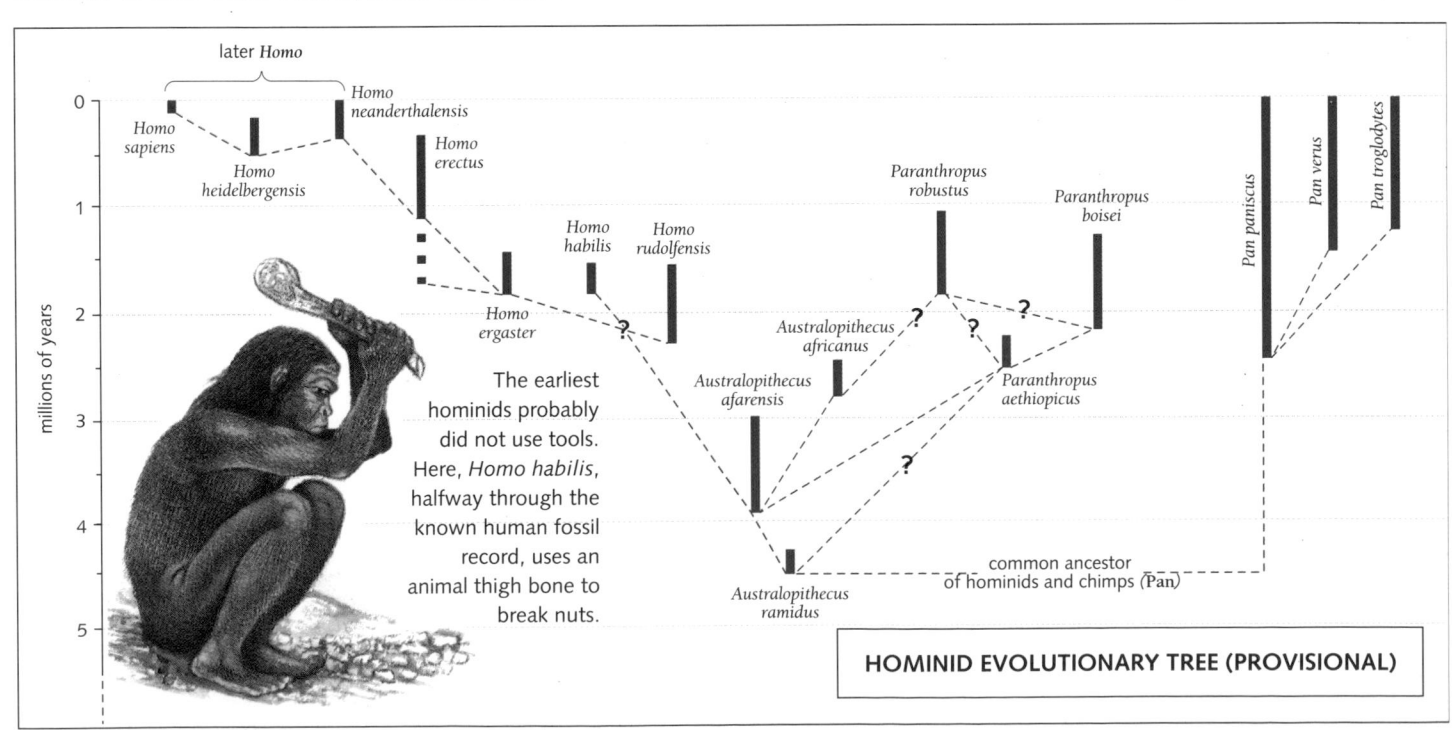

The earliest hominids probably did not use tools. Here, *Homo habilis*, halfway through the known human fossil record, uses an animal thigh bone to break nuts.

HOMINID EVOLUTIONARY TREE (PROVISIONAL)

Increasing brain size (right)
In the course of human evolution, brain size has increased. The australopithecines, such as *Australopithecus* and *Paranthropus*, were smaller than modern humans, but their brain sizes were even smaller, in the range of apes at only 400–550 cm³. This increased to 650–800 cm³ in *Homo habilis*, 850–1,100 cm³ in *Homo erectus* and 1,000–1,600 cm³ in *Homo neanderthalensis* and *Homo sapiens*. The increase in brain size with the origin of *Homo* 2 million years ago must have been associated with major increases in intelligence, and new behavioural patterns (such as the use of tools), more cooperative activities, and perhaps more advanced language use.

Australopithecus africanus

Australopithecus robustus

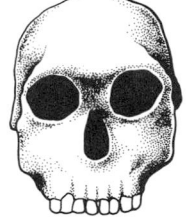

Homo habilis

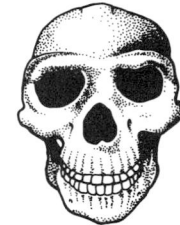

Homo erectus

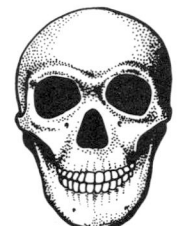

Homo sapiens neanderthalensis

Homo sapiens sapiens

Such genetic and molecular studies, as well as new fossil discoveries, show that humans split from their common ancestor with chimps about 5–7 million years ago. The first hominids were the australopithecines ('southern apes'), but the earliest steps are poorly known. The oldest hominid, *Australopithecus ramidus*, is known from isolated bones found in Ethiopia. *A. afarensis*, also from Ethiopia, is better known, mainly from the skeleton of 'Lucy', a small female hominid perhaps 1 metre tall, who walked upright but still had an ape-sized brain. *A. africanus* was a little larger, and may have given rise to some larger forms, species of *Paranthropus*.

True humans (*Homo*) arose in Africa 2–3 million years ago. They walked upright and had much larger brains than apes or the australopithecines. The first globally distributed species was *Homo erectus* about 1 million years ago, known first from Africa, and later from Europe, China and Java. This species died out, and was replaced in Europe by two 'modern' species, *H. heidelbergensis* and *H. neanderthalensis*, which died out too. Modern humans, *H. sapiens*, spread from Africa only about 150,000 years ago.

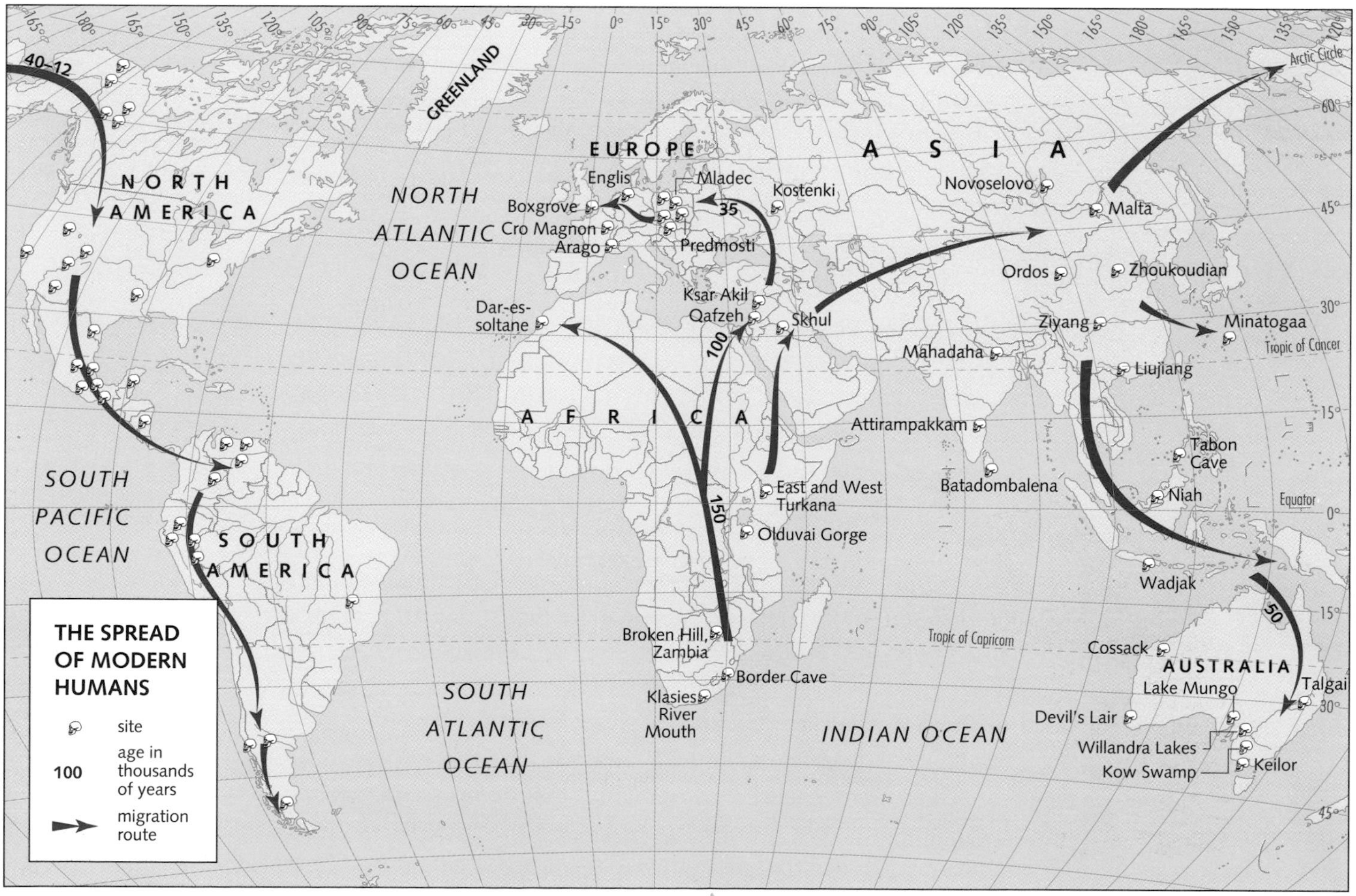

THE SPREAD OF MODERN HUMANS

🐚 site

100 age in thousands of years

➤ migration route

Section II: Natural Selection and Variation

Evolution occurs by the 'survival of the fittest'. Populations are highly variable, and generally far more offspring are produced than can survive. Natural selection is the process by which survivors are weeded out, and the way in which their superior adaptations are inherited by their offspring.

The core of Darwin's proposal in 1859 was that evolution had occurred by natural selection. He was led to his discovery of this process by a combination of circumstances, the most significant of which was his acceptance that species are not immutable and that there is a vast overproduction of young organisms.

We can divide the theory of evolution by natural selection into several propositions, each of which can be proved in nature. These propositions are: that more young organisms are produced each generation than can survive into the next generation; that individuals within populations are variable; that only certain individuals within a population, bearing certain variations, will survive into the next generation; that those individuals that have special features enabling them to survive and reproduce (the ability to find more food, to attract a mate or to escape predators) will do so; that many advantageous features of the surviving organisms are heritable and will be passed on to future generations; and that as a result of these the make-up of a population (the range of adaptations that it contains) will change, or evolve, over time.

The vast fecundity of species is well-known. Even humans, at the very low end of the reproductive spectrum, produce too many offspring. This is shown by the almost exponential present-day rate of population increase, and the fact that human populations rebound rapidly after major pestilences and wars. At the other end of the spectrum are some amazingly productive species, such as the cod, which produces millions of eggs during each reproductive cycle. Clearly, only a minute fraction of those millions can survive. Darwin quoted the example of elephants, which have a low reproductive rate like humans. He calculated that the Earth would nevertheless be filled with millions of elephants in a few centuries if all the potential offspring survived and reproduced at the earliest possible time.

Variation

Variation within populations is obvious: all humans alive today are different from one another (except for identical twins), and the same is true of all plants and animals. Variation may be dramatic, as is seen in the different groups of the single species *Homo sapiens* and in species such as the European banded snail, *Cepaea nemoralis*, which shows a great range of colours and banding patterns. These shell variations seem to relate to very local environmental conditions – the specific vegetation in a patch of land, for instance – and the patterns are adaptations that camouflage the snails from their avian predators.

Other variations are related to larger-scale regional phenomena. For example, the peppered moth, *Biston betularia*, exists in two main forms. The patchy grey morph is found in areas with clean air, where lichens grow on trees, while the black morph is especially abundant where the air is polluted and lichens are absent. The colour is strongly selected by bird predators, which pick off moths that do not match their backgrounds. Some large-scale variations are in hidden, internal features. Sickle cell anaemia is a disease in which many of the red blood cells collapse and adopt a sickle-like shape. This means that they cannot carry oxygen around the body very efficiently, causing anaemia. The control is genetic, and depends on variation in a single gene. This sickle cell gene gives protection against malaria, but can also lead to sickle cell anaemia. The balance of advantage and disadvantage helps to maintain the gene, but only where the risk of malaria is high enough to outweigh the disadvantage of sickle cell anaemia.

Adaptation
All species show variation, but the reasons for variation among the members of a species can be difficult to establish. In the case of the banded snails, the peppered moths and the sickle cell gene, good adaptive reasons can be identified, and indeed tested. In the case of mimicry, a natural experiment has already been performed. There are many cases in which palatable butterflies have evolved an astonishingly close resemblance to poisonous species. Bird predators are the controlling force, recognizing which prey are good and which are bad to eat from their colours and patterns. It takes only one sample of an evil-tasting butterfly for the bird to learn to avoid that species. Any palatable butterfly that looks even a little like a poisonous one will have a survival advantage, and the more similar it is, the greater the advantage. In other cases, several unpalatable species have evolved to look similar, reinforcing the protection afforded by a particular pattern.

In other instances, variations within species may depend on large-scale geographic and climatic phenomena. Tigers occur from tropical India to northern Siberia, and the species and subspecies vary considerably as a result of the major changes in climate and vegetation from south to north of the range. Many birds show similar variation as a result of climate. Puffins and wrens tend to be larger the further north they are from southern Europe. The larger size is probably related to the increasing cold and the need for added insulation.

Human activities have also caused variation within species. Blue tits have learnt a new behaviour in Britain as a result of the spread of towns: they sit on top of milk bottles on the doorstep and peck the foil cap to get at the milk beneath. Likewise, urban foxes in Britain have modified their diets from live prey to leftover scraps, and their hours of activity from night to day. Variation within populations can be seen in both the physical features of organisms and their behaviour.

Fecund Cod and Salmon
The overproduction of young

One of the amazing facts of nature is that most plants and animals produce far too many eggs or young. Why should a pair of frogs produce several dozen or hundred tadpoles and a cod produce 2–5 million eggs? This set Darwin thinking in 1836, just after his return from the voyage of the *Beagle* (*see pages 18–19*). Darwin had read Thomas Malthus's influential *Essay on the Principle of Population* (1798) about the economics of human societies, and Malthus's ideas set Darwin on track to develop his theory of evolution by natural selection.

Malthus noted that while human populations increase at a geometric rate (population size doubles in size with each generation), the food supply increases at only an arithmetic rate (by the addition of a fixed quantity each generation). Human populations would rapidly outstrip food sources unless there were checks and balances to the huge natural rate of increase, such as wars, famines and epidemics. If this applies to humans, Darwin reasoned, it does so much more to animal populations. There must be some principle at work that selects those animals in each generation that will survive to breed the next generation. In cases where large numbers of young are produced, there is a dramatic loss of these to predators and to starvation, and generally only the strongest, or fittest, can survive.

Plants and animals adopt one of two reproductive strategies. Many grossly overproduce young, but do not invest very much in them. As long as two of the dozens or millions produced survive to breed, the species' future is secure. Other species, including most mammals and birds, produce far fewer young (typically between one and ten at a time), and invest a great deal in them in the form of nourishment before birth and care after birth.

Fish migrations
Many species of fish migrate large distances in order to follow favourable sources of food or to avoid predators. This is especially true of those fish species with the highest rates of reproduction (*below*). In order to maximise the survival chances of their young, the different species of salmon may move enormous distances in their polar, Pacific and Atlantic habitats. The Atlantic salmon is famous for its long-distance migrations from Canada to Greenland and Europe, and the Pacific species move from Canada to Siberia.

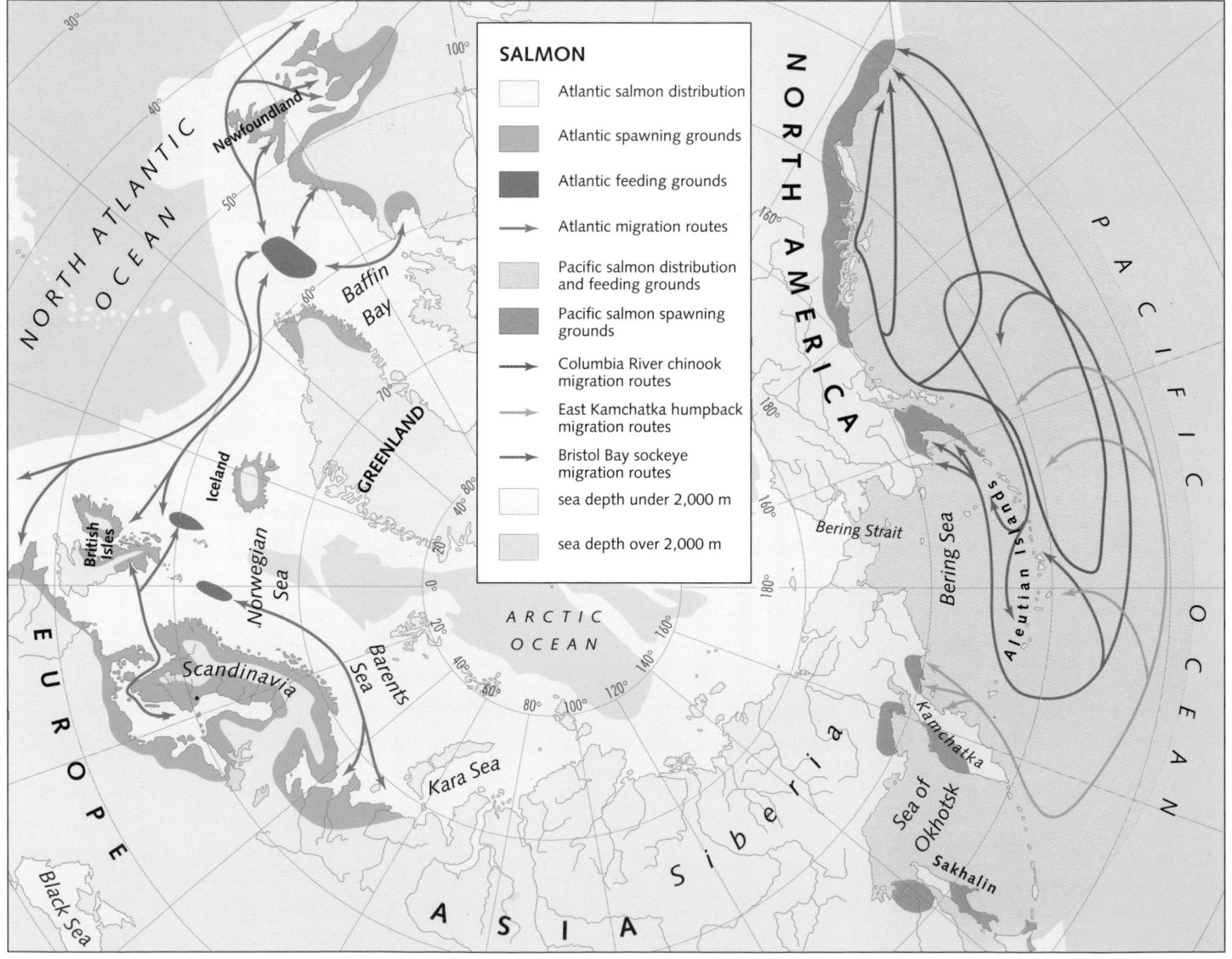

SALMON

- Atlantic salmon distribution
- Atlantic spawning grounds
- Atlantic feeding grounds
- Atlantic migration routes
- Pacific salmon distribution and feeding grounds
- Pacific salmon spawning grounds
- Columbia River chinook migration routes
- East Kamchatka humpback migration routes
- Bristol Bay sockeye migration routes
- sea depth under 2,000 m
- sea depth over 2,000 m

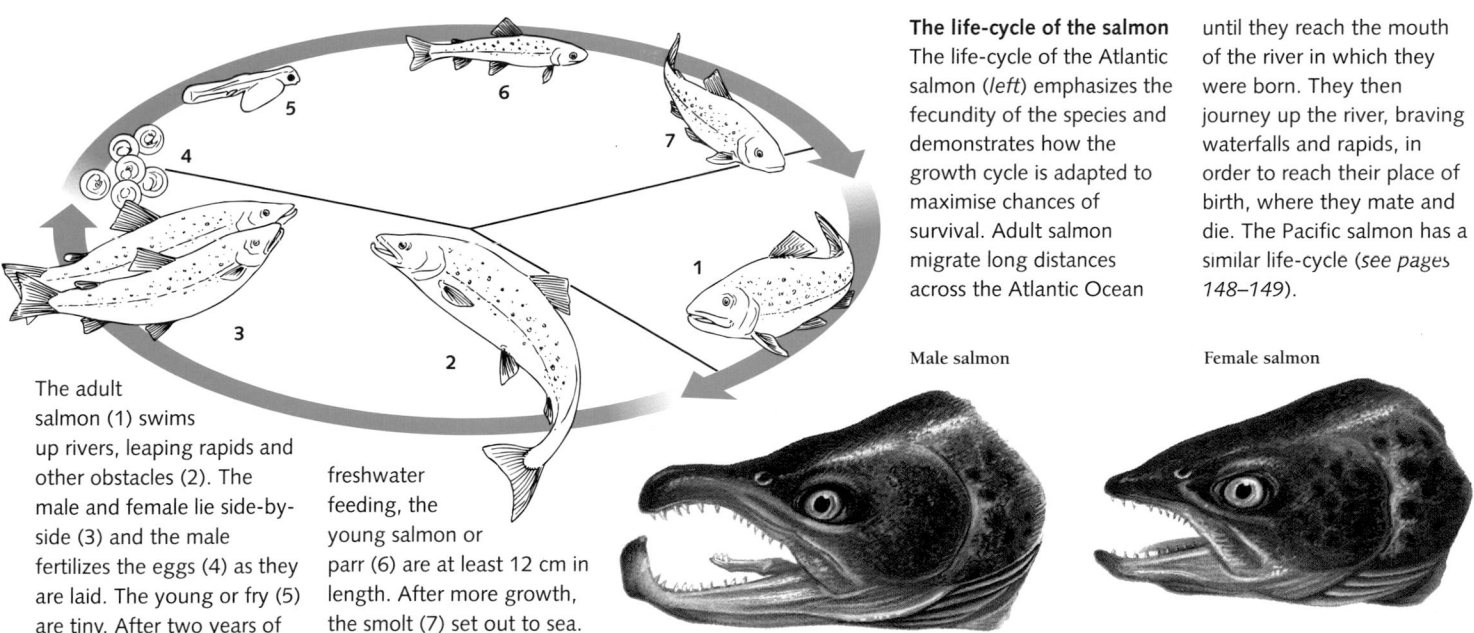

The life-cycle of the salmon

The life-cycle of the Atlantic salmon (*left*) emphasizes the fecundity of the species and demonstrates how the growth cycle is adapted to maximise chances of survival. Adult salmon migrate long distances across the Atlantic Ocean until they reach the mouth of the river in which they were born. They then journey up the river, braving waterfalls and rapids, in order to reach their place of birth, where they mate and die. The Pacific salmon has a similar life-cycle (*see pages 148–149*).

The adult salmon (1) swims up rivers, leaping rapids and other obstacles (2). The male and female lie side-by-side (3) and the male fertilizes the eggs (4) as they are laid. The young or fry (5) are tiny. After two years of freshwater feeding, the young salmon or parr (6) are at least 12 cm in length. After more growth, the smolt (7) set out to sea.

Male salmon

Female salmon

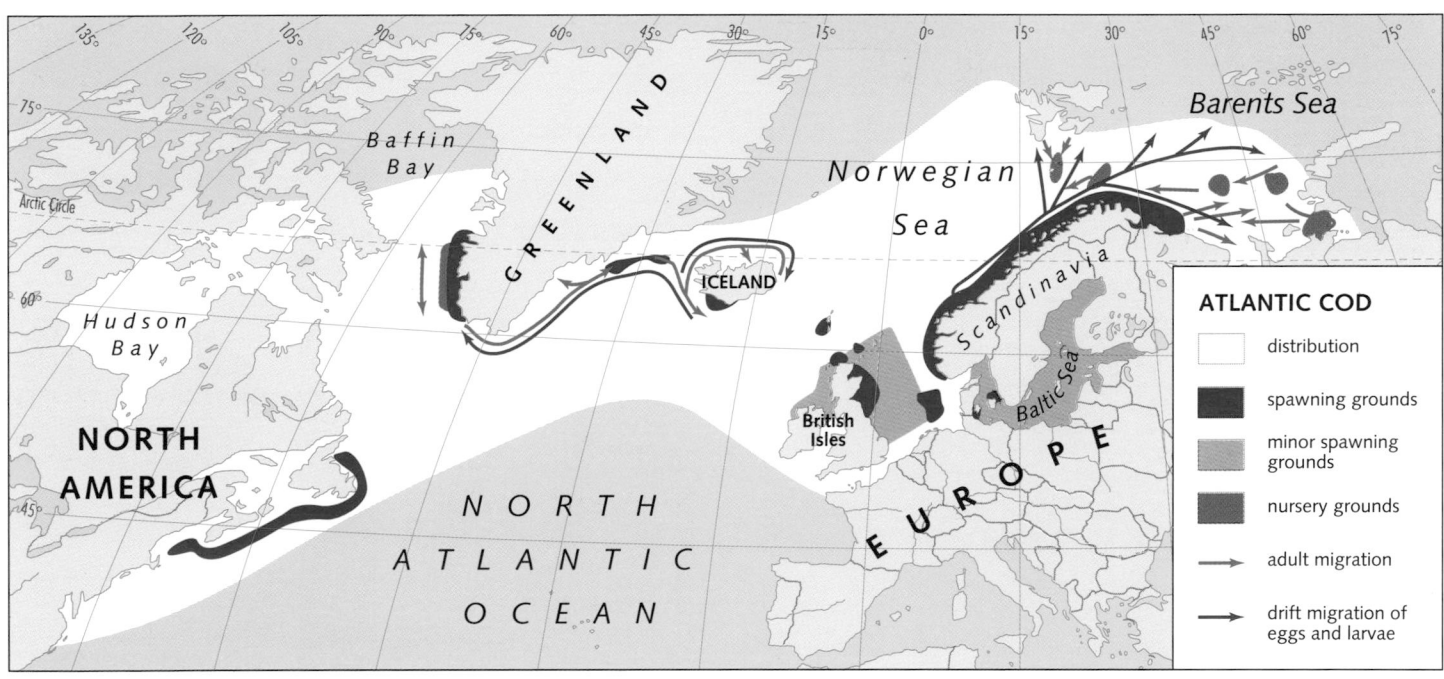

Fecundity

Fish show immense fecundity. Large salmon are impressive enough in that they can lay around 20,000 eggs in one sitting, but large female cod, which often lay in excess of 5 million eggs, are the most amazing producers of all.

Fisheries biologists have found that the body size of the female determines how many eggs she is capable of laying (*right*).

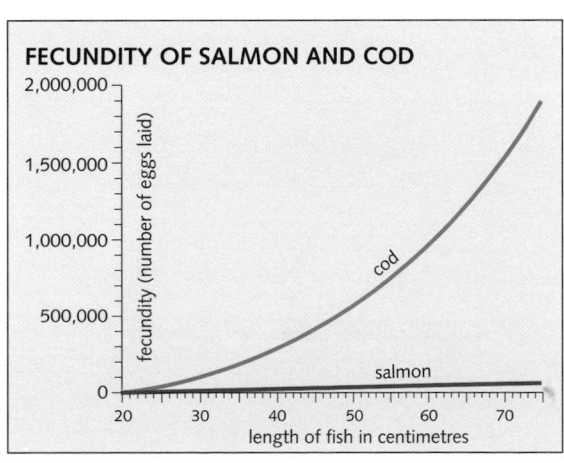

FECUNDITY OF SALMON AND COD

fecundity (number of eggs laid)

cod

salmon

length of fish in centimetres

Reproductive rates can be prodigious. A pair of mice could give rise to several million young in only a couple of years. Even slow-breeders could increase their population dramatically if there were no deaths. Darwin noted this for elephants, whose reproductive potential is even lower than that of humans. An elephant can produce about six young before it dies at the age of around 50. After 750 years, there could be nearly 19 million elephants descended from a single pair. This does not happen in nature because of the selective pressures acting against unconstrained increase. There is a 'struggle for existence' in which the weak fall and the strong survive.

Banded Snails
Natural variation in organisms

Natural selection acts on organisms engaged in the 'struggle for existence' (*see page 35*) so that the fitter individuals survive and the weaker perish. But selection does not operate unless there is variation: that is, a broad range of shapes and sizes of organisms. The natural variation of individuals within a population provides the raw material on which selection can act. The local conditions might determine that larger or smaller individuals have a survival advantage at any particular time, so those organisms will tend to survive. However, survival alone is not enough; the selection cannot be fixed in evolutionary terms unless the variations are heritable. The small or large size must be coded in the genes and passed on to the offspring. In this way, natural selection acting on the natural variations within a plant or animal population can lead to evolution.

The European land snail *Cepaea nemoralis* occurs in a great variety of colours and banding patterns. Most of them have stripes, which vary enormously in colour, intensity and width. This is an excellent example of the occurrence of variation. The proportions of different colours and of different banding patterns vary from place to place. At one time it was thought that the colours meant very little, that this was just a case of variation without adaptation. However, the patterns appear to have at least two functions: protection from predators, and thermoregulation. The stripes provide protection from the birds that feed on the snails, since they act as camouflage in the surrounding grass and leaves. The colour patterns and stripes also have a function in controlling body temperature, since darker snails retain heat and live in cooler spots.

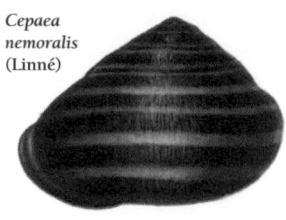

Cepaea nemoralis (Linné)

C. nemoralis var. *flavovirescens* Picard

C. nemoralis var. *olivacea* Risso

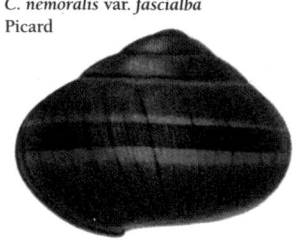

C. nemoralis var. *fascialba* Picard

The banded snail *Cepaea nemoralis* (*above*) occurs in many different colours, predominantly brown, yellow and pink. The shells typically bear five or six stripes around the circumference, which can be sharp or muted, subtle or clear, and closely or widely spaced. The relative frequency of yellow shells in western Europe increases towards the south (*left*), presumably because their value in camouflage is in proportion to the amount of dry yellow grass, which is most abundant in Mediterranean areas.

SNAIL COLOUR VARIATION IN WESTERN EUROPE

— range of *Cepaea nemoralis*

◑ frequency of yellow shells

NORTH ATLANTIC OCEAN

NORWAY

SWEDEN

FINLAND

ESTONIA

LATVIA

LITHUANIA

Baltic Sea

North Sea

DENMARK

UNITED KINGDOM

EIRE

NETHER-LANDS

BELGIUM

LUX.

GERMANY

POLAND

CZECH REPUBLIC

SLOVAKIA

FRANCE

SWITZERLAND

AUSTRIA

HUNGARY

SLOVENIA

CROATIA

BOSNIA-HERZEGOVINA

SERBIA

Bay of Biscay

PORTUGAL

SPAIN

Corsica

Sardinia

ITALY

Adriatic Sea

ALBANIA

GREECE

Mediterranean Sea

0 250 km
0 250 miles
N

These ideas have been tested by experiments in which snails with widely divergent shell patterns were mixed up and scattered in a variety of settings. It turned out that each variety is adapted to a particular background (dead leaves, fresh grass or earth, for instance), and that birds leave those snails that are best camouflaged. Yellow banded shells were found much more often in hedgerows and rough grass, where they matched the background colour. These yellow shells were much less common in woodland settings, where brown unbanded shells are predominant. The colours and patterns are genetically coded and heritable. Geneticists have made intensive studies of variation in *Cepaea nemoralis*, and have found that those genes which control shell colour vary strongly with summer temperatures but that there is no evidence for any other control of the genes by either climate or topography.

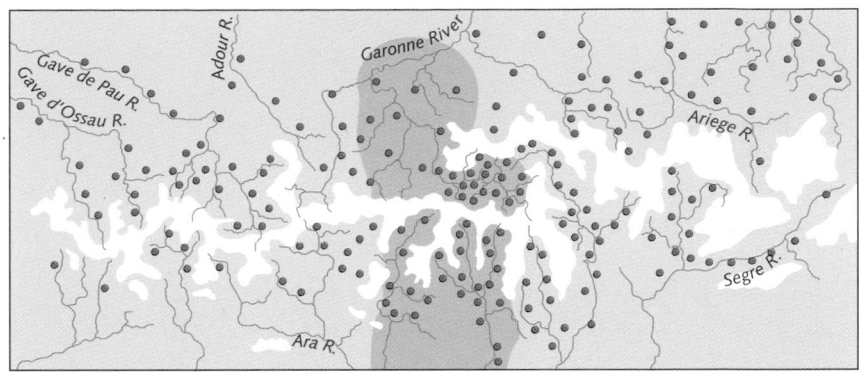

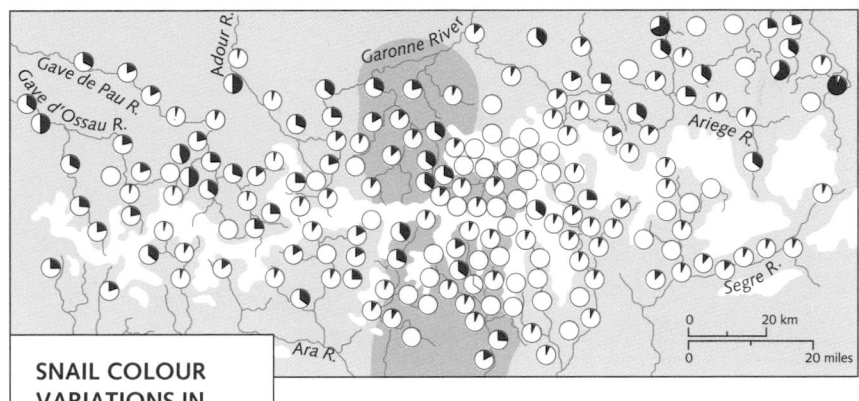

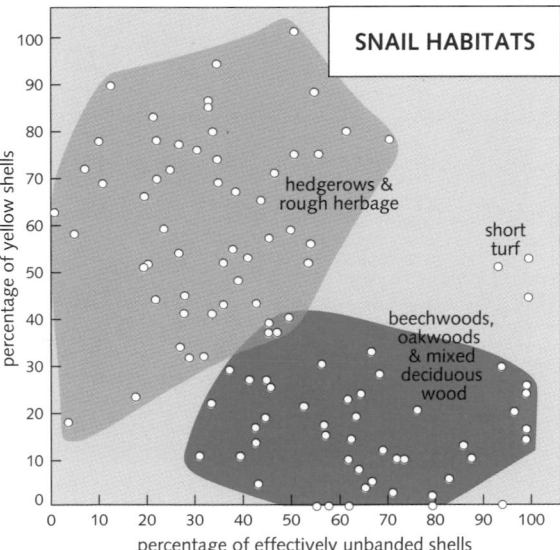

SNAIL HABITATS

hedgerows & rough herbage

short turf

beechwoods, oakwoods & mixed deciduous wood

percentage of yellow shells

percentage of effectively unbanded shells

SNAIL COLOUR VARIATIONS IN CENTRAL PYRENEES

• sites where snails were collected

◖ shell colour (yellow/pink)

Intense sampling in the Pyrenees (*above*) shows that banded snails are commonest in the valleys, rather than on high ridges. The background colour of shells, whether pink or yellow, shows little evidence of geographic variation. The same was found on the Marlborough and Berkshire Downs (*below*).

◖ frequency of five-banded shells (in red) at sampling points

Shell colours and patterns of the banded snail correspond to local vegetation (*above*).

The distributions of some colour and banding forms seem to show a large-scale pattern. For example, the proportion of yellow shells in western Europe increases dramatically from north to south, and this leads to a search for adaptive reasons. It seems that more yellow snails are found in the south of France than further north because there is more dry grass there. However, many pattern distributions make no geographic sense, since they are adapted to very local conditions. This is seen particularly when intensive collections of *Cepaea nemoralis* are made within a limited area. It is often difficult to interpret the wide variations in shell patterns that may be detected in quite a small area. Nevertheless, the great variation of shell patterns allows this species to survive predation in a broad range of habitats.

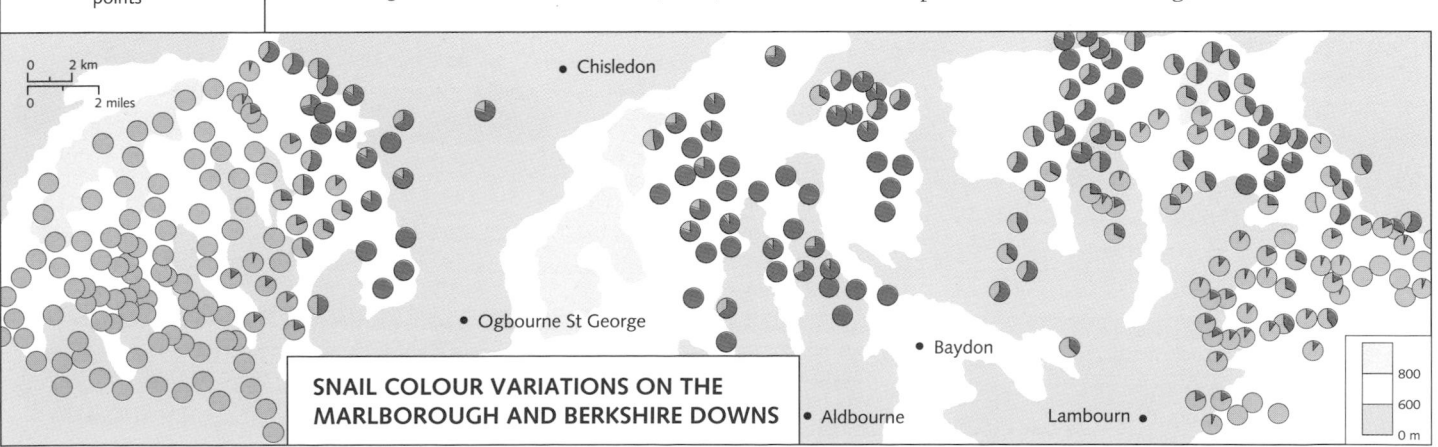

SNAIL COLOUR VARIATIONS ON THE MARLBOROUGH AND BERKSHIRE DOWNS

• Chisledon

• Ogbourne St George

• Baydon

• Aldbourne

Lambourn •

Industrial Melanism in Moths
Evolution caused by pollution

Biologists have often been able to watch evolution in action. Some of the most dramatic examples of this have been when evolution has occurred in response to human activities that result in pollution and the destruction of habitats. These cases are important because they show how evolution works, and how fast it can work when the environment changes dramatically.

One of the most famous examples of evolution in action is industrial melanism in moths. Many moths occur in two forms: a grey or pale-coloured form, and a rarer melanic, or black, form. The most impressive example of industrial melanism, that of peppered moths (*Biston betularia*), has been documented over the past 200 years. Biologists have observed major changes in the colour of the moths, from grey to black and back again.

Peppered moth populations in England have always existed in two main forms, one with a light-coloured mottled pattern and the other darker. Peppered moths escape predation by sitting on tree trunks, the mottled pattern acting as a camouflage against the irregular lichens on the bark. The light form originally dominated since it was best hidden from its predators. When the industrial revolution began, lichens on trees in the Midlands died off because the air quality became poorer. The light form became vulnerable because it now stood out against the bare and coal-blackened bark. There was a rapid increase in the proportion of the population that was composed of dark morphs in the industrial regions of England. Then, in the 1950s, with the decline of coalburning in factories and the Clean Air acts in cities, tree bark became lighter, and the trend in peppered moth populations reversed.

The quality of the camouflage has been tested with bird predators in a variety of settings. Biologists placed grey-mottle and melanic peppered moths on lichen-covered trees and bare blackened trees, and were able to prove that birds very definitely select their prey by how poorly camouflaged it is. Although birds generally have good eyesight, they still concentrate on eating moths that they see most clearly outlined against the tree bark.

The peppered moth example can only be treated as a case of true evolution if the two colour morphs are heritable. In other words, melanic forms must be shown to produce dark offspring, and grey

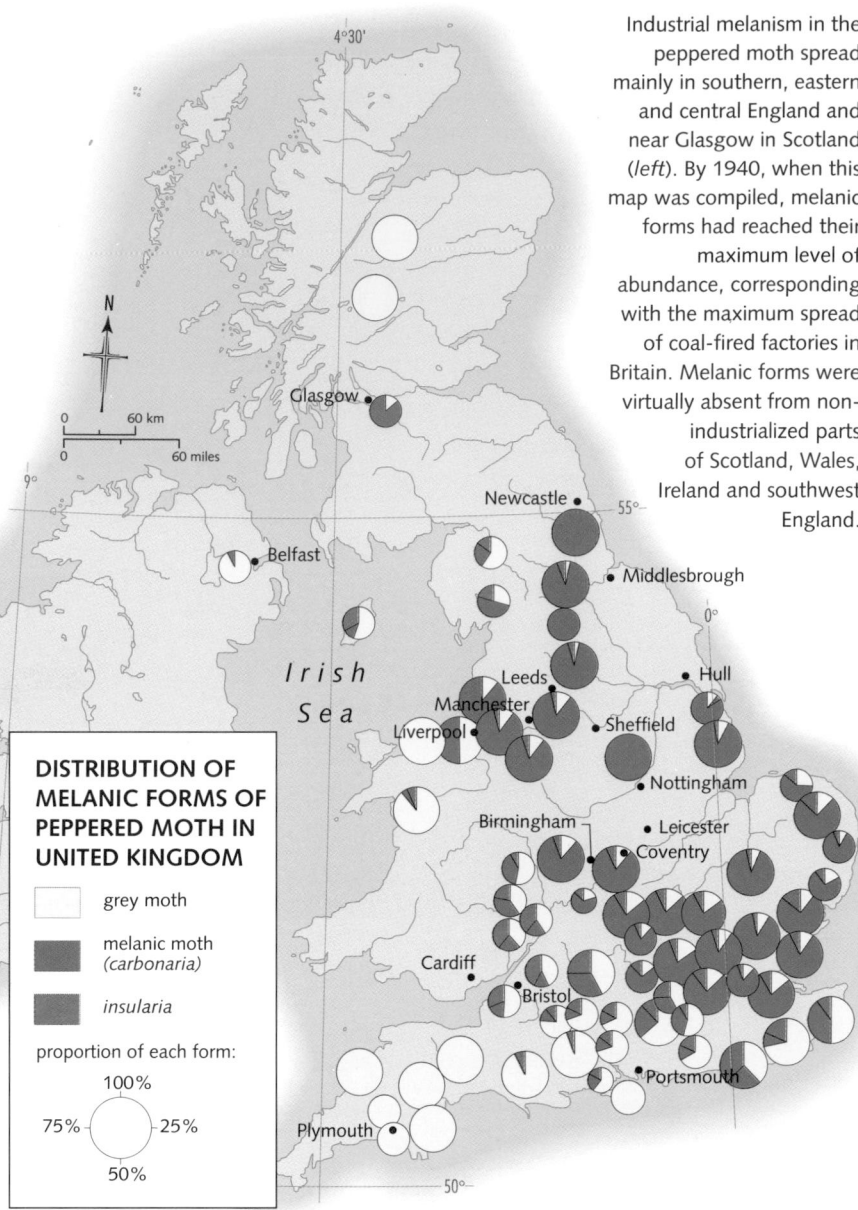

Industrial melanism in the peppered moth spread mainly in southern, eastern and central England and near Glasgow in Scotland (*left*). By 1940, when this map was compiled, melanic forms had reached their maximum level of abundance, corresponding with the maximum spread of coal-fired factories in Britain. Melanic forms were virtually absent from non-industrialized parts of Scotland, Wales, Ireland and southwest England.

DISTRIBUTION OF MELANIC FORMS OF PEPPERED MOTH IN UNITED KINGDOM

- grey moth
- melanic moth (*carbonaria*)
- *insularia*

proportion of each form:
100%
75% — 25%
50%

Peppered moths
The peppered moth, *Biston betularia*, occurs in two morphs: a grey mottled form, and a melanic (black) form (*right*). The grey morph is camouflaged from its bird predators on the lichen-covered trunks of trees, and this was the classic setting for the species. Lichens grow only in pure air. With the onset of industrialization, the quality of the air declined in extensive areas, and lichens died off. It was at this point that the melanic morph came into its own.

Melanic moth

Grey moth

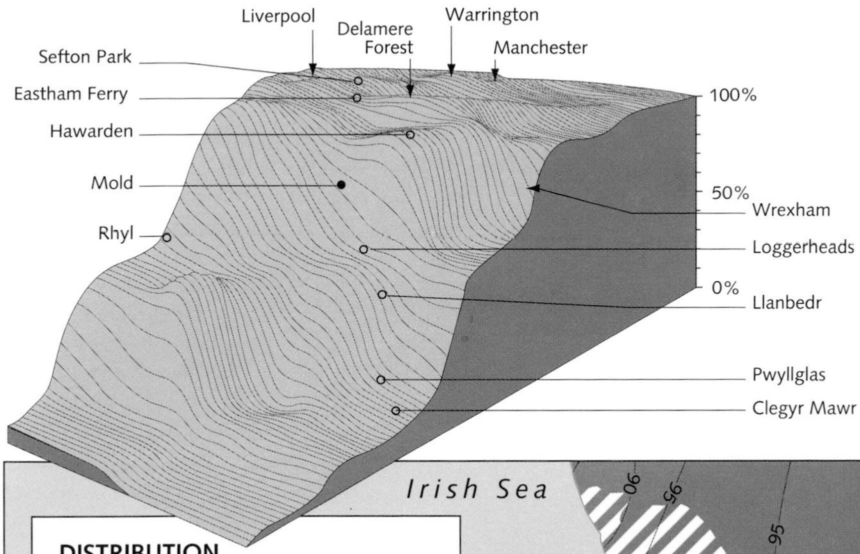

Liverpool
Sefton Park
Eastham Ferry
Hawarden
Mold
Rhyl
Delamere Forest
Warrington
Manchester
100%
50%
0%
Wrexham
Loggerheads
Llanbedr
Pwyllglas
Clegyr Mawr

Industrial melanism in the peppered moth occurred around Liverpool and Manchester. In the 1940s and the 1950s, when coalburning was still at an intense level, the melanic form was dominant around these industrial centres, but the relative abundance of grey forms increased dramatically further south and west in rural areas of Cheshire and North Wales (*left and below*).

Grey morphs of the peppered moth have made a dramatic comeback (graph, *bottom of page*). Air quality near Liverpool improved between 1960 and 1980, with levels of sulphur dioxide and smoke falling to a tenth of previous levels. Then, with a lag time of 15 years, the relative numbers of melanic peppered moths have plummetted to 50 per cent. This decline may continue.

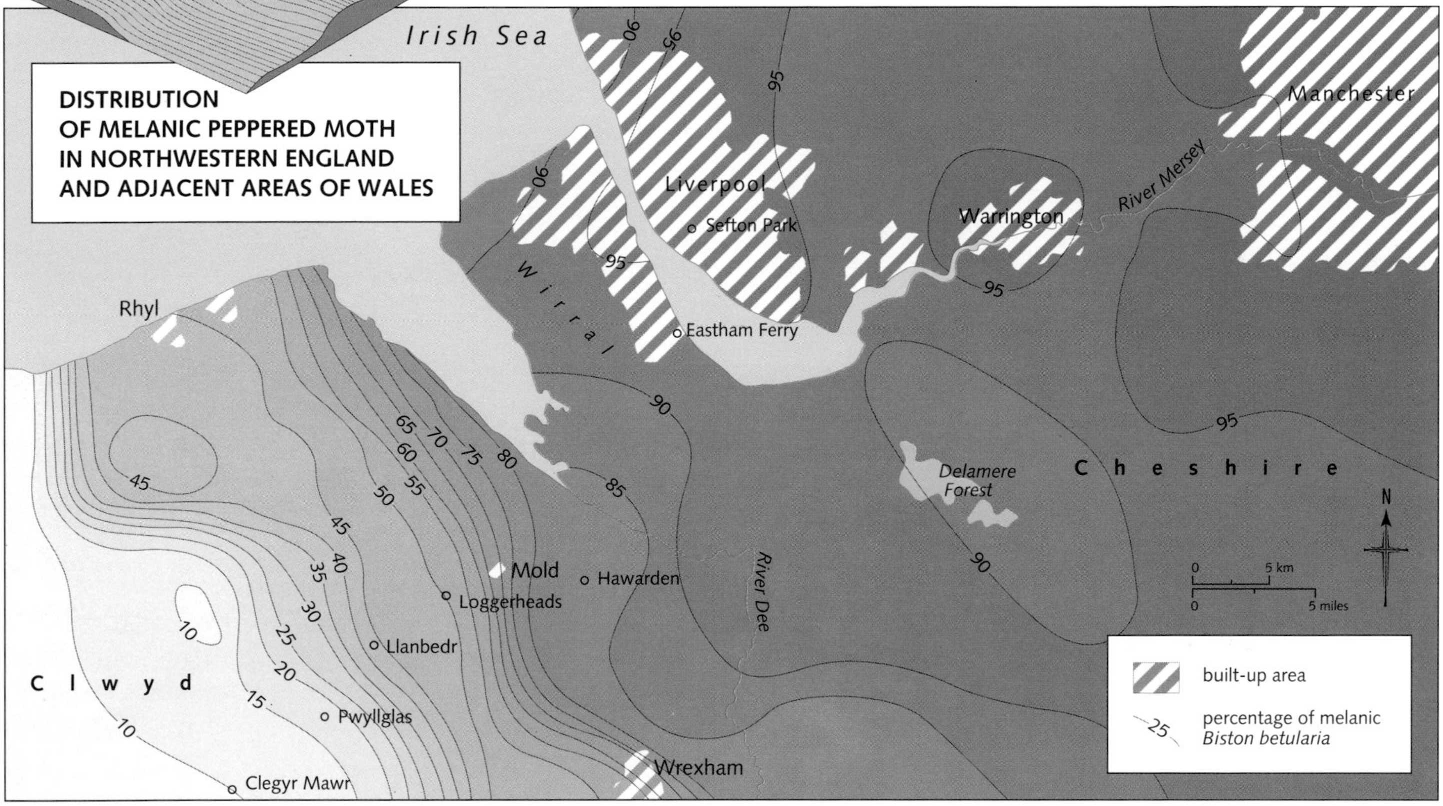

DISTRIBUTION OF MELANIC PEPPERED MOTH IN NORTHWESTERN ENGLAND AND ADJACENT AREAS OF WALES

Irish Sea

Manchester
Liverpool
Sefton Park
Warrington
River Mersey
Eastham Ferry
Wirral
Rhyl
Delamere Forest
Cheshire
River Dee
Mold
Hawarden
Loggerheads
Llanbedr
C l w y d
Pwyllglas
Clegyr Mawr
Wrexham

N

0   5 km
0   5 miles

built-up area

25   percentage of melanic *Biston betularia*

forms to produce grey offspring. If there was no heritability of colour forms, evolution to dominance by dark or light forms would not occur, even if populations shifted one way or the other. This is because such a colour shift would occur independently each generation as a result of predator pressure. When the pressure is removed, the population would revert to its normal balance of grey and black morphs. Geneticists have found the genes that control wing colour and seen that the pressure has caused selection in the moths' genome. The move from dominance by grey to dark and back to grey forms is a reversal of evolution. The changes take many generations to occur and are not simply a response within a generation to the removal of selection pressure.

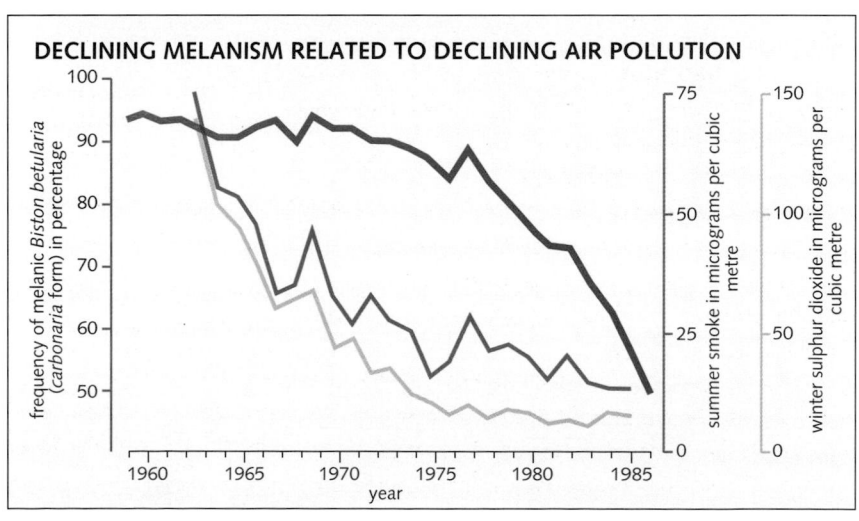

DECLINING MELANISM RELATED TO DECLINING AIR POLLUTION

Sickle Cell Anaemia
From evolutionary advantage to genetic disease

Mutations arise spontaneously in all organisms but are often of little consequence. Sometimes they are harmful, and tend to die out, and sometimes they are beneficial, and therefore tend to persist. But what is beneficial in one environment can become harmful in another. The story of sickle cell anaemia, a genetic blood disorder, shows how genetic, as well as infectious, diseases can be transported around the world, and how evolutionary advantage can turn to genetic disease.

A particularly virulent form of malaria, *Plasmodium falciparum malaria*, is prevalent in parts of the tropics and the eastern Mediterranean, and endemic in areas of tropical West Africa and India. Like other types of malaria, it is caused by a parasite that is transmitted into the blood by the *Anopheles* mosquito. This parasite attacks the red blood cells which carry haemoglobin to the body's organs. At some point during the human habitation of these regions, a mutation arose which affects the gene controlling the type of haemoglobin in the blood. Normal red blood cells are disc-shaped under a microscope, but the mutated haemoglobin makes them appear sickle-shaped. Carrying a number of such cells helps the body to resist *falciparum* malaria, so that people with sickle cells are likely to live longer in areas where malaria is common. More of them will survive to reproduce, so the gene will be passed to ever more people over the generations, and the proportion of carriers in the population will gradually increase.

But this is only the beginning of the story. Sickle cells are of benefit if present in a certain quantity, but when they rise above that level they interfere with the transport of oxygen through the blood, causing sickle cell anaemia. This disease has travelled across the world with its host population. Many of the people taken from West Africa during the slave trade would have been carriers of the sickle cell gene. Slave traders would have rejected those who were obviously sick, so it is likely that sickle cell anaemia itself did not arise until the first new generation was born in the Americas.

The same phenomenon occurred when the descendants of African slaves emigrated from the Caribbean to take up jobs in Britain in the 1950s. Although the sickle cell gene confers no advantage on people in malaria-free areas of the Caribbean, it has persisted through the generations from their African forebears, so that a significant number of the emigrants had one gene which produced sickle cells.

The incidence of sickle cell anaemia among Afro-Caribbeans in Britain was practically zero in the 1950s, but as the first immigrants married one other and had offspring, the number of cases began to rise. Immigrant populations from parts of India show the same pattern. Cases have now peaked at about 150 births per year, but are expected to fall as the effects of genetic counselling are felt.

The mutation that produces the sickle cell gene arises in all human populations, but because it is of no advantage in most regions, it does not persist in sufficient numbers to produce the double gene that causes sickle cell anaemia.

Red blood cells
Under the microscope normal red blood cells (*right*) appear spherical. Carriers of the sickle cell gene have a number of sickle-shaped cells, which provide protection against malaria. If a person has two sickle cell genes, the proportion of sickle cells rises, and this causes severe complications, one of which is a reduction in the blood's ability to transport oxygen.

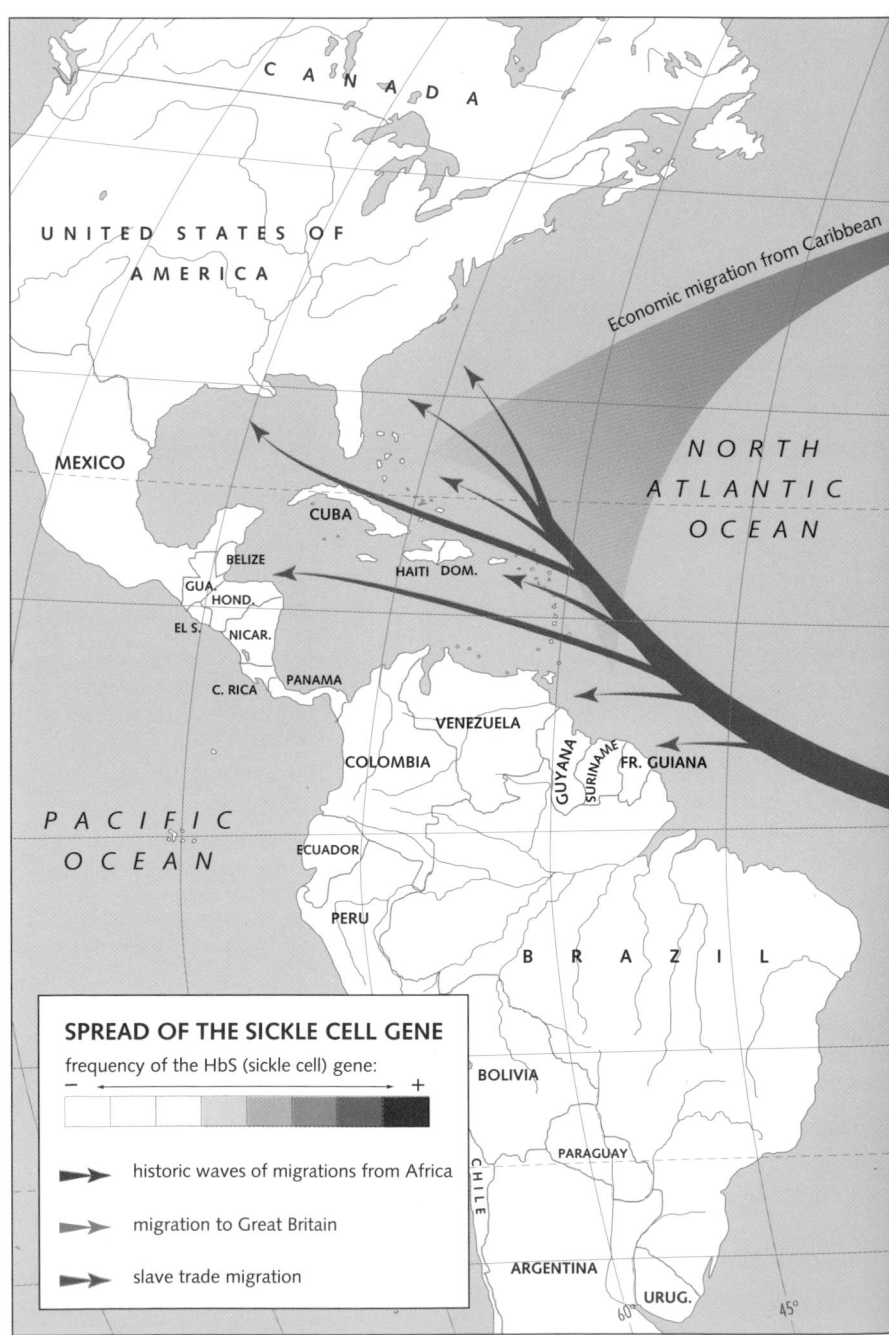

SPREAD OF THE SICKLE CELL GENE
frequency of the HbS (sickle cell) gene:

— ◄————————————► +

➤ historic waves of migrations from Africa

➤ migration to Great Britain

➤ slave trade migration

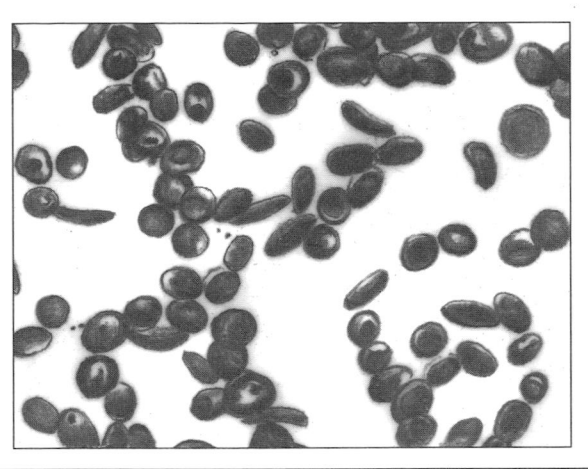

The pairing of genes

Sickle cell anaemia is controlled by a pair of genes, one inherited from each parent. If an individual is born with both genes that produce sickle cells, he or she will suffer from the disease. This can only happen if both parents are carriers of the gene, and even then there is only a one in four chance. The chart on the right shows how this works.

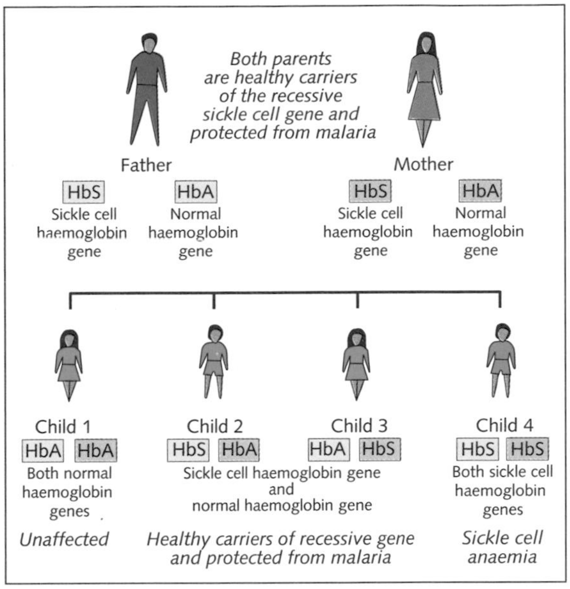

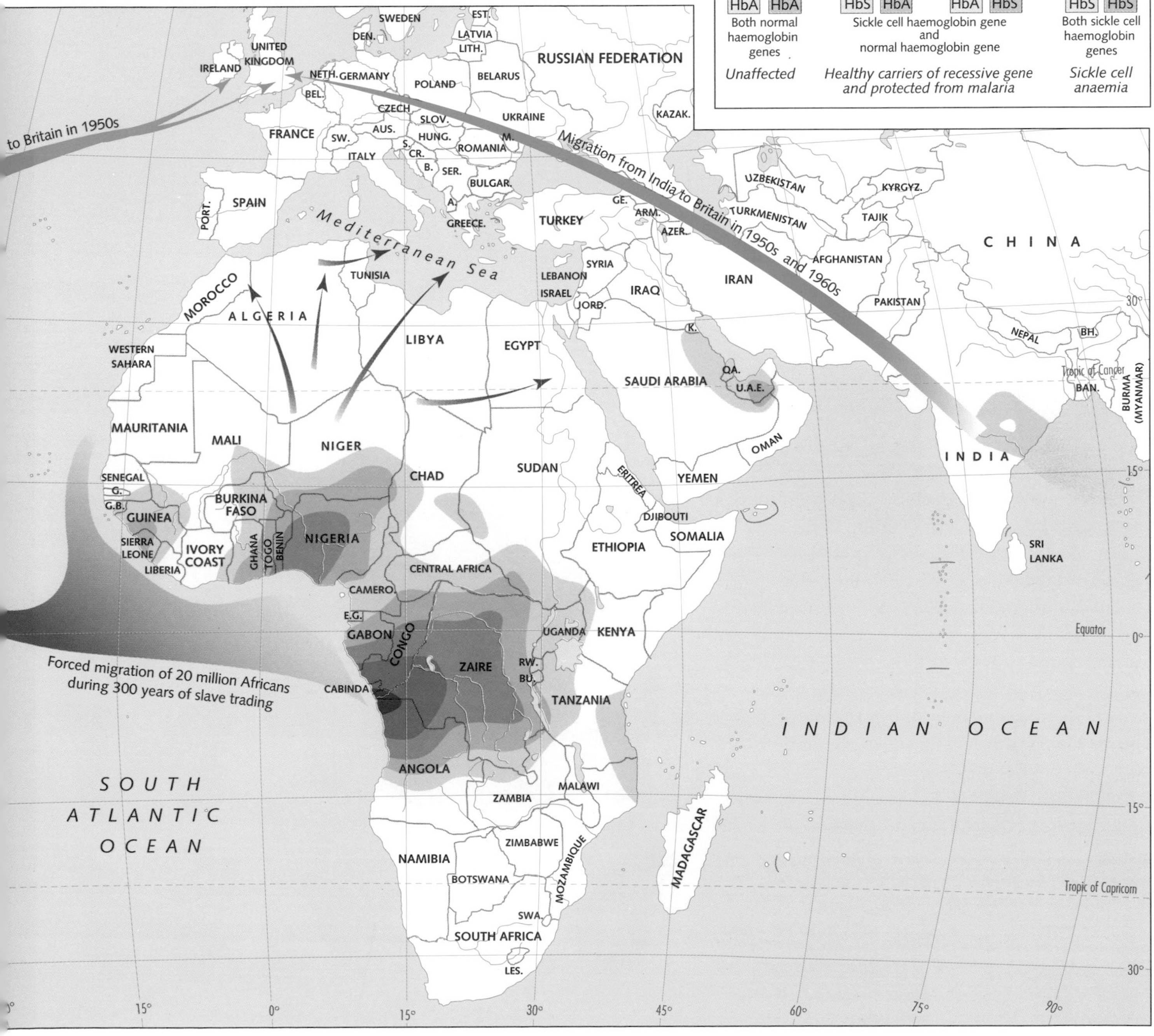

Both parents are healthy carriers of the recessive sickle cell gene and protected from malaria

Father

| HbS | HbA |

Sickle cell haemoglobin gene / Normal haemoglobin gene

Mother

| HbS | HbA |

Sickle cell haemoglobin gene / Normal haemoglobin gene

Child 1
| HbA | HbA |
Both normal haemoglobin genes

Child 2
| HbS | HbA |
Sickle cell haemoglobin gene and normal haemoglobin gene

Child 3
| HbA | HbS |
Sickle cell haemoglobin gene and normal haemoglobin gene

Child 4
| HbS | HbS |
Both sickle cell haemoglobin genes

Unaffected   *Healthy carriers of recessive gene and protected from malaria*   *Sickle cell anaemia*

to Britain in 1950s

Migration from India to Britain in 1950s and 1960s

Forced migration of 20 million Africans during 300 years of slave trading

SWEDEN
DEN.
LATVIA
EST.
LITH.
RUSSIAN FEDERATION
IRELAND
UNITED KINGDOM
NETH. GERMANY
BEL.
POLAND
BELARUS
FRANCE
SW.
CZECH
AUS.
SLOV.
HUNG.
S. CR.
B.
SER.
ROMANIA
M.
UKRAINE
KAZAK.
ITALY
A.
BULGAR.
PORT.
SPAIN
GREECE.
TURKEY
GE.
ARM.
AZER.
UZBEKISTAN
KYRGYZ.
TURKMENISTAN
TAJIK
Mediterranean Sea
SYRIA
LEBANON
ISRAEL
IRAQ
JORD.
IRAN
AFGHANISTAN
PAKISTAN
CHINA
MOROCCO
TUNISIA
NEPAL
BH.
WESTERN SAHARA
ALGERIA
LIBYA
EGYPT
SAUDI ARABIA
(K.)
QA.
U.A.E.
Tropic of Cancer
BAN.
BURMA (MYANMAR)
MAURITANIA
MALI
NIGER
CHAD
SUDAN
OMAN
INDIA
SENEGAL
G.
G.B.
GUINEA
BURKINA FASO
NIGERIA
YEMEN
ERITREA
DJIBOUTI
SRI LANKA
SIERRA LEONE
IVORY COAST
GHANA
TOGO
BENIN
LIBERIA
CENTRAL AFRICA
CAMERO
SOMALIA
ETHIOPIA
E.G.
GABON
CONGO
UGANDA
KENYA
Equator
CABINDA
ZAIRE
RW.
BU.
TANZANIA
INDIAN OCEAN
ANGOLA
ZAMBIA
MALAWI
MOZAMBIQUE
MADAGASCAR
SOUTH ATLANTIC OCEAN
NAMIBIA
ZIMBABWE
BOTSWANA
Tropic of Capricorn
SWA.
SOUTH AFRICA
LES.

15°   0°   15°   30°   45°   60°   75°   90°
30°   15°   0°   15°   30°

Mimicry in Butterflies
The evolution of warning colours and trickery in nature

One of the most dramatic pieces of evidence for evolution is mimicry, a form of convergence in which unrelated species appear to have adopted the same features for a particular purpose. Late last century, careful studies by naturalists such as Bates and Muller revealed uncanny resemblances between unrelated butterflies in South America. In some cases, several species converged on the precise colouring and patterns of a single distasteful species, while in others, several distasteful species all looked very similar.

Bates and Muller concluded that these mimicry complexes act to the advantage of the mimics and had arisen because of the strong selective advantage of either being distasteful to predators or of being thought to be so. Bird predators learn by experience not to eat unpleasant or poisonous forms. Palatable butterflies protect themselves by mimicking the distasteful form (the model).

Mimicry evolved because even a poor match may give a butterfly a chance of survival. Birds have good eyesight, but they hunt rapidly, pecking at prey that they have merely glimpsed. A hint of a warning colour or pattern may stop the bird long enough for the butterfly to escape. Through generations, predation selects ever-better mimics, until the match is nearly perfect in many cases.

There is an important trade-off in this system, which is known as Batesian mimicry. If the number of mimicking species increases in proportion to the number of distasteful species, there will come a point where the system breaks down. Birds will find that most of the butterflies they think are all the same are in fact edible. They will feed on that design of insect and the warning colours will count for nothing.

In Mullerian mimicry, several unpalatable species mimic each other. This can be seen in the parallel mimicry rings of the red and yellow heliconids, *Heliconius melpomene* and *H. erato*, found in tropical Central and South America. Variants of each species mimic each other precisely throughout their ranges. A mimicry ring has advantages for all the distasteful butterflies, since it enhances the effectiveness of the warning colour or pattern against all predators.

Batesian mimicry
The females of the swallowtail butterfly *Papilio dardanus* have an impressive ability to mimic other butterflies (*right*). They show a great range of appearances in different parts of Africa, mimicking a variety of unpalatable species. There are eight races and 55 morphs. Males are always black and yellow, and their colours and patterns vary only slightly among the races.

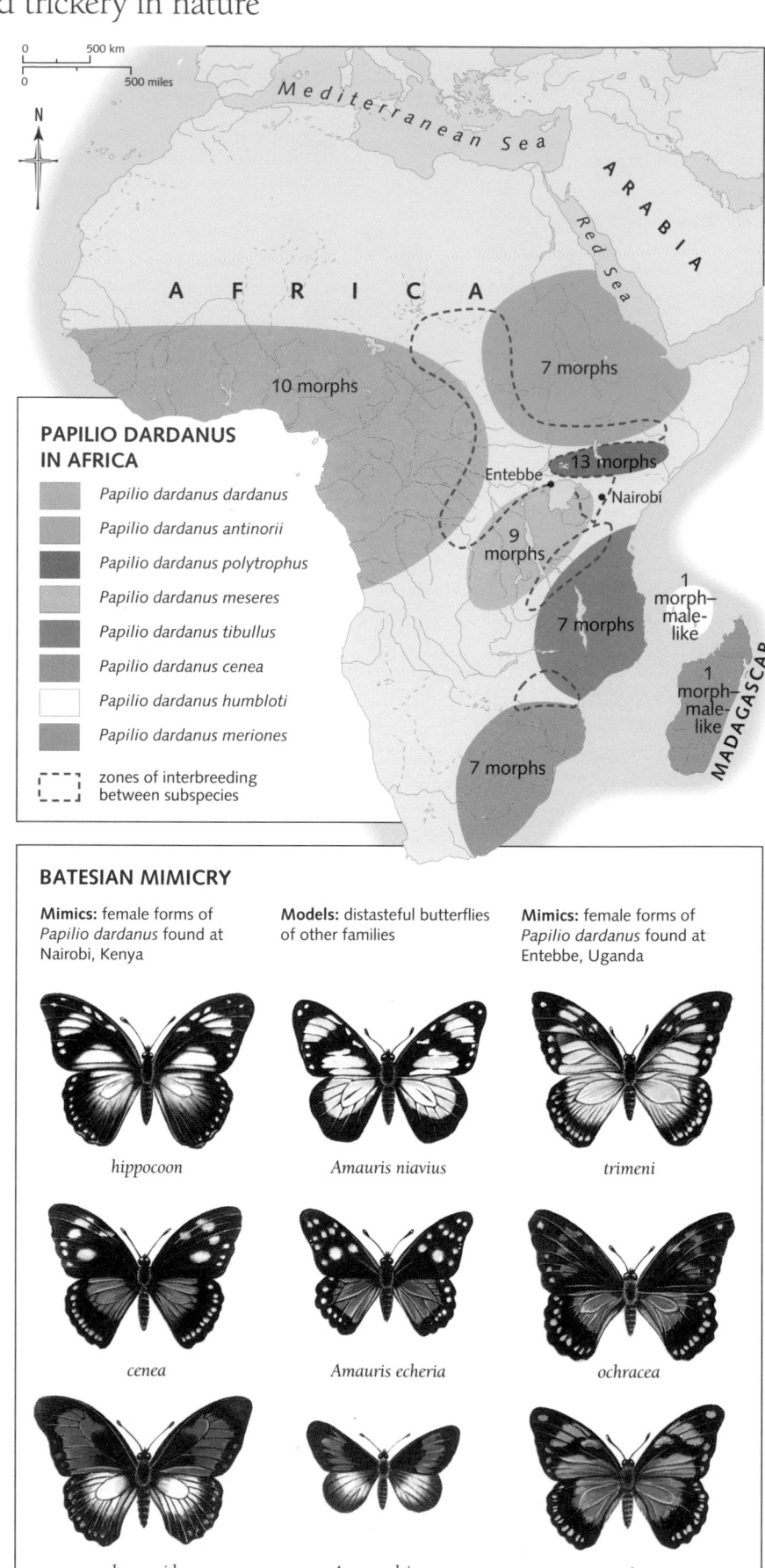

PAPILIO DARDANUS IN AFRICA

- *Papilio dardanus dardanus*
- *Papilio dardanus antinorii*
- *Papilio dardanus polytrophus*
- *Papilio dardanus meseres*
- *Papilio dardanus tibullus*
- *Papilio dardanus cenea*
- *Papilio dardanus humbloti*
- *Papilio dardanus meriones*
- - - - zones of interbreeding between subspecies

10 morphs
7 morphs
13 morphs
Entebbe
Nairobi
9 morphs
7 morphs
1 morph–male-like
1 morph–male-like
7 morphs
MADAGASCAR

BATESIAN MIMICRY

Mimics: female forms of *Papilio dardanus* found at Nairobi, Kenya

Models: distasteful butterflies of other families

Mimics: female forms of *Papilio dardanus* found at Entebbe, Uganda

hippocoon   *Amauris niavius*   *trimeni*

cenea   *Amauris echeria*   *ochracea*

planemoides   *Acraea alcioppe*   *speciosa*

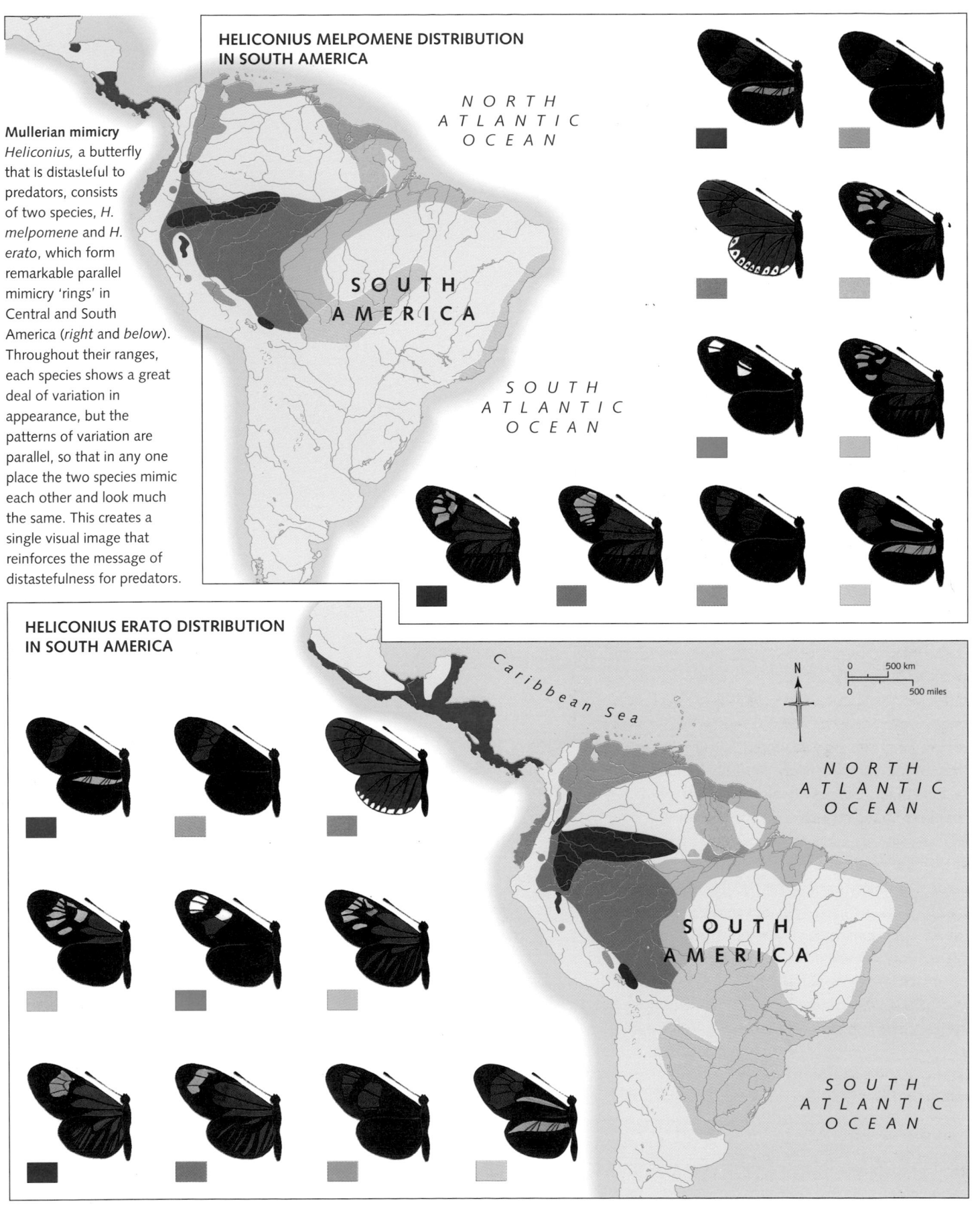

HELICONIUS MELPOMENE DISTRIBUTION IN SOUTH AMERICA

Mullerian mimicry
Heliconius, a butterfly that is distasteful to predators, consists of two species, *H. melpomene* and *H. erato,* which form remarkable parallel mimicry 'rings' in Central and South America (*right* and *below*). Throughout their ranges, each species shows a great deal of variation in appearance, but the patterns of variation are parallel, so that in any one place the two species mimic each other and look much the same. This creates a single visual image that reinforces the message of distastefulness for predators.

HELICONIUS ERATO DISTRIBUTION IN SOUTH AMERICA

Siberian Invaders

Variation within a species

The tiger provides graphic evidence of the way in which animals can adapt to changing environments while remaining a single species. In the case of the tiger, change came about when the advancing ice ages forced migration. Tigers are now found all over Asia, but they vary in important ways in each location. The species – *Panthera tigris* – has developed at least eight different races. These are still able to interbreed, as they all belong to the same species, but they are physically quite distinct.

Tigers originated 5–3 million years ago in the northern areas of the continent of Eurasia – most likely in northern Siberia, which was a region of temperate forests and grassland at that time. At the beginning of the Pleistocene, some 2 million years ago, the onset of the ice ages brought freezing conditions to these northern latitudes. Most of the tigers' food supply – grazing mammals – was forced south. Some tigers adapted to the changed conditions and stayed in Siberia, where a small number of their descendants still live. Most moved south and, over several thousand years, adapted to the new conditions that they encountered. It is these gradual adaptations in response to the increase in range that has led to the development of at least eight different races or strains within the tiger species.

In general the Siberian tigers are the largest, with males weighing over 300 kilograms (650 pounds) and reaching up to 3.5 metres (11 feet) in length, while those on the islands of Malaysia and Bali are the smallest. Tigers living in the dense jungles of Southeast Asia are darker in colour than those that live on the high plateaux farther north. The famous Bengal tigers of India come somewhere in the middle of this range, the adult male being 3 metres (9–10 feet) long and weighing 225–275 kilograms (500–600 pounds). An exception to the normal colouring of tigers is the beautiful coat of the white tigers of India. These are not albinos – they have black stripes on a creamy white background. This presumably developed as a mutation that has been preserved by interbreeding between white tigers.

As tigers moved south they fanned out in two directions around the Himalayas. The western branch reached the eastern Caucasus, Iran and Afghanistan. The eastern route led through Manchuria, Korea, China and Southeast Asia. The tiger's marked skill at swimming undoubtedly helped in its migration, particularly to outlying islands. The small size of the island tigers may be due to the original migrants being better swimmers than their larger brothers and

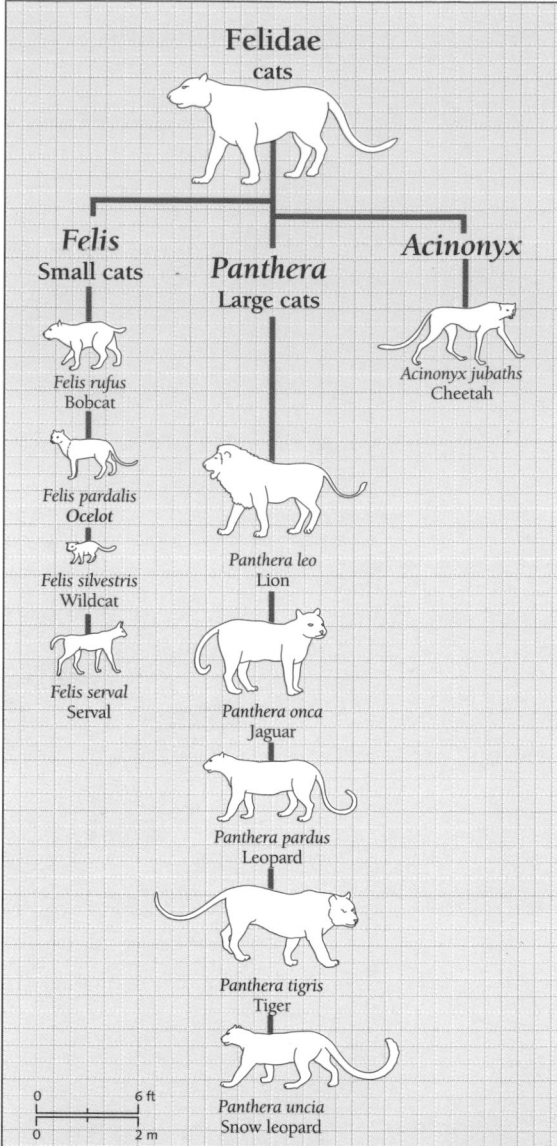

Felidae
cats

Felis
Small cats

Panthera
Large cats

Acinonyx

Felis rufus
Bobcat

Felis pardalis
Ocelot

Felis silvestris
Wildcat

Felis serval
Serval

Acinonyx jubaths
Cheetah

Panthera leo
Lion

Panthera onca
Jaguar

Panthera pardus
Leopard

Panthera tigris
Tiger

Panthera uncia
Snow leopard

0   6 ft
0   2 m

The cat family, Felidae, is found on all land masses except Australia and Madagascar. Its members vary greatly in shape and size and are generally divided into two groups: *Felis*, the small cats, and *Panthera*, the large cats. The cheetah has distinct characteristics and is generally classified in a separate group.

The tiger is the largest living member of the family. Bigger cats, such as the sabre-toothed cat *Smilodon*, existed in the past, but cats' prey has become smaller and the modern tiger is large enough to kill its biggest potential prey, the elephant. The *Panthera* group includes lions, leopards and jaguars as well as tigers. The third zoological name on the map opposite indicates that the races are a variation within the species. They do not differ enough to form separate species, presumably because their environments have not yet required a huge adaptation in their genetic make-up. Lions and leopards, on the other hand, have developed into separate species from tigers and from each another.

The Siberian tiger, *Panthera tigris altaica* (*left*), is the largest tiger race. Tigers originated in Siberia, but conditions have changed dramatically since their emergence over 3 million years ago, when Siberia was a warm region with lush vegetation. The ice ages drove most tigers south; those remaining in Siberia are confined to the extreme east. The range of tigers has decreased over the whole of Asia because of human settlement, but hunting has been reduced and tigers are now a preserved species in many countries.

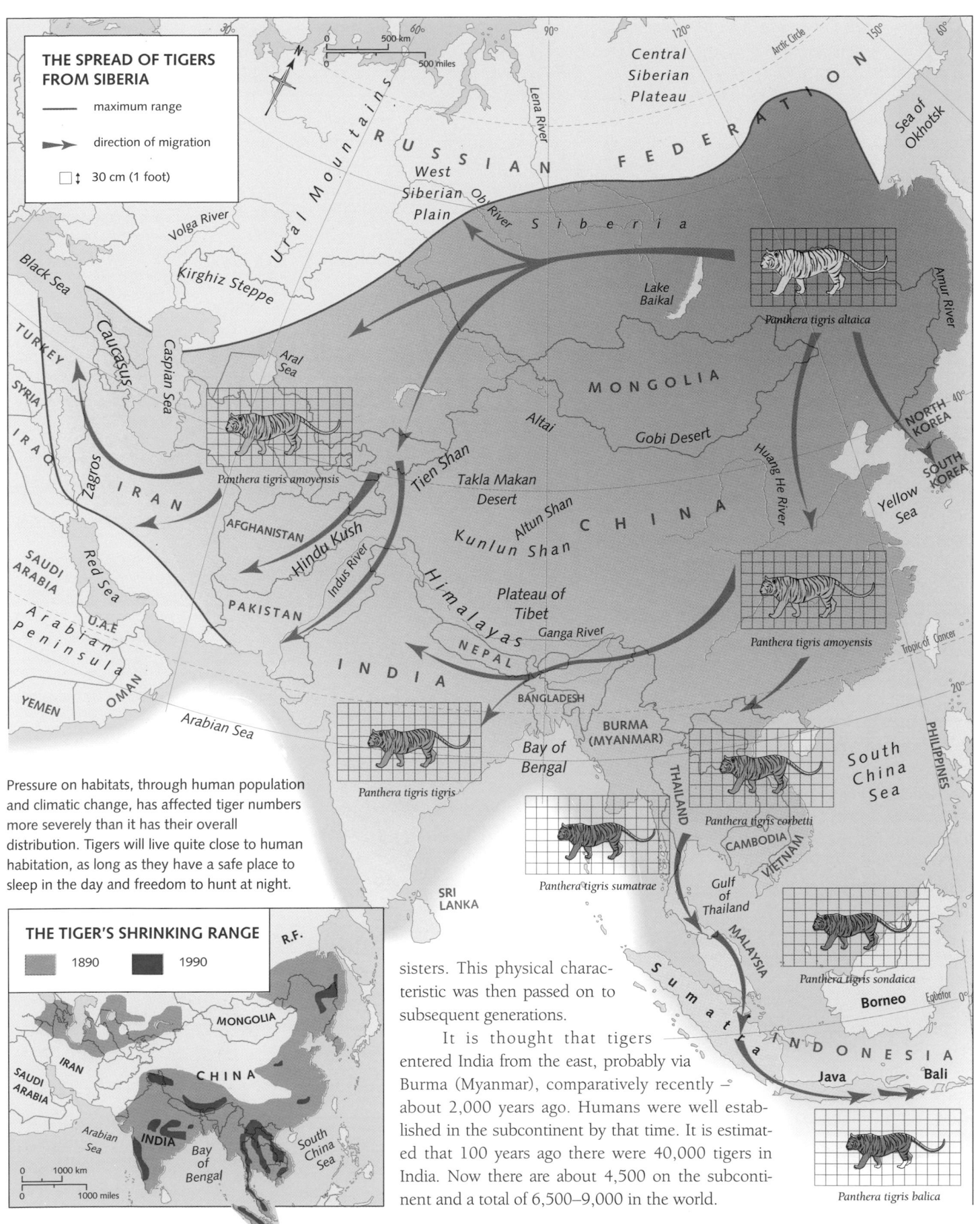

THE SPREAD OF TIGERS FROM SIBERIA

— maximum range

➤ direction of migration

☐ ↕ 30 cm (1 foot)

Panthera tigris altaica

Panthera tigris amoyensis

Panthera tigris amoyensis

Panthera tigris tigris

Panthera tigris corbetti

Panthera tigris sumatrae

Panthera tigris sondaica

Panthera tigris balica

Pressure on habitats, through human population and climatic change, has affected tiger numbers more severely than it has their overall distribution. Tigers will live quite close to human habitation, as long as they have a safe place to sleep in the day and freedom to hunt at night.

THE TIGER'S SHRINKING RANGE

☐ 1890   ☐ 1990

sisters. This physical characteristic was then passed on to subsequent generations.

It is thought that tigers entered India from the east, probably via Burma (Myanmar), comparatively recently – about 2,000 years ago. Humans were well established in the subcontinent by that time. It is estimated that 100 years ago there were 40,000 tigers in India. Now there are about 4,500 on the subcontinent and a total of 6,500–9,000 in the world.

Bergmann's Rule
The dependence of body size on temperature

Biologists have known for a long time that there may be predictable distributions of particular colours or body sizes across the total range of a species. One of the best-established of such large-scale patterns of geographic variation can be seen in warm-blooded animals such as birds and mammals, which seem to be larger in cold climates. This was first noticed by German biologist Bergmann in the 19th century, and has since proved to be generally applicable.

Why should Arctic mammals and birds be larger than their temperate zone relatives? The classic Pleistocene faunas consist of large mammals such as mammoths, mastodons and woolly rhinoceroses (*see pages 114–115*), and the later natural faunas of northern parts of Europe and North America consist of bison, the aurochs (giant wild ox), giant Irish deer and the like. The explanation seems to be that cold polar climates can be faced only by medium-sized and large birds and mammals.

Warm-bloodedness has many advantages, not least that birds and mammals can live in cold conditions that cold-blooded reptiles cannot survive. A disadvantage is that warm-blooded birds and mammals must eat ten times as much as cold-blooded animals of the same body weight, and the colder the climate, the more they must eat to keep warm. Small animals lose body heat faster than large ones, simply because their surface to volume ratio is very high. A large animal is insulated to some extent by its large size. Hence, very small mammals and birds are found only in tropical regions, and small birds and mammals are not found at polar latitudes.

Within species, latitudinal size change can be dramatic. Many North American birds, such as the downy woodpecker, increase vastly in size from south to north. Biologists measured the wing length of this bird, since it is quicker to record than body weight but is a precise indication of body weight, as wing size is determined by weight. It increased from an average of 86 millimetres in southern Florida to 96 millimetres in northern Wisconsin. Similar extraordinarily regular patterns of increase in wing length, and hence body size, have been observed in puffins and wrens in Europe. The most striking feature of these patterns is their predictable regularity, which is a clearly adaptive aspect of geographic variation.

Downy woodpecker
Geographic variation in the wing length of the downy woodpecker in North America (*below*) is used as an indication of variation in body weight. Studies throughout the eastern states allowed biologists to plot the geographic pattern of change in some detail. From south to north, wing length increases by ten millimetres. This is a good example of Bergmann's rule, according to which warm-blooded animals are larger in colder latitudes.

WING LENGTH VARIATION IN THE DOWNY WOODPECKER

90 — wing length in millimetres

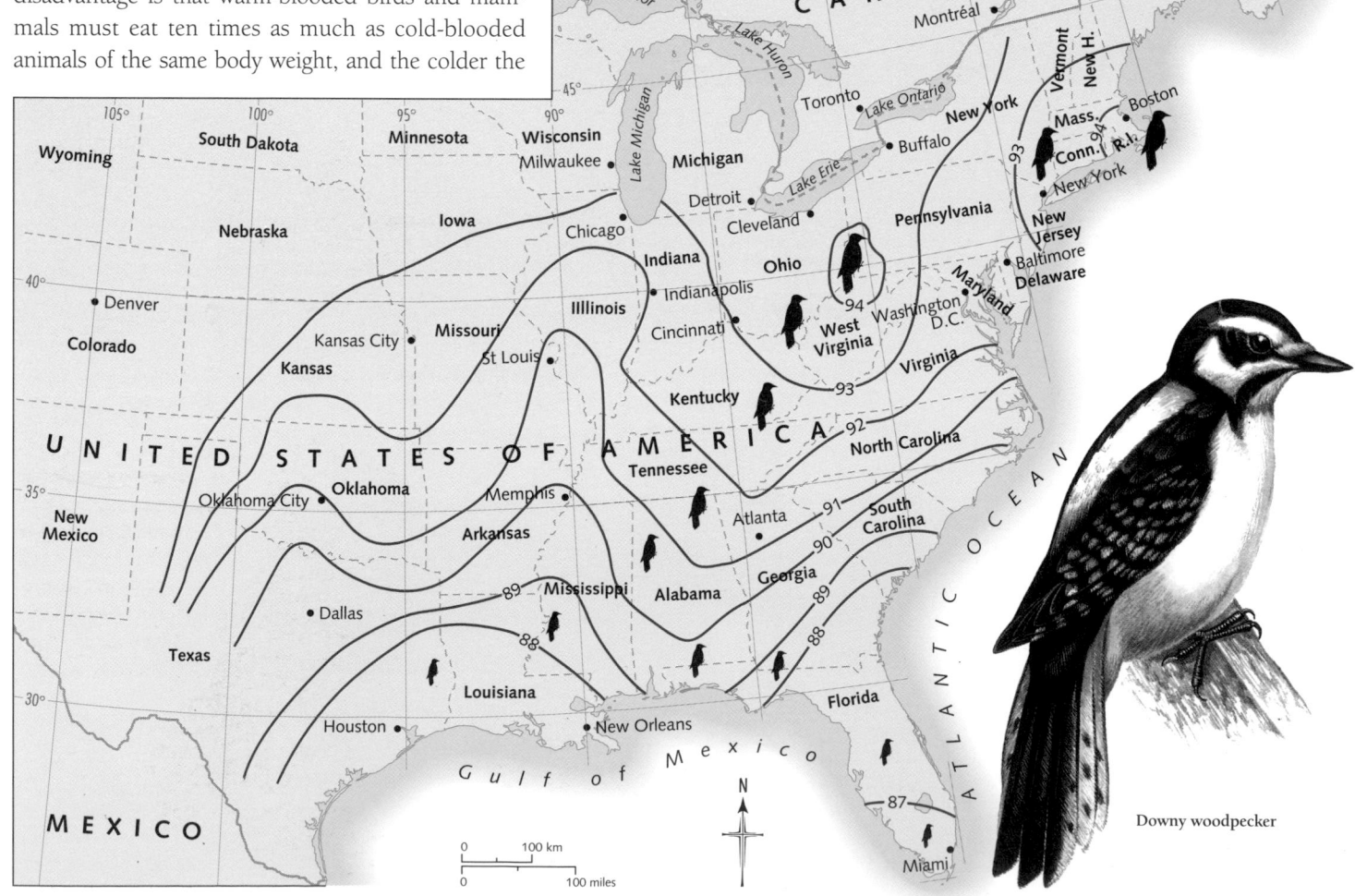

Downy woodpecker

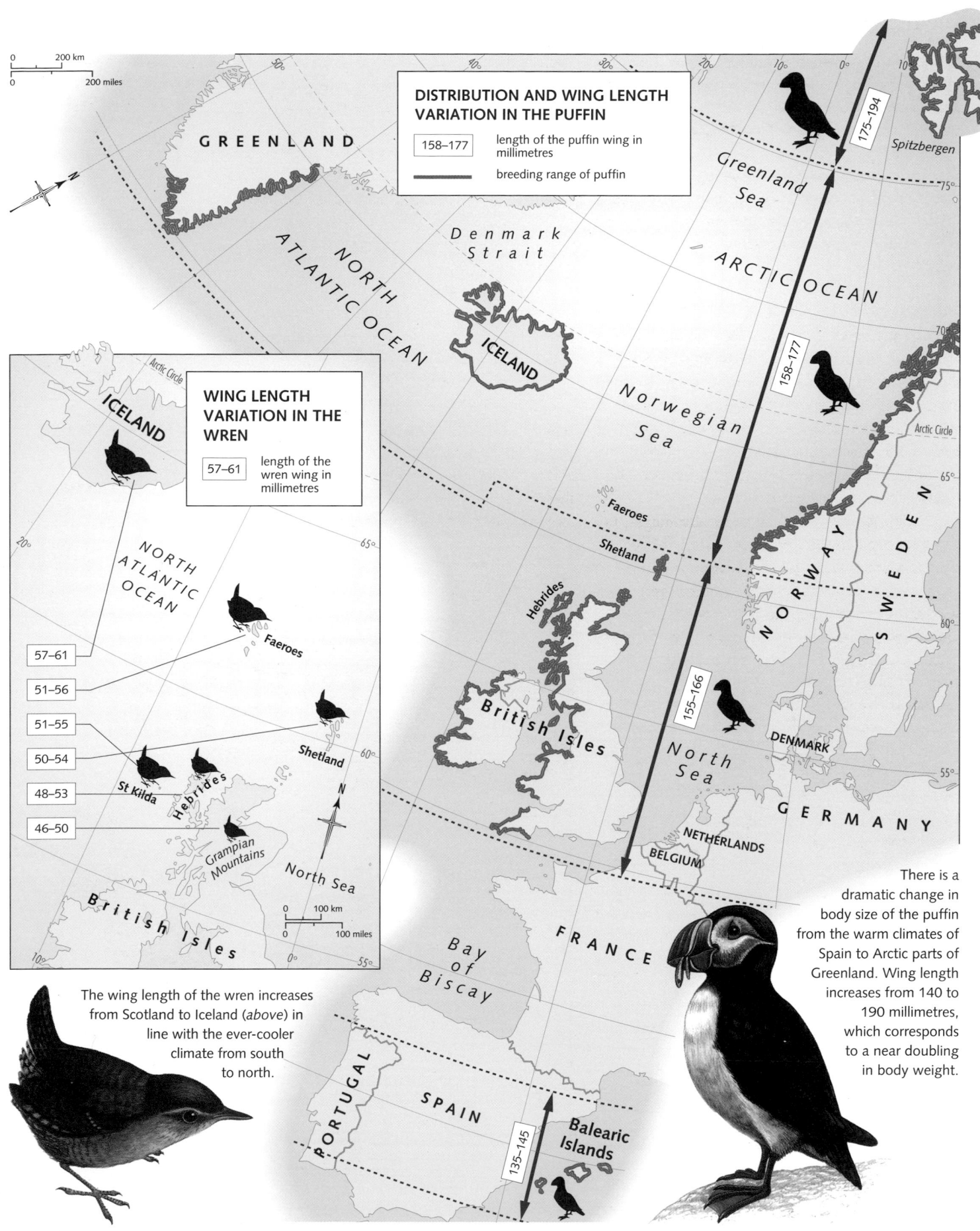

DISTRIBUTION AND WING LENGTH VARIATION IN THE PUFFIN

158–177 length of the puffin wing in millimetres

breeding range of puffin

WING LENGTH VARIATION IN THE WREN

57–61 length of the wren wing in millimetres

57–61
51–56
51–55
50–54
48–53
46–50

175–194

158–177

155–166

135–145

The wing length of the wren increases from Scotland to Iceland (*above*) in line with the ever-cooler climate from south to north.

There is a dramatic change in body size of the puffin from the warm climates of Spain to Arctic parts of Greenland. Wing length increases from 140 to 190 millimetres, which corresponds to a near doubling in body weight.

Learned Behaviour

Changes in animal habits within human environments

Human activities have caused evolution in the form and appearance of plants and animals within historical time (*see pages 38–39*). Many animals have also modified their behaviour when they have come into contact with humans. This has been particularly common in the past two or three centuries, in line with the dramatic increase in human populations and the spread of cities.

Animals have invaded the urban habitat in a variety of ways. Parasites were the first to move in and are particularly prevalent where humans live in crowded conditions. Buildings provide a special set of living places for animals. Ledges on tall buildings are like cliffs on which birds can perch and nest. Birds and bats live inside the roof spaces of buildings, mice live in wall spaces, and rats inhabit sewers and drains. Hoards of insects and small mammals find all the warmth and food they need in houses. Cockroaches, silverfish, ants, flies, beetles, moth caterpillars, mites, rats and mice may spend all their lives inside houses, and generation after generation survives in this human landscape.

Other plants and animals have adapted to life in cities. Many plants, insects, birds and mammals specialize in occupying patches of urban wasteland. City parks provide flowering plants, often exotic imports, for insects to feed on. Bees in particular profit from these plants. Suburban gardens provide succulent vegetables for insects, slugs, birds and mammals, while rivers, canals and ornamental lakes become aquatic habitats for fish, water plants, insects and water birds.

Behaviour modifications among birds in cities have been studied in detail. Blue tits in Britain are a classic example of the rapid adaptation of behaviour to new opportunities. In Britain, milk has been delivered to the doorsteps of homes in the growing suburban areas since the 1920s. Blue tits rapidly learned to exploit this new source of nourishment by perching on the edge of the milk bottle, pecking a hole in the aluminium foil cap and drinking the cream. This behaviour was noted early enough for biologists to track its spread throughout the country. Presumably the behaviour is not genetically coded but rather a result of birds in different regions learning by observation. The habit is now passed from parents to offspring throughout the country.

Garbage is the other recent innovation that has attracted new urban animals. Throughout the world, medium-sized carnivorous mammals have modified their behaviour to become urban dwellers – coyotes,

Bottle opening
Blue tits generally feed on insects. In the 1920s, they discovered an easy new food supply in the bottles of milk left on people's doorsteps throughout Britain. They pierced the foil caps and drank the rich cream. Biologists followed the spread of the behaviour, and found it had been transmitted widely within only 20 years.

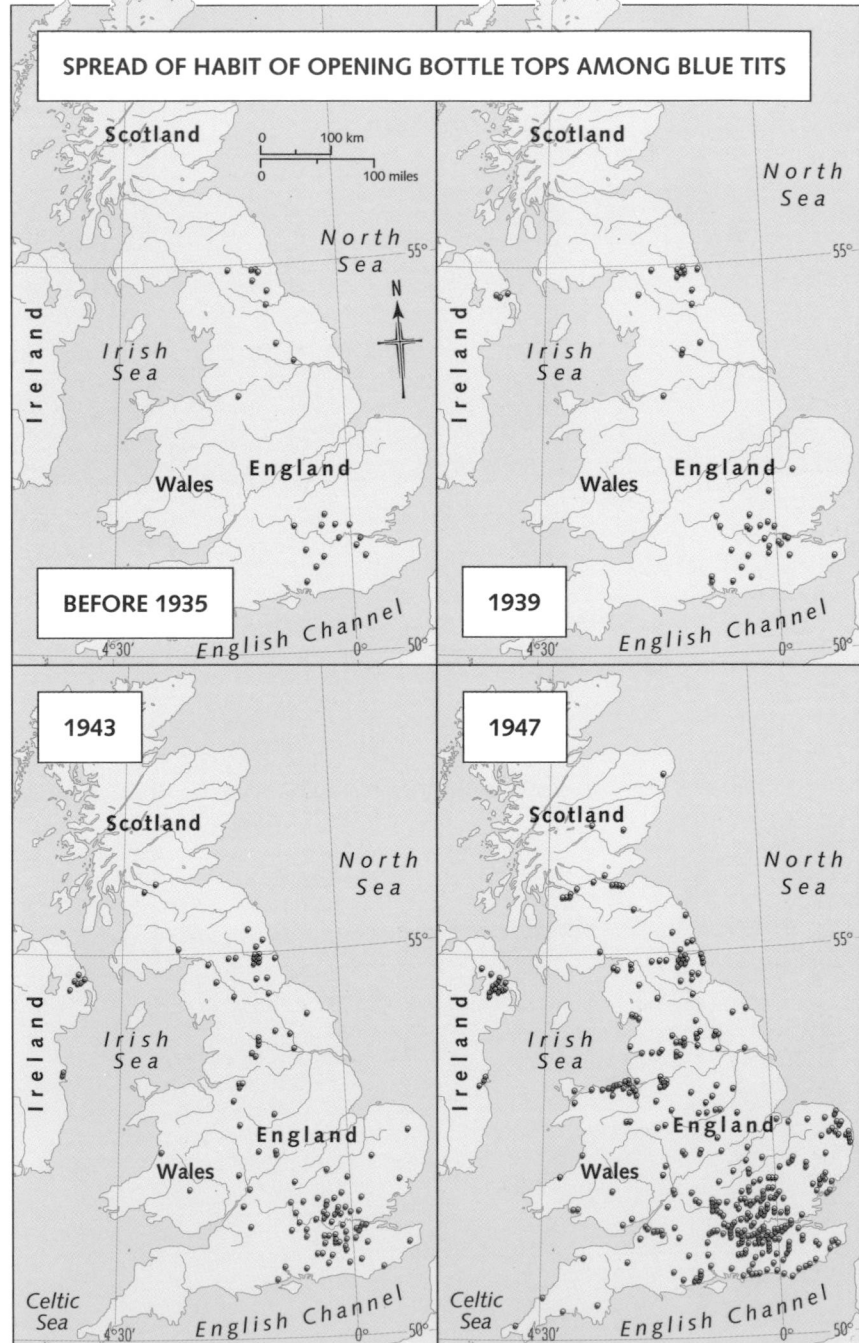

SPREAD OF HABIT OF OPENING BOTTLE TOPS AMONG BLUE TITS

BEFORE 1935

1939

1943

1947

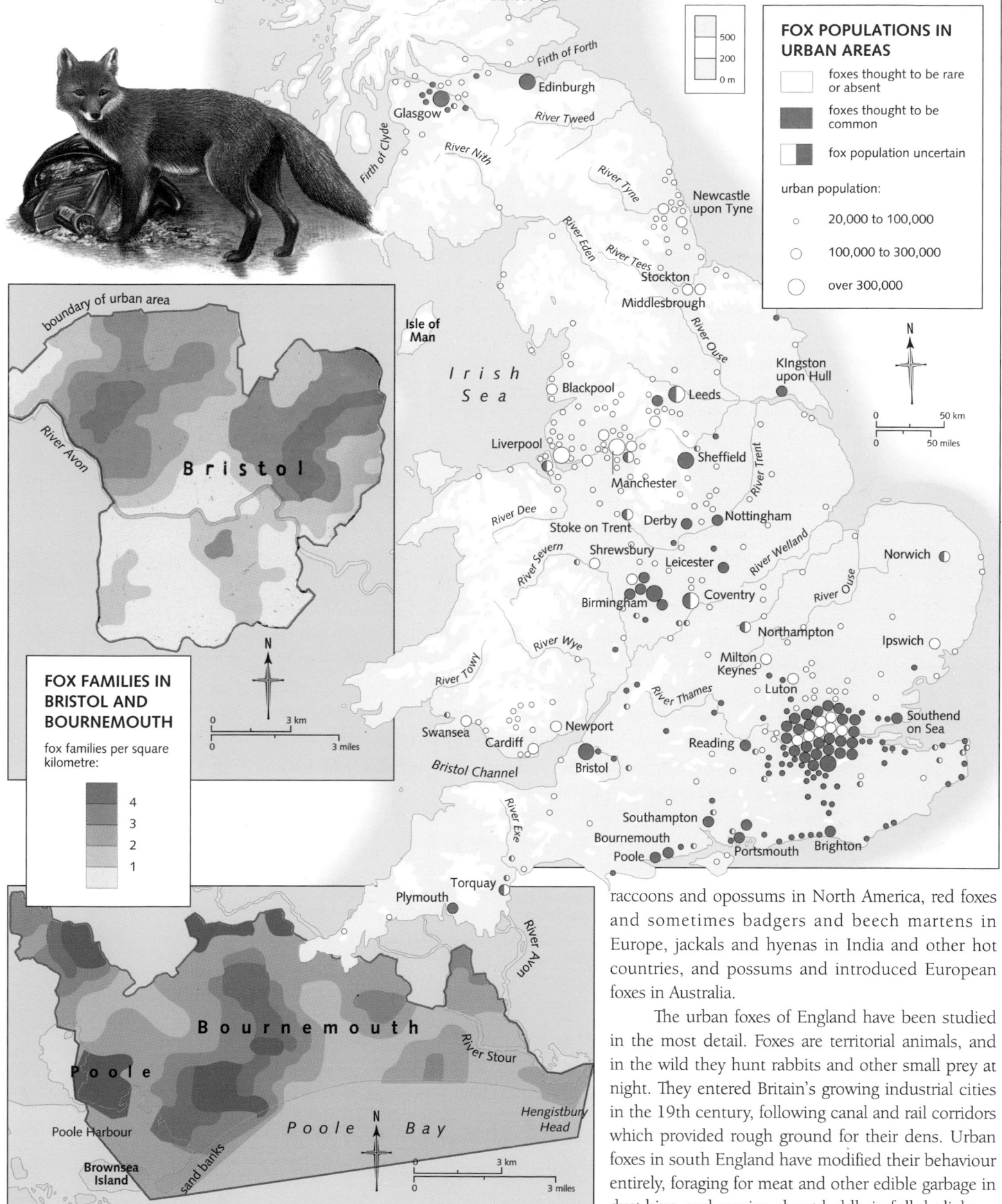

FOX POPULATIONS IN URBAN AREAS

foxes thought to be rare or absent

foxes thought to be common

fox population uncertain

urban population:

○ 20,000 to 100,000

○ 100,000 to 300,000

○ over 300,000

FOX FAMILIES IN BRISTOL AND BOURNEMOUTH

fox families per square kilometre:

4
3
2
1

boundary of urban area

raccoons and opossums in North America, red foxes and sometimes badgers and beech martens in Europe, jackals and hyenas in India and other hot countries, and possums and introduced European foxes in Australia.

The urban foxes of England have been studied in the most detail. Foxes are territorial animals, and in the wild they hunt rabbits and other small prey at night. They entered Britain's growing industrial cities in the 19th century, following canal and rail corridors which provided rough ground for their dens. Urban foxes in south England have modified their behaviour entirely, foraging for meat and other edible garbage in dust bins, and moving about boldly in full daylight.

Section III: The Origin of Species

Species originate when a pre-existing species splits. The commonest mode of speciation is geographic, in which natural barriers separate populations. Through time, and lack of contact, each half of the species evolves apart, until they are sufficiently distinct to be classified as two independent species.

Species may be defined in many ways. In ancient Greece, Plato saw that plants and animals do not form a meaningless continuum but can be divided into clear groups such as dogs, cats, horses and oak trees. He also noted that there is a great deal of variation in the basic design within each group. Plato thought that each species was focused around an ideal or perfect form and that actual cats and dogs were imperfect versions of that ideal animal. This was an essentialist view of species, in which each had been formed or created as a perfect type, or essence, that was presumably ideally designed to do its job in nature. Any changes that subsequently occurred led to a falling-off in the degree of perfection.

By Darwin's time, this essentialist view of species had been modified somewhat into a typological view. In Christian terms, each species was based around a central 'type' created by God. Again, variations around that type were regarded as imperfect mimics of the original. This view was codified in terms of immutable species and evolving varieties – the species would remain the same, but over time varieties arose, either naturally or as a result of human intervention. These varieties could look very different from one another, as do the breeds of dogs and the different populations of *Homo sapiens*. We need not explore here the antiquated views of some white European scientists as to which modern human groups were closest to, and which were furthest from, the ideal human species as conceived by God.

Darwin cut through this comfortable mythology when he realized that species are not immutable (*see pages 16–19*). He understood that varieties, subspecies and other divisions of species are no different from species except in terms of scale. Just as subspecies and varieties can diverge from one another through time, so too can species.

Populations

Darwin initiated the modern populational view of species, according to which each species consists of numerous populations, all restricted to a particular area. Generally there is occasional or frequent interbreeding, which maintains a mixing of genetic material, or a gene flow, between the populations within a species. Individual animals move from time to time and bring their genes from one population to another. Seeds may be blown further than normal, and may travel from one habitable patch to another. Darwin realized that when these corridors of gene flow were broken, speciation could occur. This, he surmised, is what had happened on the Galapagos Islands, where each island, although separated by only a short stretch of water from neighbouring islands, may be characterized by its own endemic species of finch, and variety of tortoise.

This concept of speciation was formalized 100 years later as the geographic, or allopatric, model of speciation. According to this model, the key step in speciation is the introduction of a barrier to gene flow. This may be a large geographic barrier, such as a sea or mountain range, or it may be some seemingly insignificant feature, such as a narrow stream or a patch of treeless ground. The subsets of the species isolated on either side of the barrier will evolve differently for two reasons. First, each population begins with a different selection of the gene pool (the available genetic diversity) of the parent species. Second, each population may be subject to different selective pressures. The rate of change may be rapid. This is seen where species have been moved to new habitats in the last few hundred years, as in the case of the European house sparrows in North America.

If the subpopulations come back into contact, they may be able to reestablish the gene flow. If it is reestablished only to a certain extent, a hybrid zone may arise, where two distinctive species, or subspecies, occasionally cross. An example is the zone between the carrion crow and the hooded crow in central and northern Europe.

The biological species concept

The geographic speciation concept was developed in the 1930s and 1940s by the distinguished evolutionary biologist Ernst Mayr. Hand-in-hand with that concept, he reestablished a useful field-based definition of a biological species as consisting of all those individuals that can interbreed and produce viable offspring. This is a simple definition that depends to a large extent on the realities of how organisms recognize other members of their own species. All members of a species must be potentially interfertile, and this proves that all dogs, from tiny chihuahuas to great danes, belong to one species, since they can all breed together. The 'viable offspring' caveat is essential because it excludes interspecies crosses. The horse and the ass, for instance, can breed, but their offspring, mules and hinnies, cannot.

There are numerous modern cases where speciation may be seen in progress, and where the biological species concept comes into play. The diversification of European sparrows in their new habitats and the hybrid zone between hooded and carrion crows have been mentioned. Ring species are another superb example of speciation in action. In the case of arctic gulls, two distinctive species in one part of the world may intergrade through various intermediates elsewhere. Fruit flies on the Hawaiian Islands, cichlid fish in the lakes of the East African rift valley and Australian tree creepers show geographically controlled species distributions, where it is possible to reconstruct precisely when and how the species split into their present high diversities.

House Sparrows
Evolution in historical time

Most large evolutionary changes in the history of life took place over thousands or millions of years. However, it is also possible to observe evolution in action at the present time and in the recent past. Human activities have sometimes created natural experiments, and the spread of the house sparrow is an excellent example of this.

House sparrows (*Passer domesticus*) originally occupied much of Europe, extending in a belt as far east as the Moscow basin, and south into parts of North Africa. A further portion of the range ran down to India and parts of Southeast Asia. The geographic range of sparrows has been greatly extended in the past 200 years by human intervention, and unique evidence exists about the precise dating of much of this expansion. Strikingly, there is evidence for evolutionary change even during the short time span of 100 to 200 years.

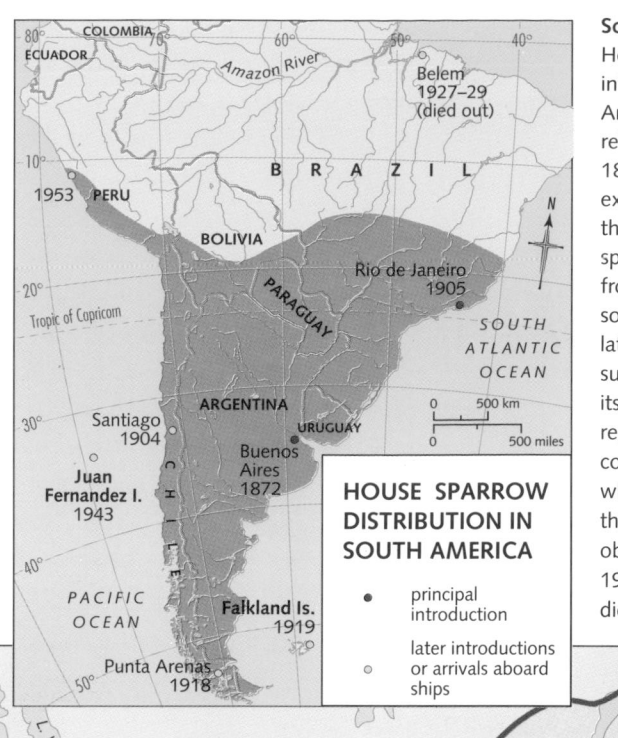

HOUSE SPARROW DISTRIBUTION IN SOUTH AMERICA

- • principal introduction
- ○ later introductions or arrivals aboard ships

South America (*left*)
House sparrows were first introduced into South America when 20 pairs were released in Buenos Aires in 1872. At first their range expansion was slow, but it then exploded, and house sparrows are now known from most of the continent south of the 15° line of latitude. The bird was not successful in establishing itself in more tropical regions, however, and a colony at the city of Belem, which lies near the mouth of the Amazon River, was observed only from 1927 to 1929, when it apparently died out.

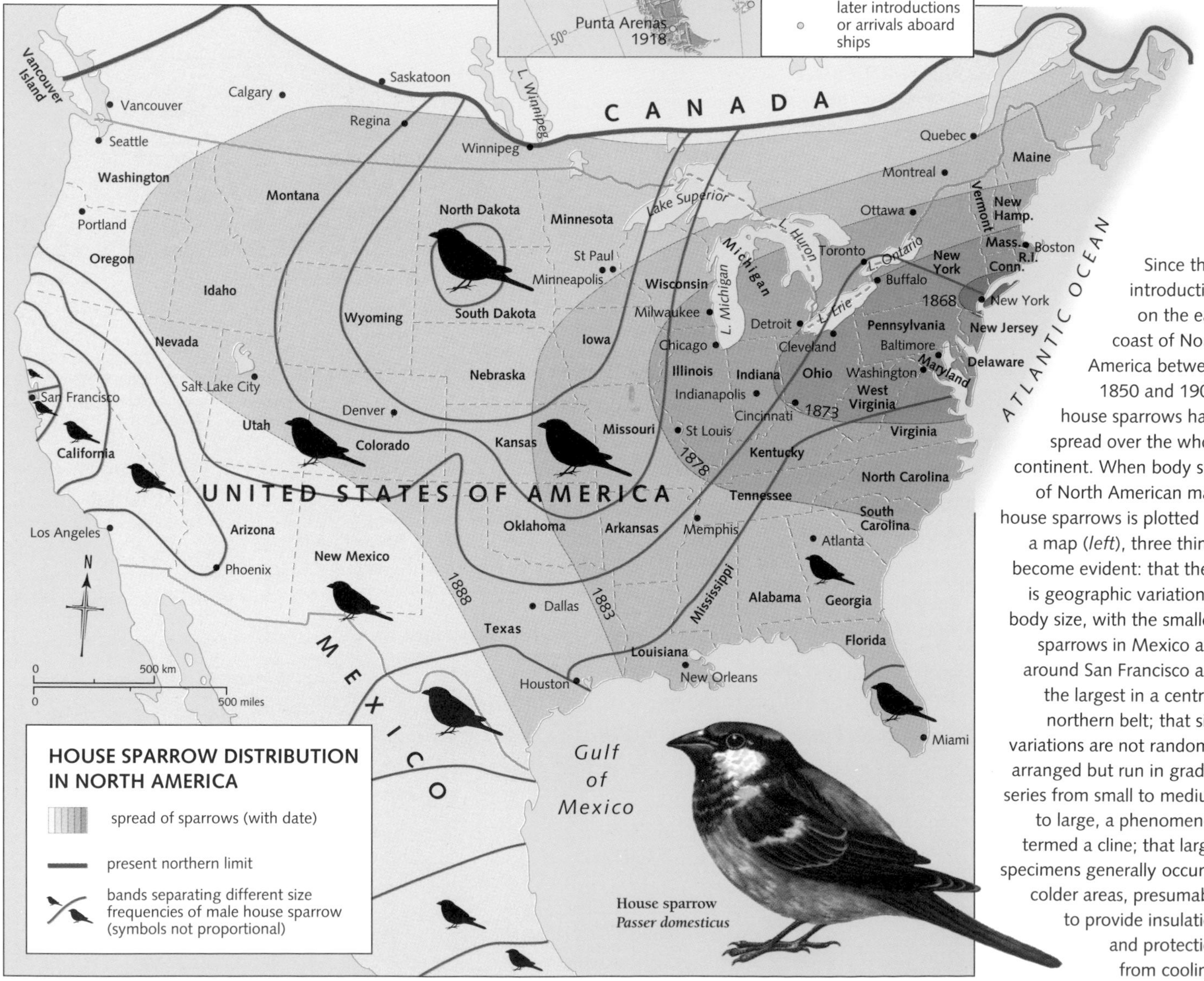

HOUSE SPARROW DISTRIBUTION IN NORTH AMERICA

- ░ spread of sparrows (with date)
- ── present northern limit
- ✕ bands separating different size frequencies of male house sparrow (symbols not proportional)

House sparrow
Passer domesticus

Since their introduction on the east coast of North America between 1850 and 1900, house sparrows have spread over the whole continent. When body size of North American male house sparrows is plotted on a map (*left*), three things become evident: that there is geographic variation in body size, with the smallest sparrows in Mexico and around San Francisco and the largest in a central-northern belt; that size variations are not randomly arranged but run in graded series from small to medium to large, a phenomenon termed a cline; that larger specimens generally occur in colder areas, presumably to provide insulation and protection from cooling.

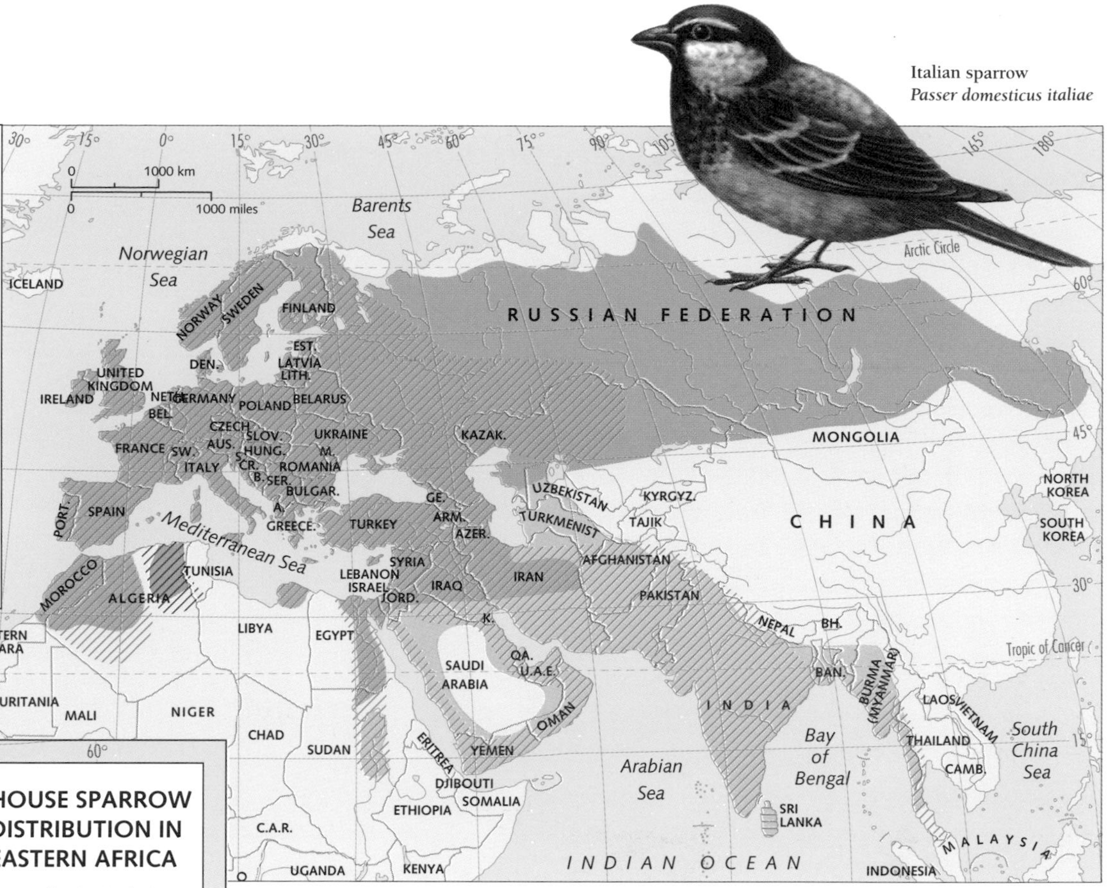

Italian sparrow
Passer domesticus italiae

HOUSE SPARROW DISTRIBUTION IN PALEARCTIC AND ORIENTAL REGIONS

present-day:

- *domesticus* subspecies
- *indicus* subspecies
- *domesticus x hispaniolensis* hybrid

early 19th-century:

- house sparrow
- *domesticus x hispaniolensis* hybrid

Right: *Passer domesticus* falls into two subspecies – *domesticus* and *indicus*.

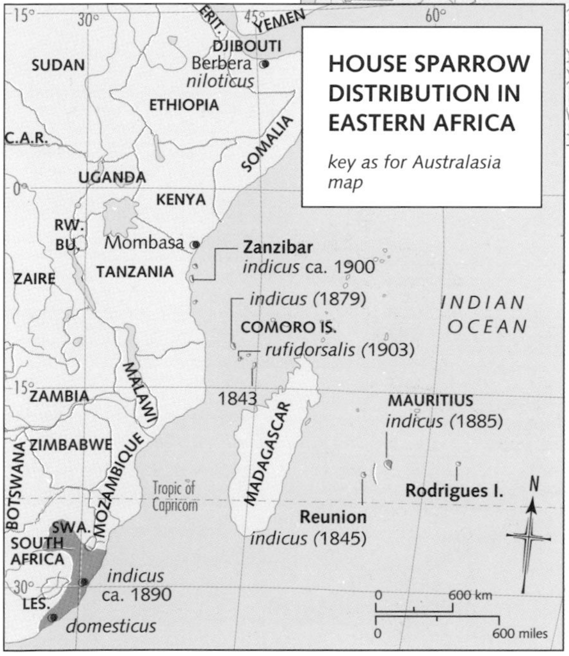

HOUSE SPARROW DISTRIBUTION IN EASTERN AFRICA

key as for Australasia map

House sparrows extended their range enormously in the 19th century, over much of Russia (*above*). They were introduced on many separate occasions to islands in the Indian Ocean and to mainland Africa (*left*), as well as to Australia, New Zealand and many of the offshore islands (*below*).

Further range extensions were entirely artificial. As British colonists occupied far-flung lands, they deliberately introduced house sparrows, for sentimental reasons or in attempts to control insect pests. House sparrows were brought from Europe to North America between 1852 and 1900. Sparrows were introduced in South America from 1872 to 1953, in South Africa in about 1890, in Australia from 1861 to 1871, and in New Zealand from 1866 to 1948.

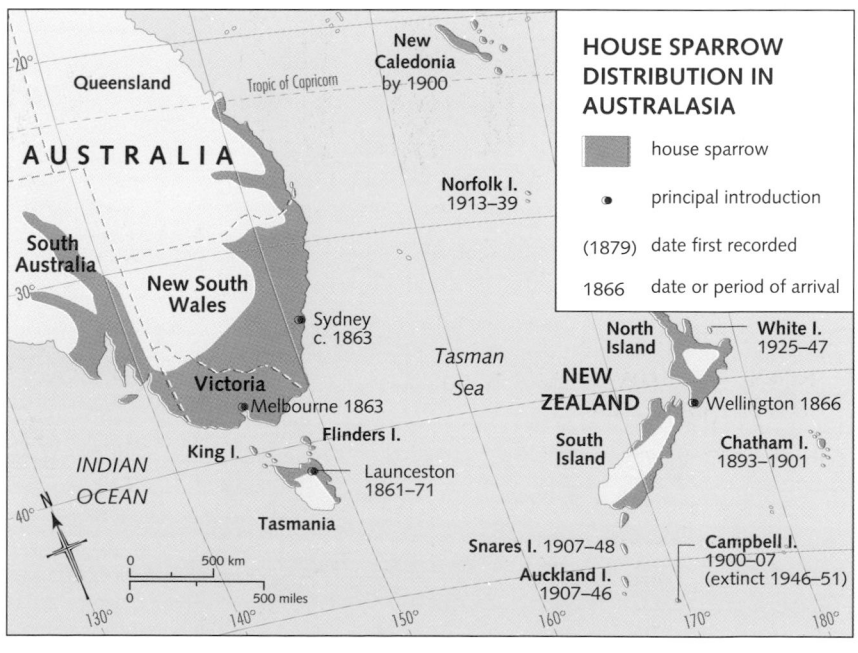

HOUSE SPARROW DISTRIBUTION IN AUSTRALASIA

- house sparrow
- principal introduction
- (1879) date first recorded
- 1866 date or period of arrival

There are nine or ten species of sparrow, with distributions covering parts of Africa, Europe and Asia. *Passer domesticus*, which is closely related to *Passer hispaniolensis*, the Spanish sparrow, falls into two subspecies: *domesticus*, which is restricted to Europe and Africa, and *indicus*, which is found in parts of the Middle East, Arabia, the Indian subcontinent and parts of Southeast Asia. *Passer domesticus* seems to have had a tendency to expand its range. Particularly dramatic has been the enormous eastwards expansion of *Passer domesticus* over Asiatic Russia to the Pacific coast – a distance of some 3,000 kilometres, achieved in only 100 years, largely as steppe lands came under the plough.

The European Crows
Evidence for the splitting of species

One of the keys to an understanding of evolution, as Charles Darwin realized, is the origin of species. The separation of two species from a single existing species must have happened millions of times in the past, yet the likelihood of seeing such an event in progress is small. Darwin argued that the processes of speciation probably took their course from the division of a single species into distinctive races or subspecies, living in different parts of the geographic range. Over time, some of the races might become so distinct that they would cease to interbreed, and they would hence be termed independent species. At the time of transition from subspecies to species, the incipient species might still interbreed to some extent along a hybrid zone – a narrow band of territory where the two populations meet. An excellent example of such a hybrid zone is seen in the distribution of the European crows.

There are two closely related species of crow in Europe: the hooded crow (*Corvus corone*) and the carrion crow (*Corvus cornix*). The former is found in western Europe, while the latter occurs in eastern and northern Europe and much of Asia. The two species meet along a broad hybrid zone, up to 10 kilometres wide, in which they interbreed and produce hybrid young. This zone can be interpreted in two ways: it may represent the beginnings of full isolation between two populations, and in the future all hybridization may cease as the two forms evolve full genetic isolation and become fully independent species; or it may represent a secondary contact zone, where the two independent species had evolved in geographic isolation, and later came together and interbred to a limited extent.

Close relatives of the crows are the jackdaws. Again, there are two species: the jackdaw, *Corvus monedula*, and the Daurian jackdaw, *Corvus dauuricus*. Their ranges meet, but there is no evidence of a hybrid zone. The jackdaw's breeding range extends over most of Europe south of Scandinavia, parts of North Africa, and eastwards over the Middle East and Central Asia. Its feeding range extends further east into Asia and northwards into Russia. The Daurian jackdaw's resident and breeding range, in China, eastern Russia and parts of Southeast Asia, is quite distinct, although the extended non-breeding ranges of the species do meet. Perhaps the two species of jackdaw were fully established before the European crow split into two species, and in the future the hybrid zone between the two crow species will perhaps disappear as the species evolve apart.

CROW DISTRIBUTION
- hooded crow
- carrion crow
- carrion crow additional winter range
- hybrid zone

Carrion crow
Corvus cornix

Hooded crow
Corvus corone

Jackdaws (*right*)
There are two species: the jackdaw (*Corvus monedula*), found from Europe to Central Asia, and the Daurian jackdaw (*Corvus dauuricus*), typical of China and eastern Asia. The breeding ranges of the two species are separate, and there is no hybrid zone, although their extended feeding ranges overlap.

Daurian jackdaw
Corvus dauuricus

The carrion crow, *Corvus cornix* (*top*), is black all over, while the hooded crow, *Corvus corone* (*bottom*), is grey with a black head, wings and tail. The Daurian jackdaw, *Corvus dauuricus* (*right*), is mainly black but is greyish on the back and sides of the head.

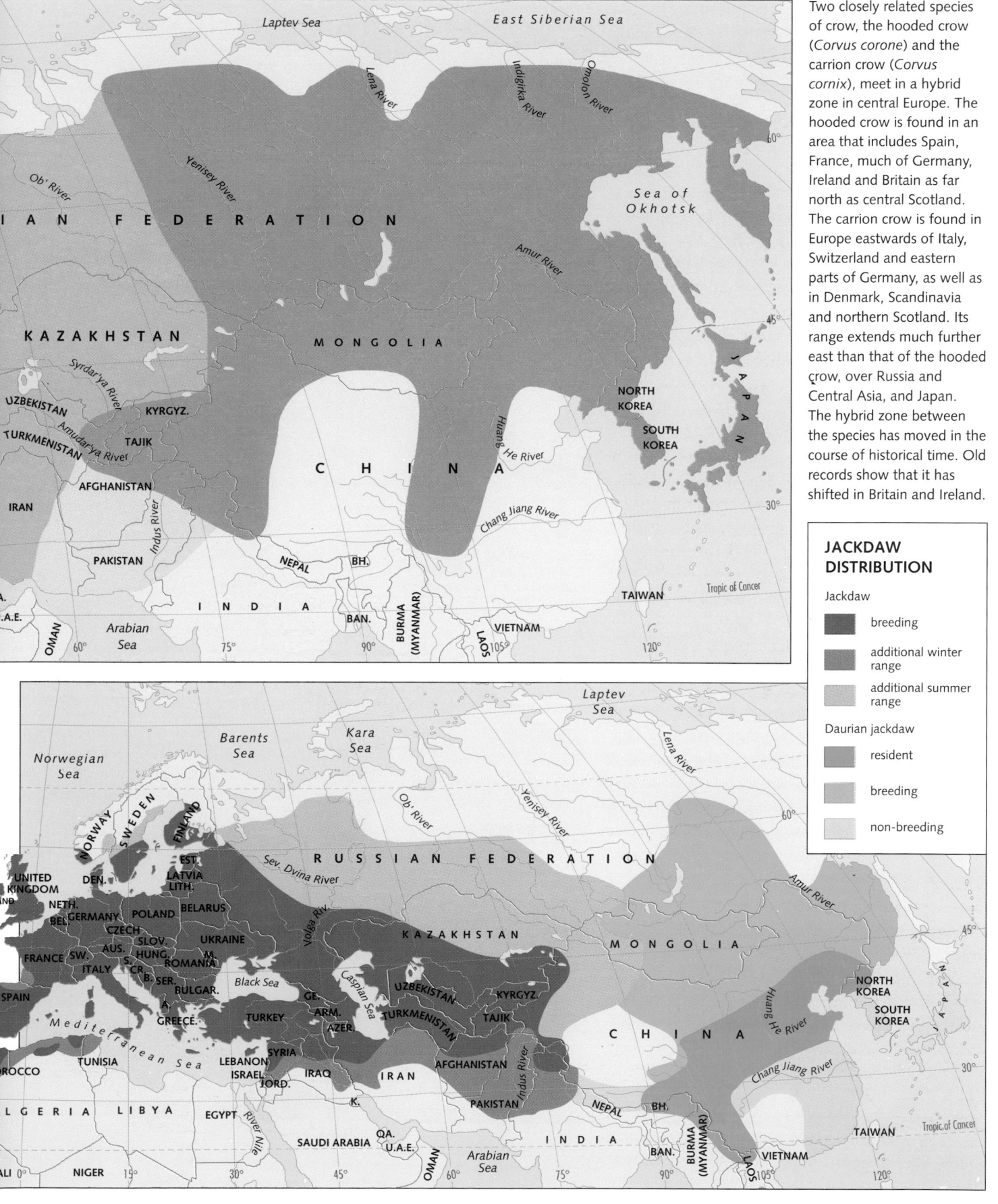

Two closely related species of crow, the hooded crow (*Corvus corone*) and the carrion crow (*Corvus cornix*), meet in a hybrid zone in central Europe. The hooded crow is found in an area that includes Spain, France, much of Germany, Ireland and Britain as far north as central Scotland. The carrion crow is found in Europe eastwards of Italy, Switzerland and eastern parts of Germany, as well as in Denmark, Scandinavia and northern Scotland. Its range extends much further east than that of the hooded crow, over Russia and Central Asia, and Japan. The hybrid zone between the species has moved in the course of historical time. Old records show that it has shifted in Britain and Ireland.

JACKDAW DISTRIBUTION

Jackdaw

- breeding
- additional winter range
- additional summer range

Daurian jackdaw

- resident
- breeding
- non-breeding

Arctic Gulls
A ring species as evidence for recent speciation

Problems in identifying some species of sea gulls have led biologists to recognise an important example of speciation in progress over a time-scale of about 10,000 years. Ornithologists easily distinguished between the herring gull (*Larus argentatus argentatus*) and the lesser black-backed gull (*Larus fuscus fuscus*) in western Europe, but problems arose when they extended their studies over Russia and North America. It seemed that the lesser black-backed gull became more and more like the herring gull the further east one went, while the herring gull became increasingly like the lesser black-backed gull as one traversed North America and eastern Siberia.

In northwestern Europe, the herring gull and the lesser black-backed gull are distinct species that do not interbreed. The Siberian lesser black-backed gull (*Larus fuscus antelius*) in northern Siberia differs only subtly from the lesser black-backed gull. However, the next subspecies, Heuglin's gull, found over much of central Russia, has been classified with both the lesser black-backed and the herring gull. It is currently generally classified as a herring gull, *Larus argentatus heuglini*. In North America only herring gulls are found, all assigned to *Larus argentatus smithsonianus*, a subspecies that differs only a little from the European form. This subspecies grades imperceptibly into the easternmost Siberian form, the Vega gull (*Larus argentatus vegae*). The next subspecies west, Birula's gull (*Larus fuscus birulae*), can also interbreed with its neighbours. The final link is Heuglin's gull, noted above as intermediate between herring gulls and lesser black-backed gulls.

This sequence of gradual changes between subspecies is a classic ring species, where subspecies blend insensibly along a path of distribution but where the end points are so distinct from each other that they cannot interbreed, and are hence independent species. The ring here was produced by changes in gull distribution caused by ice advances during the last great ice age. Gulls in northern Europe and Asia seem to have been forced southwards into central Asian and Mediterranean refuges. When the ice sheets withdrew northwards, the Mediterranean populations moved into northern Europe as the lesser black-backed gull. Central Asian gulls moved into northern Siberia as Birula's gull, and gulls from eastern Siberia extended their range across Alaska and Canada, then over the Atlantic to western Europe, as the herring gulls. Over the past 10,000 years, these populations have become more distinct, and intermediates have evolved to fill the gaps in the range.

The gull species
In northwestern Europe, the herring gull (*Larus argentatus argentatus*) is large, white and grey in colour, and has pink legs. The lesser black-backed gull (*Larus fuscus fuscus*) is smaller, has darker feathers in the wings and over the back, and has yellow legs. Five subspecies form a complete ring between these, with lesser black-backed gulls heading east and herring gulls heading west.

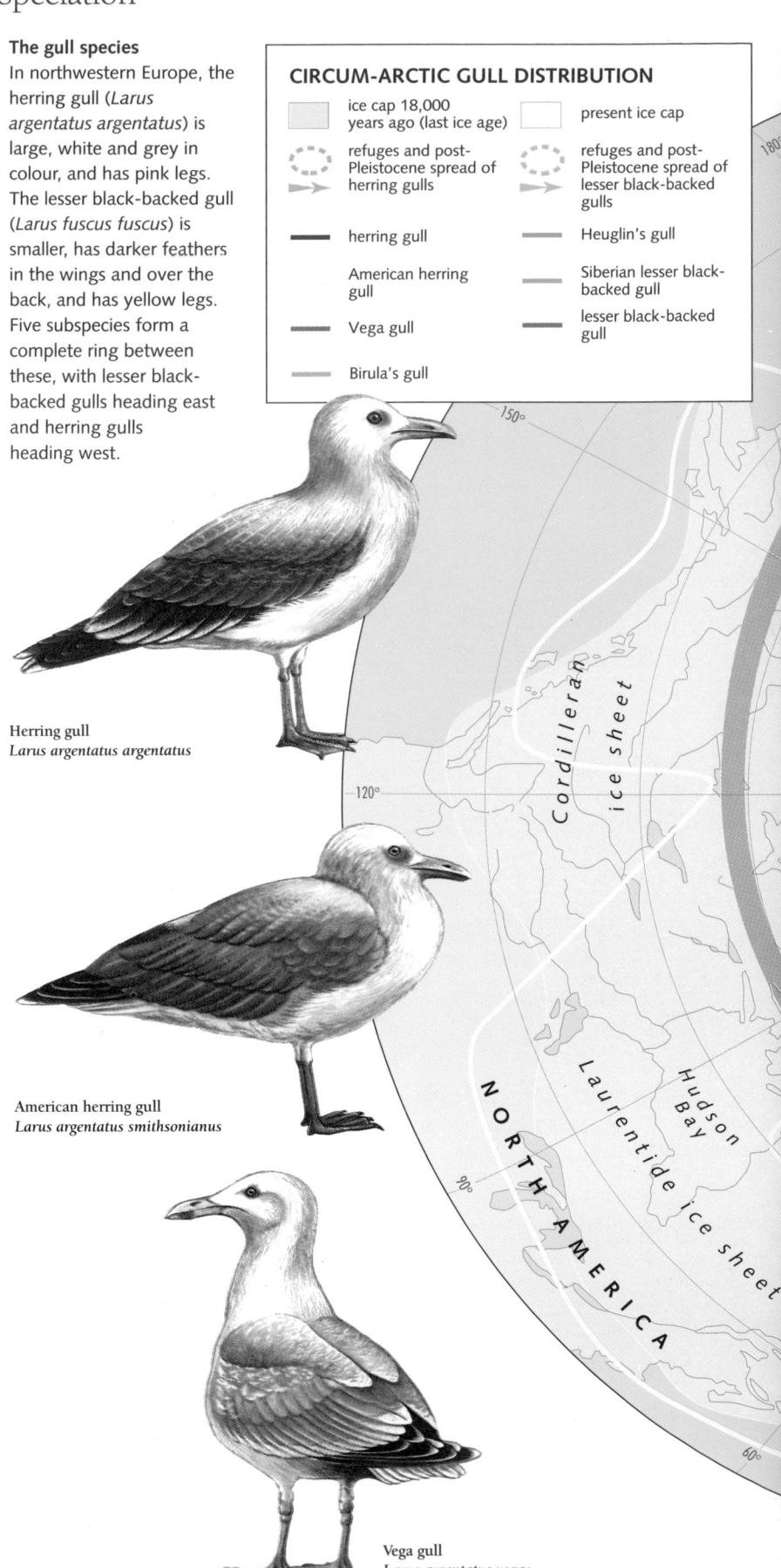

CIRCUM-ARCTIC GULL DISTRIBUTION

	ice cap 18,000 years ago (last ice age)		present ice cap
	refuges and post-Pleistocene spread of herring gulls		refuges and post-Pleistocene spread of lesser black-backed gulls
	herring gull		Heuglin's gull
	American herring gull		Siberian lesser black-backed gull
	Vega gull		lesser black-backed gull
	Birula's gull		

Herring gull
Larus argentatus argentatus

American herring gull
Larus argentatus smithsonianus

Vega gull
Larus argentatus vegae

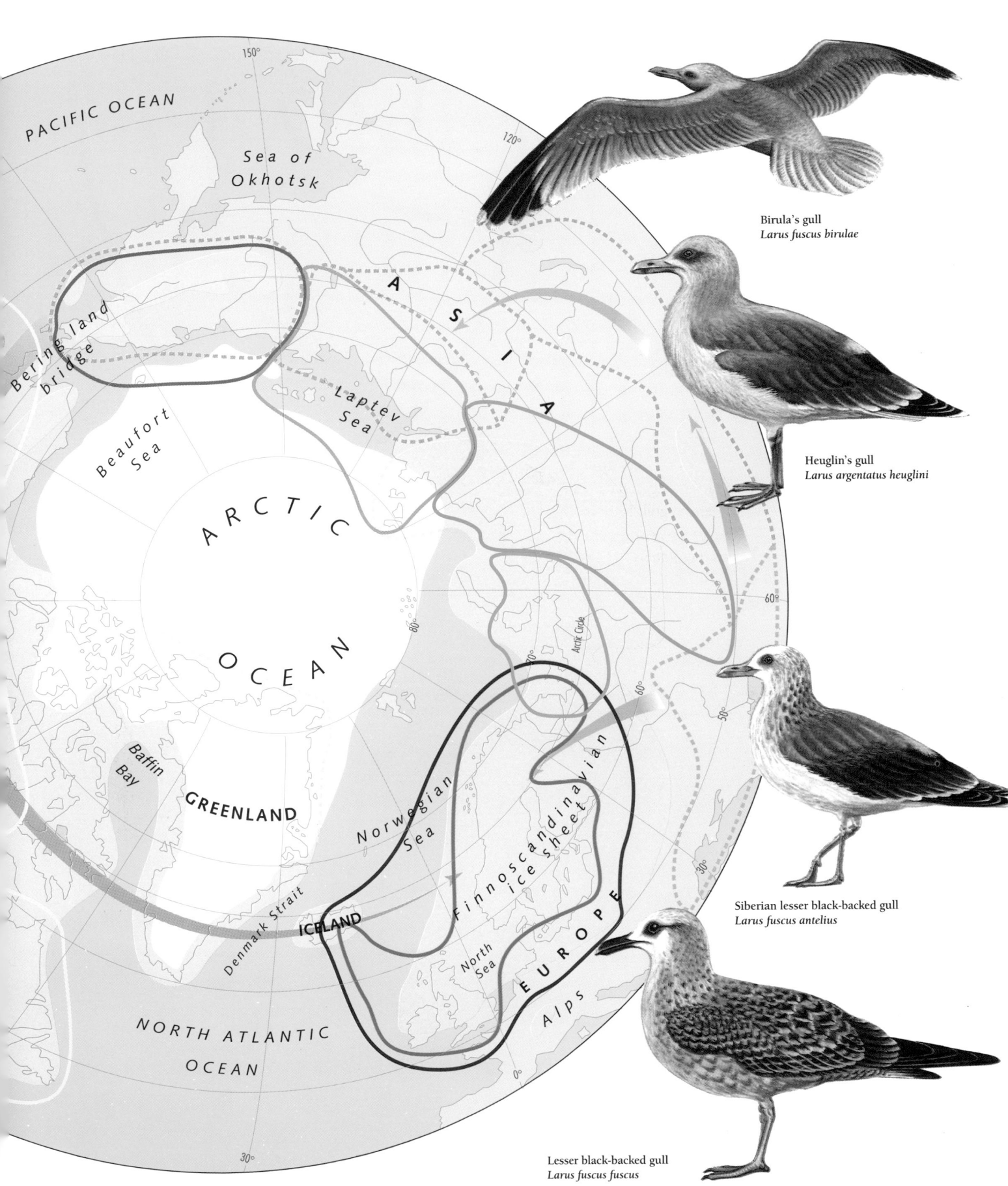

PACIFIC OCEAN

Sea of Okhotsk

Bering land bridge

Beaufort Sea

Laptev Sea

ASIA

ARCTIC OCEAN

Baffin Bay

GREENLAND

Norwegian Sea

Finnoscandinavian ice sheet

Denmark Strait

ICELAND

North Sea

EUROPE

Alps

NORTH ATLANTIC OCEAN

Arctic Circle

Birula's gull
Larus fuscus birulae

Heuglin's gull
Larus argentatus heuglini

Siberian lesser black-backed gull
Larus fuscus antelius

Lesser black-backed gull
Larus fuscus fuscus

Fruit Flies in Hawaii
Speciation by geographic isolation

Perhaps the commonest cause of speciation has been geographic isolation. According to this theory, a split in the distribution range of a species causes the populations on either side of the barrier to evolve away from each other. The differences may be minor at first, equivalent to the level of varieties or subspecies. In time, however, the populations will become so different that they can no longer interbreed and are therefore two distinct species. Hawaii, with its fruit fly populations, provides a superb natural laboratory in which these processes can be reconstructed.

Fruit flies are a diverse group consisting of about 1,500 species of the genus *Drosophila*, 350 of which are restricted to the Hawaiian Islands. They range from small forms to large ones with wingspans of 22 millimetres, and show great variation in wing shapes and patterns, leg shapes and eye patterns. Each of the major Hawaiian islands has its own set of unique species, and it seems clear that the 350 species diversified during the history of the islands.

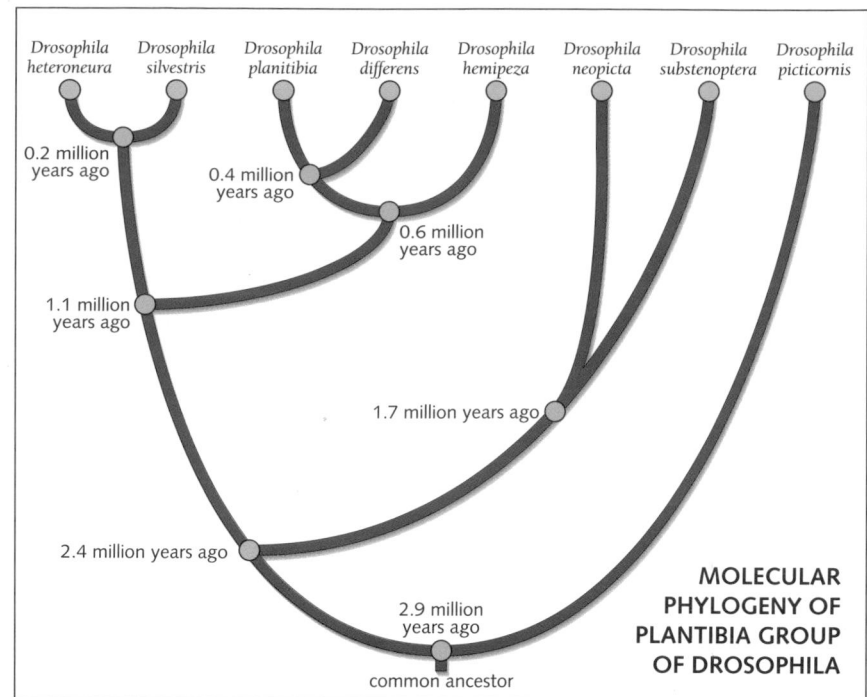

MOLECULAR PHYLOGENY OF PLANTIBIA GROUP OF DROSOPHILA

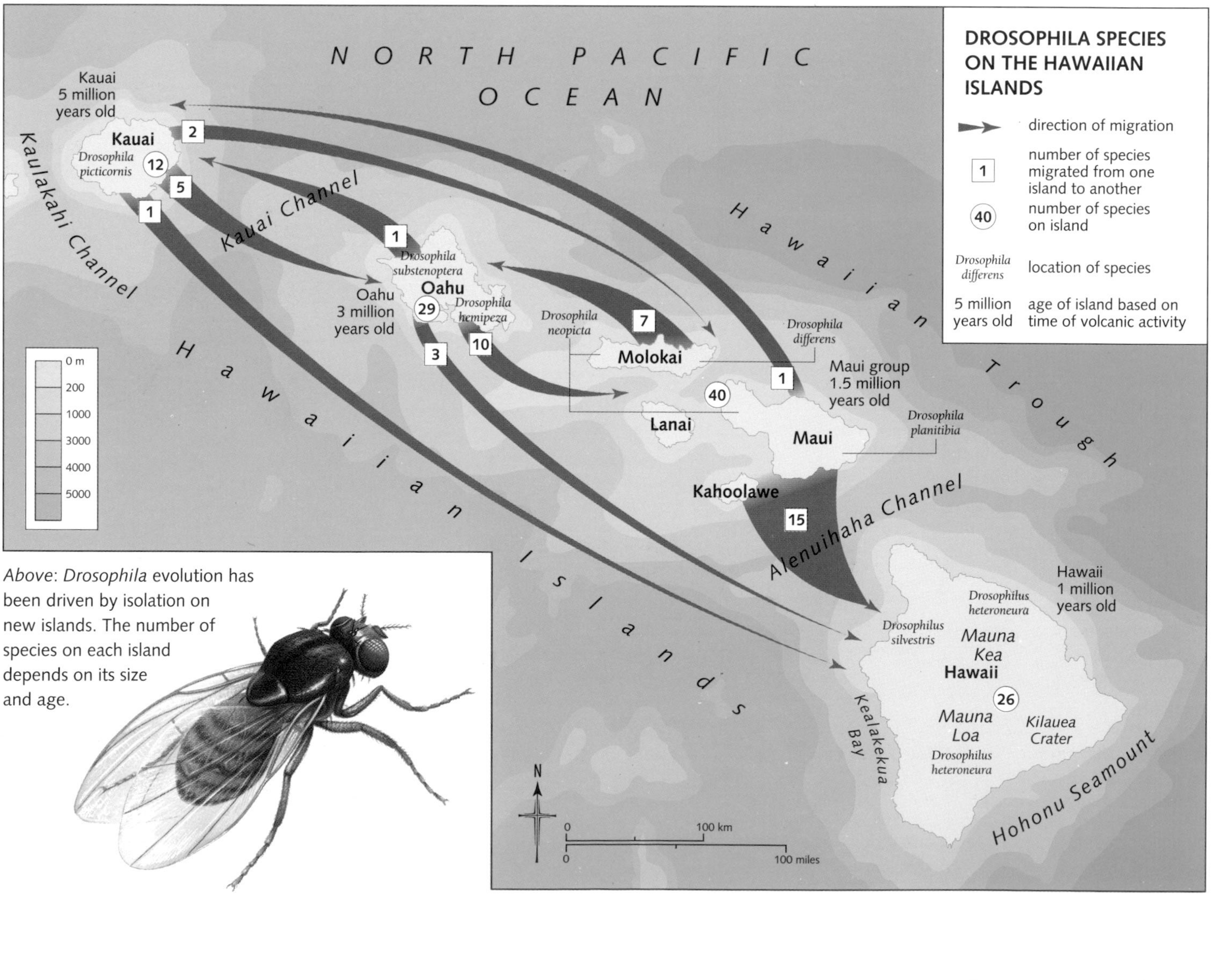

Drosophila heteroneura · *Drosophila silvestris* · *Drosophila planitibia* · *Drosophila differens* · *Drosophila hemipeza* · *Drosophila neopicta* · *Drosophila substenoptera* · *Drosophila picticornis*

0.2 million years ago
0.4 million years ago
0.6 million years ago
1.1 million years ago
1.7 million years ago
2.4 million years ago
2.9 million years ago
common ancestor

DROSOPHILA SPECIES ON THE HAWAIIAN ISLANDS

➤ direction of migration

[1] number of species migrated from one island to another

(40) number of species on island

Drosophila differens location of species

5 million years old age of island based on time of volcanic activity

NORTH PACIFIC OCEAN

Kauai 5 million years old

Kauai
Drosophila picticornis (12)

[2]
[5]
[1]

Kaulakahi Channel

Kauai Channel

Hawaiian Islands

[1]
Drosophila substenoptera
Oahu (29)
Oahu 3 million years old
Drosophila hemipeza
[3]
[10]

Drosophila neopicta

[7]

Molokai

Drosophila differens

(40)

Lanai

Maui group 1.5 million years old

Drosophila planitibia

Maui

Kahoolawe

[15]

Alenuihaha Channel

Hawaiian Trough

Hawaii 1 million years old

Drosophilus heteroneura

Mauna Kea
Hawaii
(26)
Mauna Loa
Kilauea Crater
Drosophilus heteroneura

Drosophilus silvestris

Kealakekua Bay

Hohonu Seamount

0 m
200
1000
3000
4000
5000

N

0   100 km

0   100 miles

Above: Drosophila evolution has been driven by isolation on new islands. The number of species on each island depends on its size and age.

The Hawaiian Islands formed over the last 6 million years as volcanoes rising from the Pacific floor. The origin of each island and hence the maximum date of arrival of *Drosophila* to each can be dated. The splitting of species occurred when barriers of various kinds prevented populations from interbreeding. These are classic examples of geographic (allopatric) speciation, in which peripheral isolates diverge rapidly and form separate species.

The isolating geographical barrier was usually the ocean. Stray individuals wandered onto new islands as they became habitable and evolved in isolation from the parent stocks. In other cases, lava flows split populations. This has been observed on the main island, Hawaii, in the last 150 years.

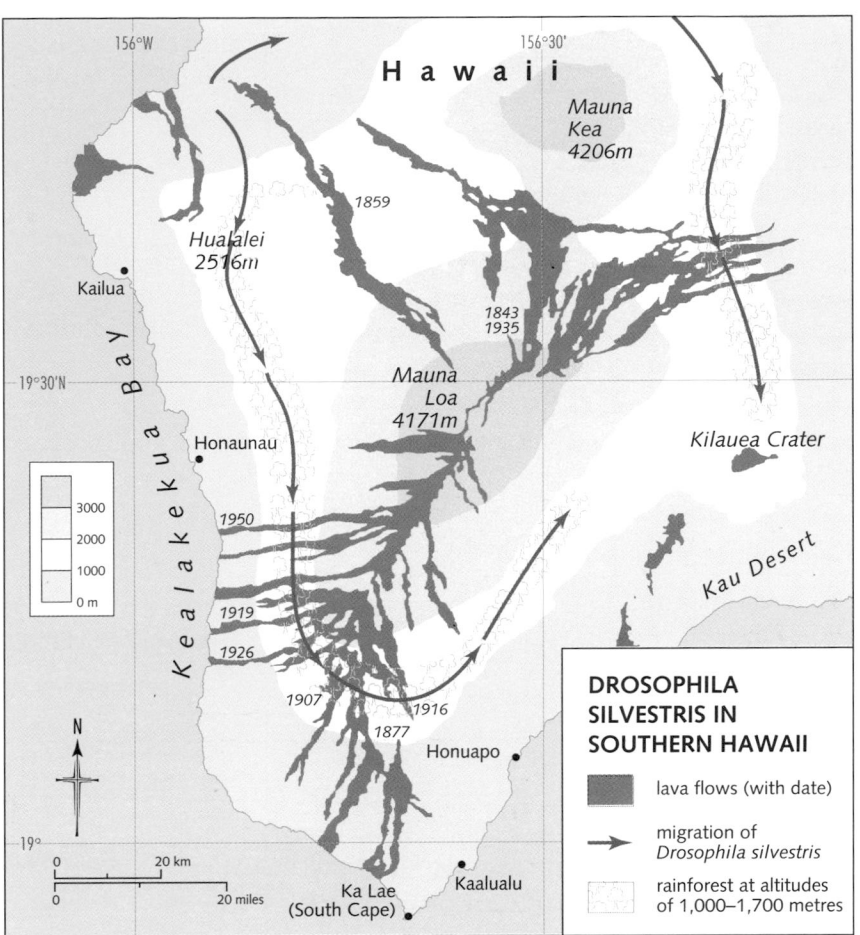

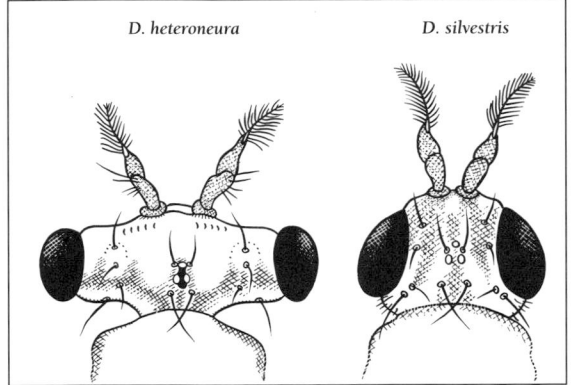

DROSOPHILA
SILVESTRIS IN
SOUTHERN HAWAII

- lava flows (with date)
- migration of *Drosophila silvestris*
- rainforest at altitudes of 1,000–1,700 metres

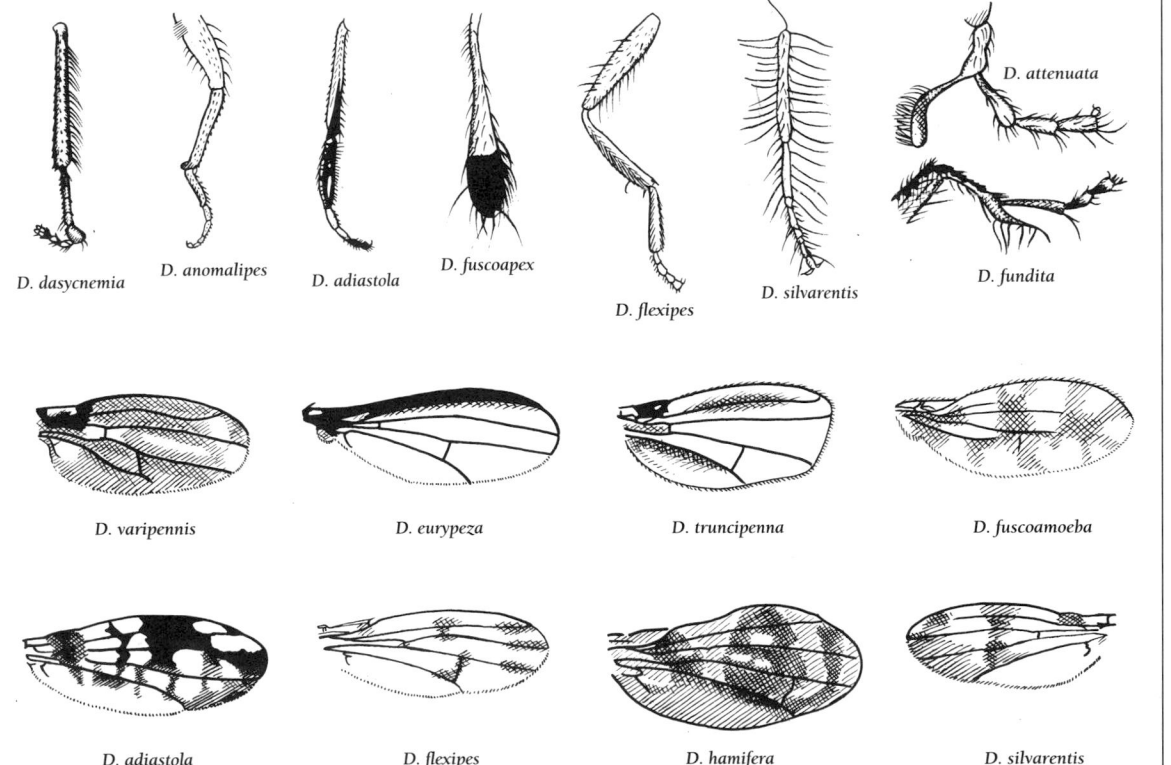

Species of the fruit fly *Drosophila* show enormous variation in head and eye shape (*above*), and in leg and wing shape (*right*). Wing patterns are particularly distinctive (*right*). Much of the rapid speciation of *Drosophila* on the Hawaiian Islands has been geographically controlled. New volcanic islands provide testing grounds for evolution. Also, habitats are fragmented by lava flows and topographic restrictions (*top right*), and species ranges may be split. Here, the range of *Drosophila silvestris* may divide into three or four as a result of recent lava flows. This will result in several new species.

D. dasycnemia   D. anomalipes   D. adiastola   D. fuscoapex   D. flexipes   D. silvarentis   D. attenuata   D. fundita

D. varipennis   D. eurypeza   D. truncipenna   D. fuscoamoeba

D. adiastola   D. flexipes   D. hamifera   D. silvarentis

Cichlid Fish in Lake Victoria

Rapid evolution in a species flock

One of the most spectacular examples of rapid evolution is the cichlid fish of Lake Victoria in Tanzania and the neighbouring African Great Lakes. Here, 700 species have diversified in only 20 million years. Individual lakes are packed with unique species. Lake Victoria, for example, has more than 170 species, of which all but six are unique. The most diverse genus of cichlid in the lakes, *Haplochromis*, may contain as many as 200 species, although recent taxonomic revisions have led to the division of *Haplochromis* into several genera.

The African Great Lakes began to form 20 million years ago. Some fossil cichlids from Rusinga Island, dated at 18 million years old, may be forms of *Haplochromis*, but fossils are rare and offer little guide to evolution. Much of the speciation seems a lot more recent – probably Pleistocene and hence restricted to the last 2 million years. Within the large lakes, many species have very localized ranges and highly specialized habitat and feeding preferences. Perhaps a combination of major physical changes induced isolation and rapid speciation in small pockets around the lake margins. Lake levels rose and fell several times, isolating patches of water, and certain habitats, such as rocky shores, are patchily distributed. The cichlid species flocks appear to show extreme cases of localized speciation by isolation both geographically and in terms of diet and habitat.

Recent studies of the DNA of cichlid fish show how rapid their evolution has been. It has not been resolved whether all cichlids in each of the lakes arose from a single ancestral form or from several. In Lake Victoria, for example, there may have been three ancestral stocks, one in each of the major river systems underlying the present lake. In any case, molecular and geological evidence suggests that the *Haplochromis* species flock of Lake Victoria arose in the very recent geological past, although the range of dates is still wide, from 750,000 to 14,000 years ago.

Present-day species distributions on the fine scale are, however, complex. It is not possible to show the isolated distributions of each species because many species arose when the lakes were shallower and more subdivided into narrow prongs. Populations became isolated in the lake branches and evolved into new species. Mixing occurred when the lakes filled up and the isolated tributaries joined together. The species distinctions are very clear in terms of overall body shape and head shape, and this variation has been produced by the wide range of diets and lifestyles of modern cichlid fish.

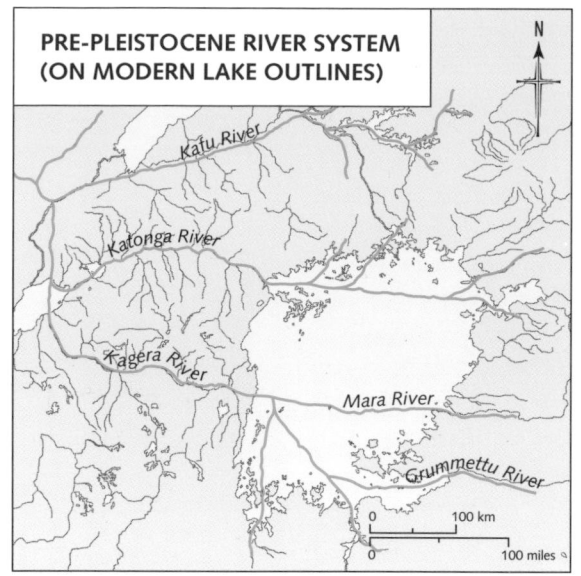

PRE-PLEISTOCENE RIVER SYSTEM (ON MODERN LAKE OUTLINES)

In Pleistocene times there was no lake, merely a system of major rivers draining to the west (*left*). These have been reconstructed from studies of modern topography and river patterns. Uplift west of the lake caused westward-flowing rivers to drain to the east. The Katonga and Kagera rivers reversed their flow, and the basin where the lake now lies began to fill. Cichlid fish evolution was most intense early in the filling of the lake, when there were many isolated lake branches (*below*).

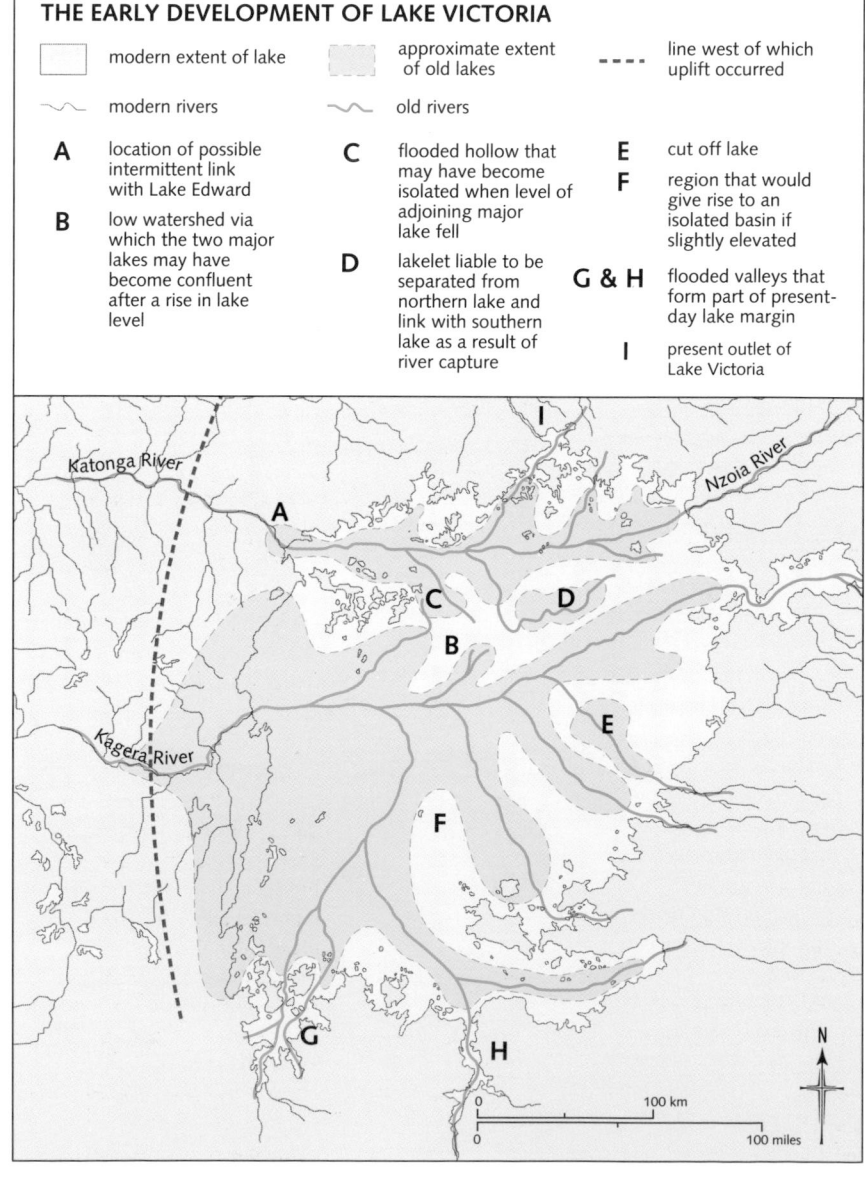

THE EARLY DEVELOPMENT OF LAKE VICTORIA

- modern extent of lake
- modern rivers
- approximate extent of old lakes
- old rivers
- line west of which uplift occurred

A location of possible intermittent link with Lake Edward

B low watershed via which the two major lakes may have become confluent after a rise in lake level

C flooded hollow that may have become isolated when level of adjoining major lake fell

D lakelet liable to be separated from northern lake and link with southern lake as a result of river capture

E cut off lake

F region that would give rise to an isolated basin if slightly elevated

G & H flooded valleys that form part of present-day lake margin

I present outlet of Lake Victoria

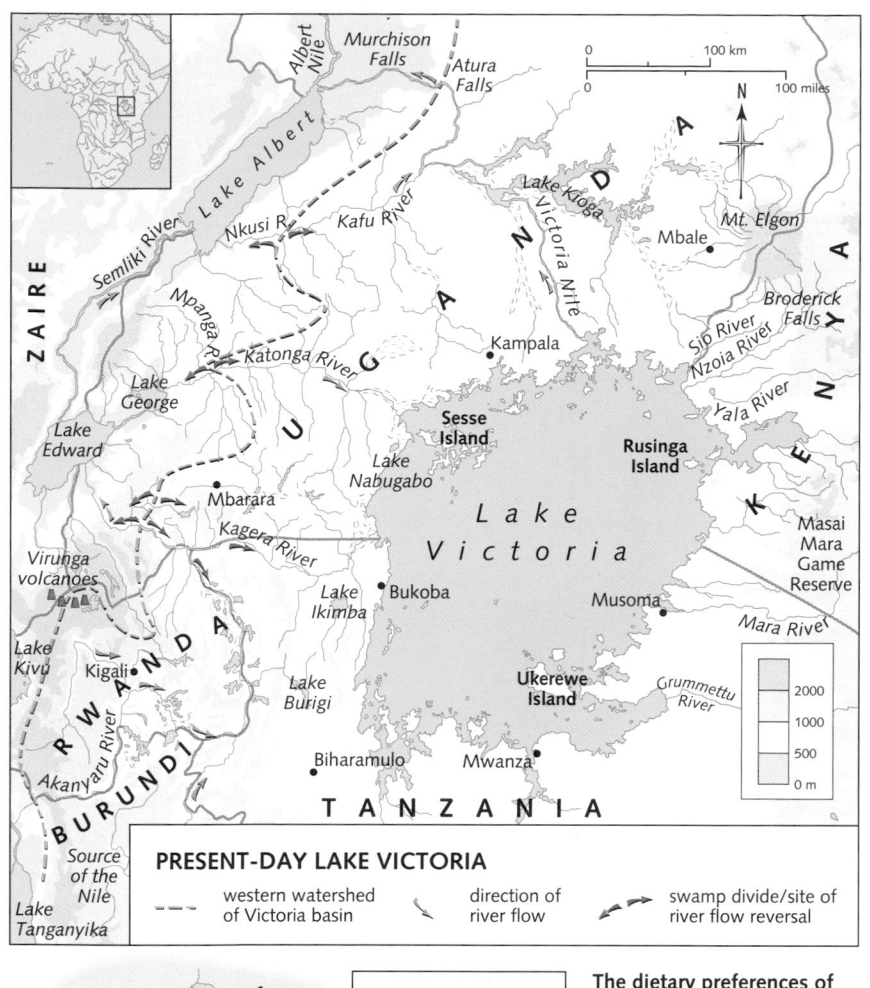

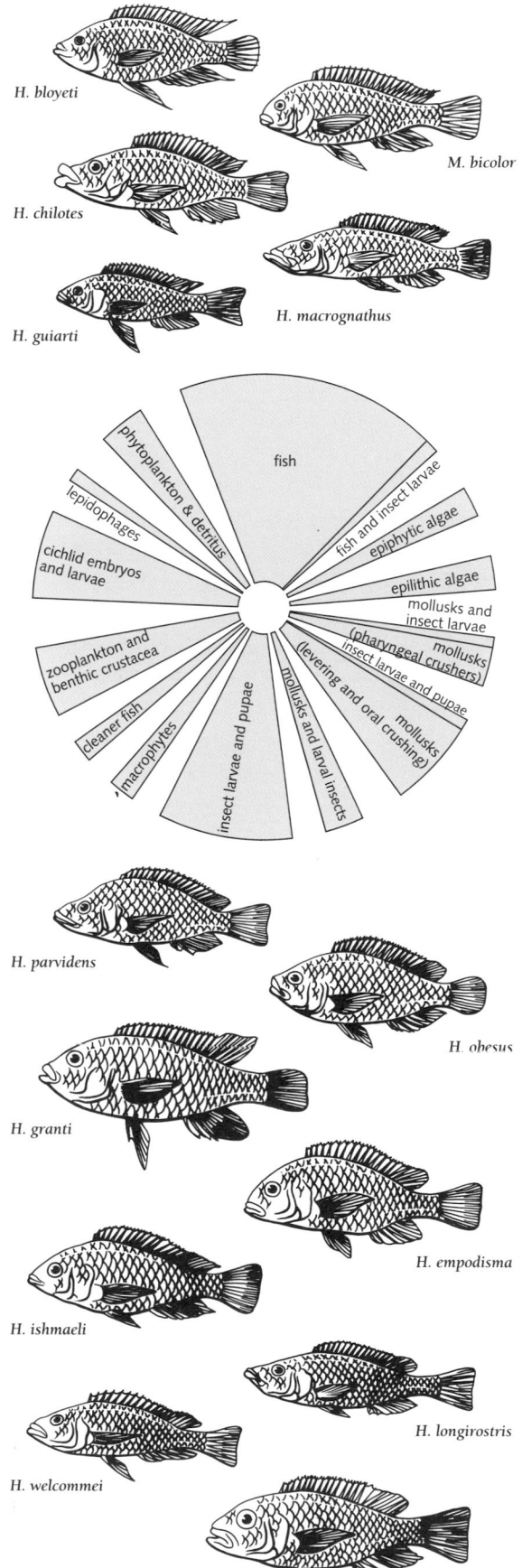

The dietary preferences of cichlid fish

Cichlid fish eat a wide variety of food types (*right*), including other fish, fish larvae and eggs, algae growing on rocks, floating plant plankton, detritus on the lake floor, floating animal plankton, insect larvae and pupae, scale-scraping from other fish, snails and bivalves, cleaning parasites from other fish, and crabs. Fish and insect-eating are the commonest dietary types. Overall body shape is no guide to diet, but the nature of the teeth is highly variable. The pie diagram on the right shows the relative numbers of cichlid species that are adapted to each dietary mode. A random selection of species of *Haplochromis* and one species of *Macropleurodus* are shown.

Right: the detailed distribution of species of the cichlid fish *Tropheus* in Lake Tanganyika. Six species show discontinuous distributions around the lake margins. The patterns have been rendered complex by lake flooding, and by the meeting of previously isolated populations.

Australian Tree Creepers
Speciation driven by habitat fragmentation

Species distributions are often closely linked to habitat. Many species have highly specialized preferences in terms of the topography, rainfall, temperature range or vegetation with which they are associated. Even quite minor differences in any of these factors may make a particular location uninhabitable for a species. This close association of species with habitats is a ready source of new species, as habitats change and as barriers are established. A classic example of close adaptation to specific habitats is seen in the Australian tree creepers.

The Australian tree creepers are members of the family Climacteridae and show species distributions that correspond closely to particular vegetation types. Five species of the *Climacteris picumnus* group, *C. picumnus*, *C. rufa*, *C. wellsi*, *C. melanura* and *C. melanota*, have distributions that are tied to savannah and sclerophyll (dry forest with thick-leaved plants) habitats in peripheral zones around Australia. The group also occurs in eucalyptus trees along watercourses, which explains their extensive distribution in western New South Wales and the mid west. Most *Climacteris picumnus* species have isolated distributions, separated by tracts of unsuitable terrain. Species distributions meet only in eastern Australia, and these are connected by a steep colour cline; in other words, the members of two species meet over a hybrid zone in which specific colours change rapidly over a relatively short distance on the ground.

Other tree creeper species include *Climacteris leucophaea*, which is associated with wet sclerophyll forest and montane forest and extends its distribution inland along rivers. It is replaced in the wet forests of the southwest by *C. rufa* of the *picumnus* group. *C. leucophaea* has also extended its range to New Guinea, where it occurs in highland forests. The two final *Climacteris* species, *C. affinis* and *C. erythrops*, are also associated with particular vegetation types. *C. affinis* is found in areas of low scrubland termed mallee and mulga, while *C. erythrops* is found in a narrow strip of hill forest in the southeast.

This shows how apparently subtle differences in the ways tree creepers exploit the trees in which they feed and live can have profound effects on their distribution. Presumably, as habitats have shifted with climatic change, so too have the birds' habitats. Changes in woodland habitats will directly affect them. It is likely that the different species originated in restricted forest patches near the coasts, after the Pleistocene, and expanded along the coastal strips and inland as the woodlands and scrub extended.

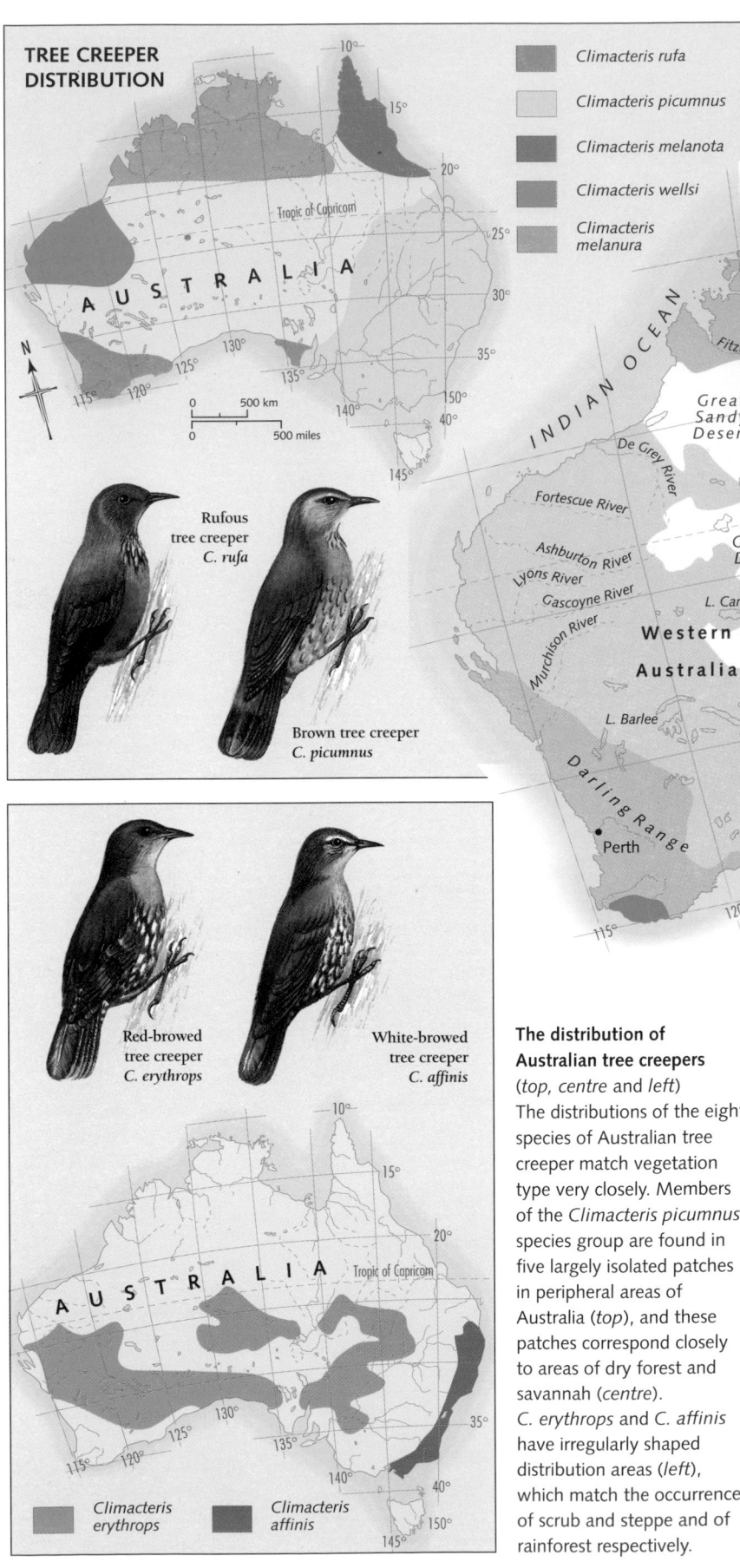

The distribution of Australian tree creepers (*top, centre* and *left*) The distributions of the eight species of Australian tree creeper match vegetation type very closely. Members of the *Climacteris picumnus* species group are found in five largely isolated patches in peripheral areas of Australia (*top*), and these patches correspond closely to areas of dry forest and savannah (*centre*). *C. erythrops* and *C. affinis* have irregularly shaped distribution areas (*left*), which match the occurrence of scrub and steppe and of rainforest respectively.

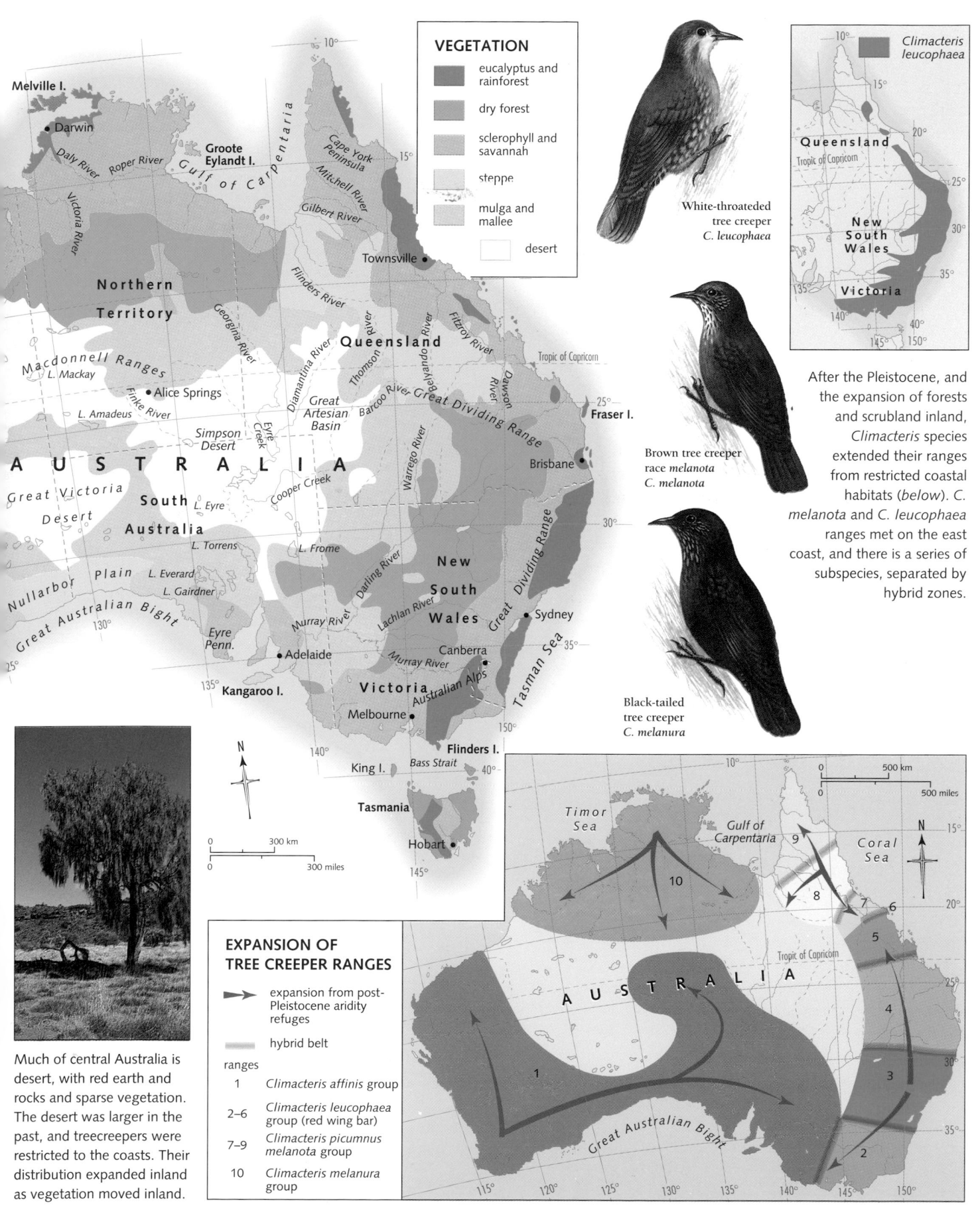

VEGETATION

- eucalyptus and rainforest
- dry forest
- sclerophyll and savannah
- steppe
- mulga and mallee
- desert

White-throateded tree creeper
C. leucophaea

Brown tree creeper
race *melanota*
C. melanota

Black-tailed tree creeper
C. melanura

Climacteris leucophaea

After the Pleistocene, and the expansion of forests and scrubland inland, *Climacteris* species extended their ranges from restricted coastal habitats (*below*). *C. melanota* and *C. leucophaea* ranges met on the east coast, and there is a series of subspecies, separated by hybrid zones.

Much of central Australia is desert, with red earth and rocks and sparse vegetation. The desert was larger in the past, and treecreepers were restricted to the coasts. Their distribution expanded inland as vegetation moved inland.

EXPANSION OF TREE CREEPER RANGES

→ expansion from post-Pleistocene aridity refuges

hybrid belt

ranges

1 *Climacteris affinis* group

2–6 *Climacteris leucophaea* group (red wing bar)

7–9 *Climacteris picumnus melanota* group

10 *Climacteris melanura* group

Part Two
Life on Earth,
The Geography of Evolution

A Brief History of Life

Continental Movement

The Legacy of the Ice Ages

Section IV: A Brief History of Life

Since life originated 3,500 million years ago, the number of species on Earth has increased dramatically. Studies of the fossil record show how that diversification may have occurred. They also reveal major extinction events and phases of recovery and provide detailed information about evolution.

Large-scale aspects of evolution – macroevolution – are studied mainly from the fossil record. In Darwin's time, much less was known about fossils and the age of the Earth, although paleontologists knew that there was a vast array of extinct plants and animals, that these ancient organisms occurred in predictable associations (faunas and floras), and that there were sequences of fossils as one passed from ancient to more modern rocks. Darwin showed how all species, living and extinct, were related by patterns of common descent, and hinted strongly that these ever-diversifying lineages could probably be traced back to a single common ancestor of all living things.

Darwin's intuition was almost certainly correct. Techniques to determine the precise ages of rocks have since been developed or improved, and it is possible to estimate the time-scale of any long-term evolutionary pattern or process. Improvements in dating have come partly from the invention of the radiometric technique, in which natural rates of radioactive decay are measured in rock samples. In addition, the techniques of relative dating using fossils have advanced to such an extent that it is now possible, for some periods at least, to divide time up into packets of a quarter of a million years or less.

Improvements in dating have gone hand-in-hand with a huge increase in the number of named fossil species and in precise records of their occurrence. Until 1859, the date of publication of Darwin's *Origin*, several hundred species had been found and described by 30 or 40 paleontologists operating mainly in the British Isles, France and Germany. Today that stock of knowledge has probably increased a thousandfold. Tests of the quality of the fossil record have shown that it is good enough to show the broad patterns of the diversification of life and the timing and magnitude of major extinction events.

Diversification and extinction

The fossil record of Precambrian times, the first 4 billion years of the Earth's history, has been worked out only this century, and it reveals the key early stages in the history of life. Evidence suggests that life originated 3,500 million years ago, and that the first life forms were single-celled organisms resembling modern viruses and blue-green algae. Life remained microscopic until 1,000 million years ago, when the first multi-celled organisms occur. These simple chains and balls of cells mark a change to the kinds of plants and animals that we can recognize, and left a reasonable fossil record. The problem is that although we can detect the beginnings of diversification between 600 and 500 million years ago, this is probably long after many earlier major pulses of diversification, which are concealed from us by the small size of the organisms involved.

The fossil record, read literally, seems to show that life diversified in fits and starts. There were dramatic increases in diversity, perhaps when new habitats were occupied or a major new adaptation had arisen. There were leaps in diversity when shells and hard skeletons evolved in marine animals, when animals moved onto land, when insects and birds began to fly and when flowering plants were established. After these pulses of radiation, some lasting tens of millions of years, the rate of diversification slows down until the next burst.

The pattern of increasing diversity has suffered a number of setbacks, when unusually large numbers of species disappeared in a relatively short time span. Five of these extinction events were larger than the others, accounting for losses of 50 per cent or more of species. The biggest of these mass extinctions, 250 million years ago, was an almost total annihilation of life, when perhaps only 5 per cent of species survived. The most famous, 65 million years ago, was smaller, and was marked by the loss of 50 per cent of species, including the dinosaurs. Surviving species seem to have evolved quickly to fill empty ecospace, and the study of recovery after mass extinction is a fascinating window onto the true potential of evolution unconstrained.

Rates

Rates of evolution may be assessed from laboratory experiments and historical observations of change. The fossil record offers a view of longer-term rates. Evidently, different species and larger groups evolve at widely different rates, which may depend on circumstances. Rates may be held down by competitive interactions with other species, in which the range of permissible adaptations is constrained by species with very similar modes of life. They may also be taxon-dependent: that is, some groups may have a propensity to evolve quickly and become species-rich, while others (the 'living fossils') may evolve more slowly and never achieve high diversity.

At a more detailed level, some fossil records, particularly those covering the past 10 million years, are good enough to reveal small-scale rates of change. One of the most dramatic discoveries from such records is that evolution may operate in two modes, fast and slow. The norm is slow evolution, when species do not change through time. From time to time, this background stasis is punctuated by bursts of evolution, when flurries of new species may originate. This pattern of evolution, termed punctuated equilibrium, had not been predicted from studies of modern plants and animals, yet it might be typical. It may be that speciation is the key to evolutionary diversification, as Darwin realized, and to evolutionary change. If this is indeed the case, the evidence for it came from the fossils.

Patterns of Change
Long-term evolutionary trends

Charles Darwin was particularly interested in the origins of diversity – one of the biggest questions to be answered about the evolution of life. He was reasonably sure that all living organisms and fossils had evolved from a single common ancestor, and modern studies of the most ancient fossils show that this ancestor probably lived about 3,500 million years ago. Paleontologists and evolutionary biologists still debate the question of how life diversified from its distant origins.

Today there are 5–50 million species on Earth (*see pages 28–29*), and life must have diversified from one to many millions of species. There are a number of possible ways in which this expansion might have taken place. Perhaps the pattern was additive – essentially a straight line showing a regular addition of several thousand new species to the overall total every million years. Or perhaps the pattern of the diversification of life followed an exponential curve, in which the number of species doubled every few million years. Alternatively, the pattern might have shown a rapid rise to modern diversity levels very early on, suggesting that there is some kind of fixed global carrying capacity of species. After that rise, the level might have remained roughly constant.

Paleontologists have attempted to document

Critics of these endeavours have pointed out that the fossil record is incomplete. There are many gaps in the rock record, and many kinds of plants and animals are not well preserved as fossils, particularly those which do not have a hard skeleton of any kind. Perhaps the curves simply show a general measure of the quality of a poor fossil record. In other words, the apparent diversity levels are low in earlier parts of geological time simply because the older parts of the record are poorer than more recent geological time intervals. This could be because rocks and fossils are lost from ancient rocks by erosion and major plate tectonic movements (*see pages 86–87*), and fossils are more difficult to collect. But paleontologists have tested these ideas and found that the curves do show something real. Even after all sources of error are factored out, the patterns shown here seem to survive. Life has diversified in fits and starts, roughly along a straight line in most cases.

Another major feature of the history of life, mass extinction, is clear from the diversificaton graphs. Major extinction events show up as dramatic drops in diversity. There have been five major mass extinctions (the 'big five'), of which one, at the end of the Permian 250 million years ago (*see pages 70–71*) was vast, and four others (late Ordovician,

Diversification

Life has diversified in an irregular fashion, with numerous fits and starts – at least according to the fossil evidence (*below*). Information has been collected on the number of families of marine animals and land vertebrates and on the number of species of land plants through the past 550 million years, when the fossil record is reasonably good. Marine animals seem to have reached a plateau level of 450 families by about 500 million years ago. They were hard hit by the mass extinction that occurred 250 million years ago, and have risen steadily in diversity since then. Land plants arose 400 million years ago, and they too have followed a pattern of rapid rises, plateaux and mass extinctions. The pattern for land vertebrates is rather different – it follows an exponential curve, with low levels up to 100 million years ago, followed by a curve of ever-increasing diversity since that time.

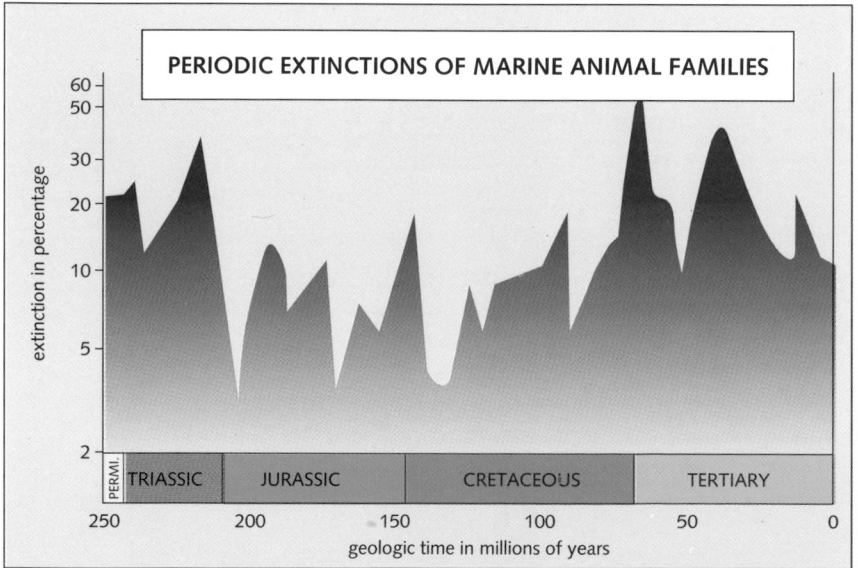

the precise patterns of the expansion of life since the 1960s. Different workers have made huge lists of all the fossil species, genera and families known to date and plotted graphs of the diversity of marine animals, vertebrates and land plants through time. These graphs all show a pattern somewhere between a straight line and an exponential curve.

A recent suggestion has been that mass extinctions have occurred periodically (*above*), peaking every 26 million years. This would imply a single extraterrestrial cause of extinction, presumably impact.

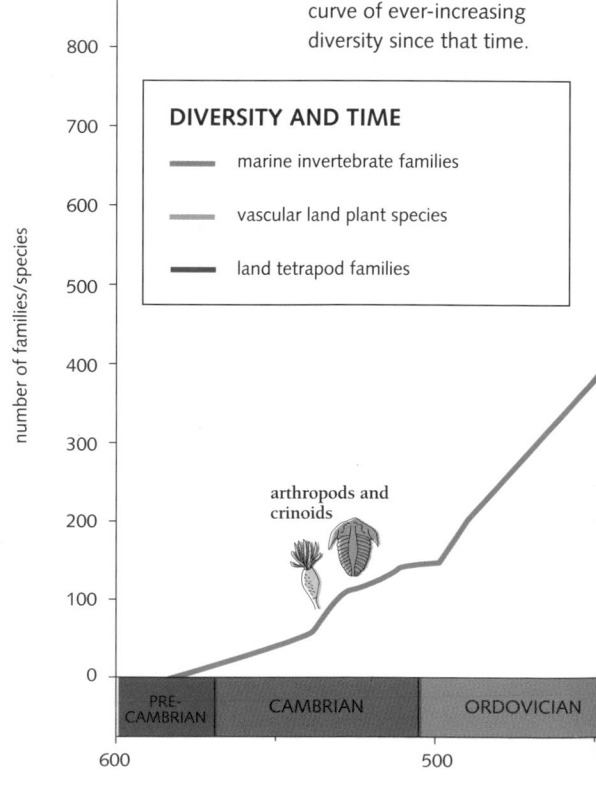

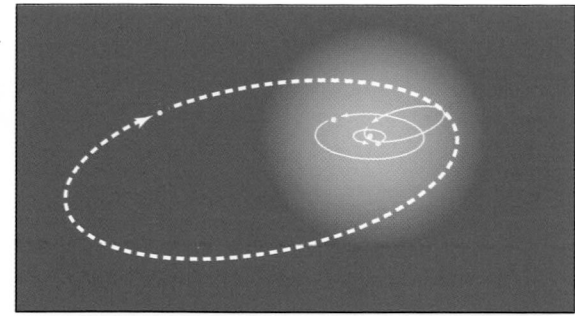

Nemesis

Three dramatic theories have been proposed for the astronomical causes of mass extinctions. One of the suggestions is that the Sun has a sister star, which has been dubbed Nemesis. This sister star disturbs the outer comet cloud once every 26 million years.

Tilting galaxy

The second theory is that our galaxy – the Milky Way – tilts up and down like a giant saucer balanced on a pencil once every 26 million years. As the galaxy changes its tilt, comets from the outer ring of comets hurtle into the solar system and hit the planets.

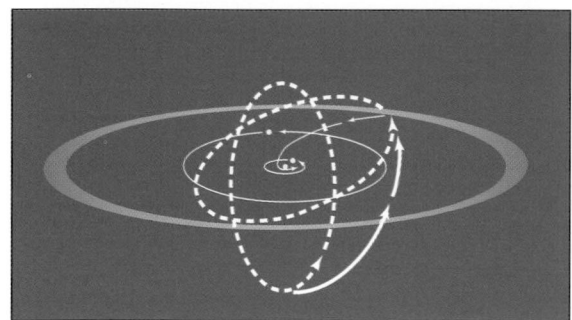

Planet X

The third theory for the astronomical cause of mass extinctions is that there is a tenth unknown planet, called Planet X, lying beyond Pluto. Perhaps every 26 million years its orbit strays into the outer comet cloud. None of these ideas has yet been proved.

late Devonian, end-Triassic, and end-Cretaceous) were smaller. The end-Permian event wiped out 50 per cent of families (scaling to 95 per cent of species), and the other four 15–20 per cent of families (40–60 per cent of species). The best-known mass extinction, although not the biggest, was at the end of the Cretaceous, when the dinosaurs died out (*see pages 72–73*). Mass extinctions were rapid, wiped out great numbers of species, and seem to have been global in extent and to have affected all kinds of organisms indiscriminately.

There have also been dozens of smaller extinction events through the history of life, each accounting for the loss of 1–5 per cent of families. Some were regional, while others were selective, affecting only certain kinds of organisms. The latest, at the end of the Pleistocene 10,000 years ago, saw the end of large mammals in many parts of the world, but other groups were unaffected (*see pages 114–115*). Present-day human-induced extinctions may eventually rank as a mass extinction.

An intriguing, but controversial, recent suggestion is that extinction events have been periodic over the past 250 million years at least. This is a proposal of regular occurrence that would demand an astronomical explanation.

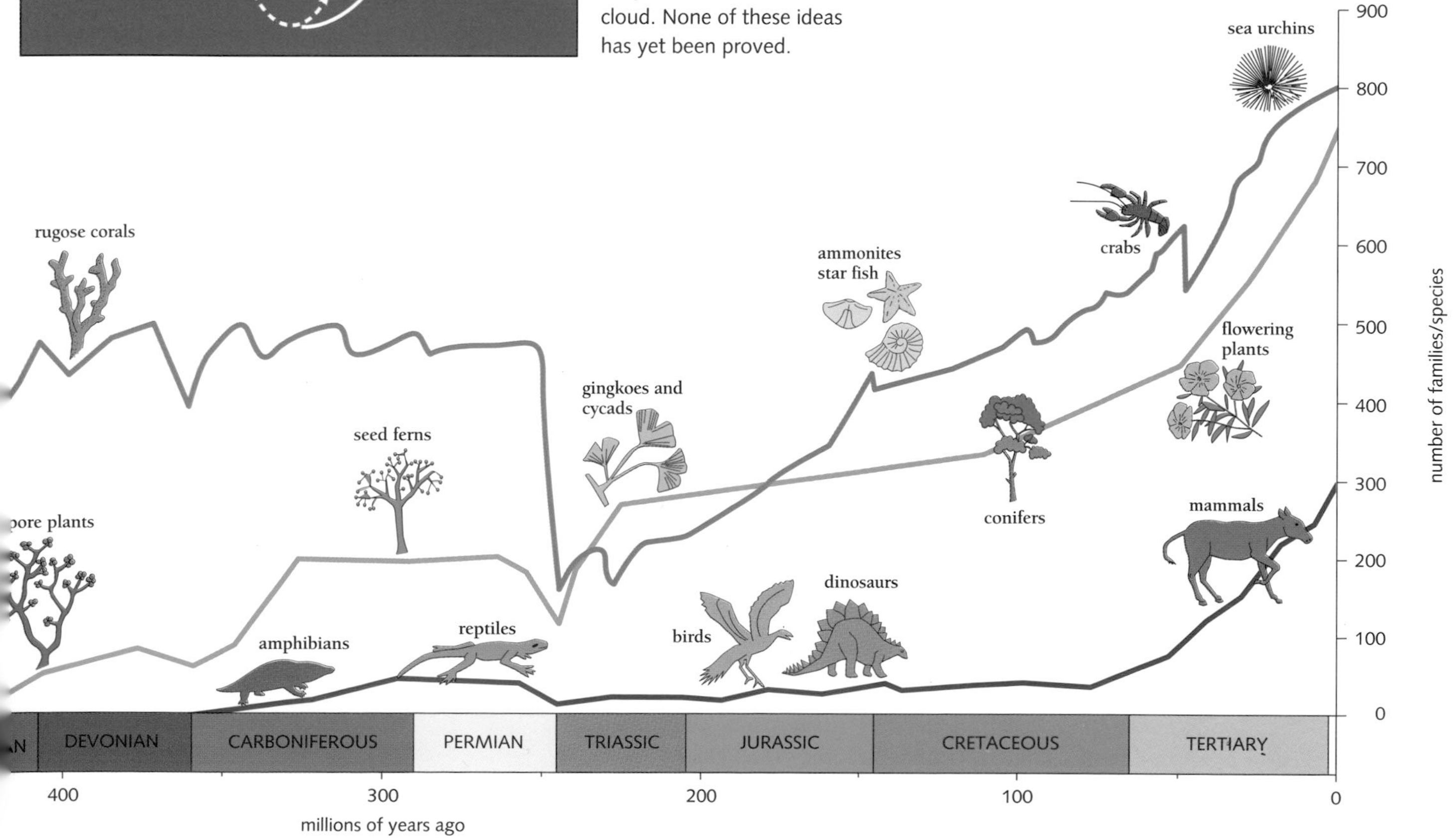

The Biggest Mass Extinction

How life was almost completely wiped out 250 million years ago

The biggest mass extinction of all time happened 250 million years ago, at the end of the Permian period. It has been estimated that 95 per cent of species became extinct at this time, in what was the greatest crisis for life yet known. Surprisingly, however, for such an extraordinarily dramatic event, scientists still know relatively little about what happened. It is not certain whether the catastrophe lasted only a few years, or whether it took between 5 and 10 million years in all. The cause is also uncertain.

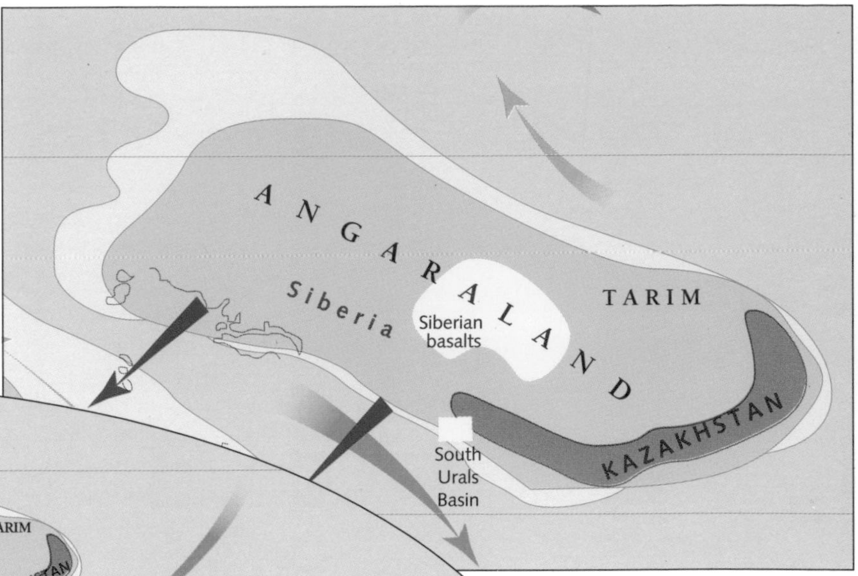

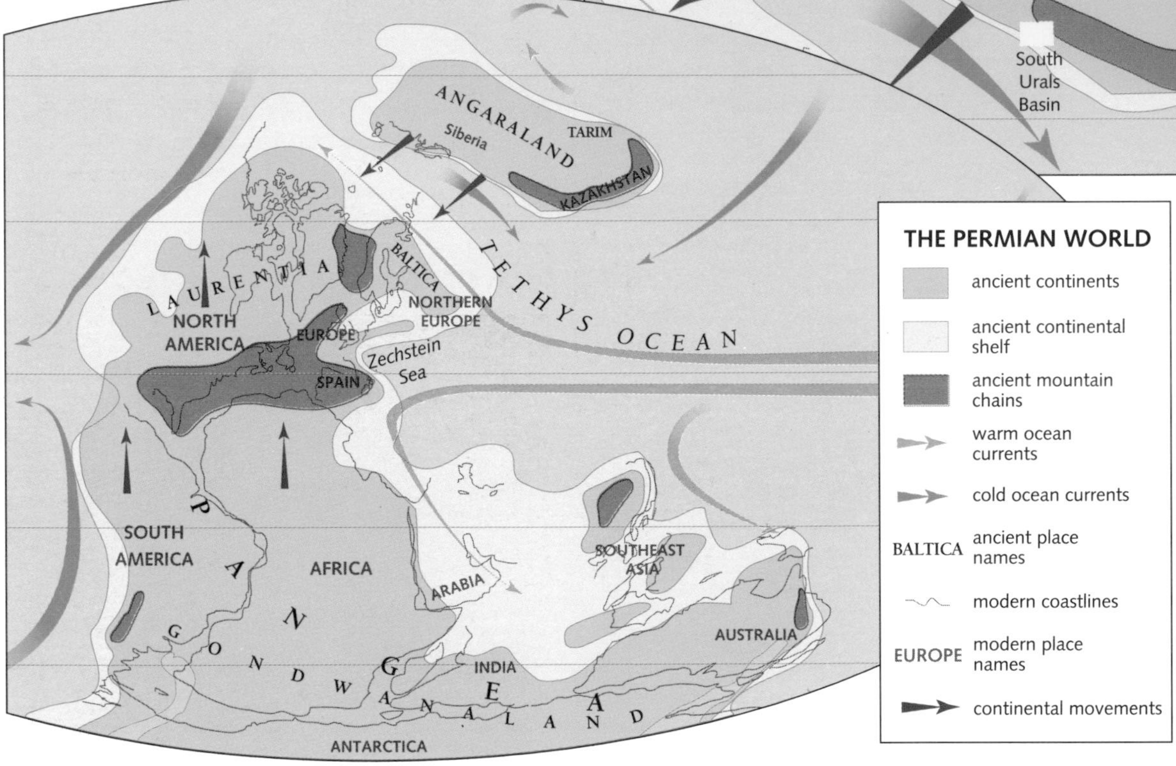

During the Permian, the continents drifted together and fused, forming Pangea (*left*). This map shows the Earth in the Early Permian. During the next 30 million years, Angaraland, consisting of Siberia and surrounding areas, was the last to join Pangea. The Ural Mountains were then uplifted. Huge amounts of sediment were eroded from the rising mountain chain (*right*) into lowland basins in the Late Permian, and thousands of amphibians and reptiles were buried.

THE PERMIAN WORLD

- ancient continents
- ancient continental shelf
- ancient mountain chains
- warm ocean currents
- cold ocean currents
- BALTICA — ancient place names
- modern coastlines
- EUROPE — modern place names
- continental movements

The extinctions affected all groups of plants and animals, in the sea and on land. Some major groups disappeared in the Late Permian – the trilobites, the rugose and tabulate corals, the majority of the brachiopods, and other forms that had been characteristic of the seabed throughout the previous 300 million years. On land too, most of the extraordinary mammal-like reptiles became extinct, as did the *Glossopteris* flora (*see pages 86–87*). The loss of life amounted to 50 per cent of all families, which scales to 95 per cent of species – the nearest life has ever come to complete annihilation.

There are two main views about the cause of this mass extinction: long-term climatic change, and massive volcanism. Throughout much of the Permian period, there was long-term climatic and topographic change as the various continents moved together and fused to form the supercontinent Pangea. This process of continental amalgamation reached its height at about the time of the mass extinction. Perhaps the uniting of the continents cut down dramatically on restricted seaways and lowland areas. Also, plants and animals that had previously been isolated now came into contact, and it may be that a massive loss of diversity took place.

A carnivorous Late Permian reptile, *Titanosuchus* (*below*) is known from South Africa but had many close relatives in Russia at the same time. This rather mammal-like reptile belongs to a group wiped out by the end-Permian mass extinction.

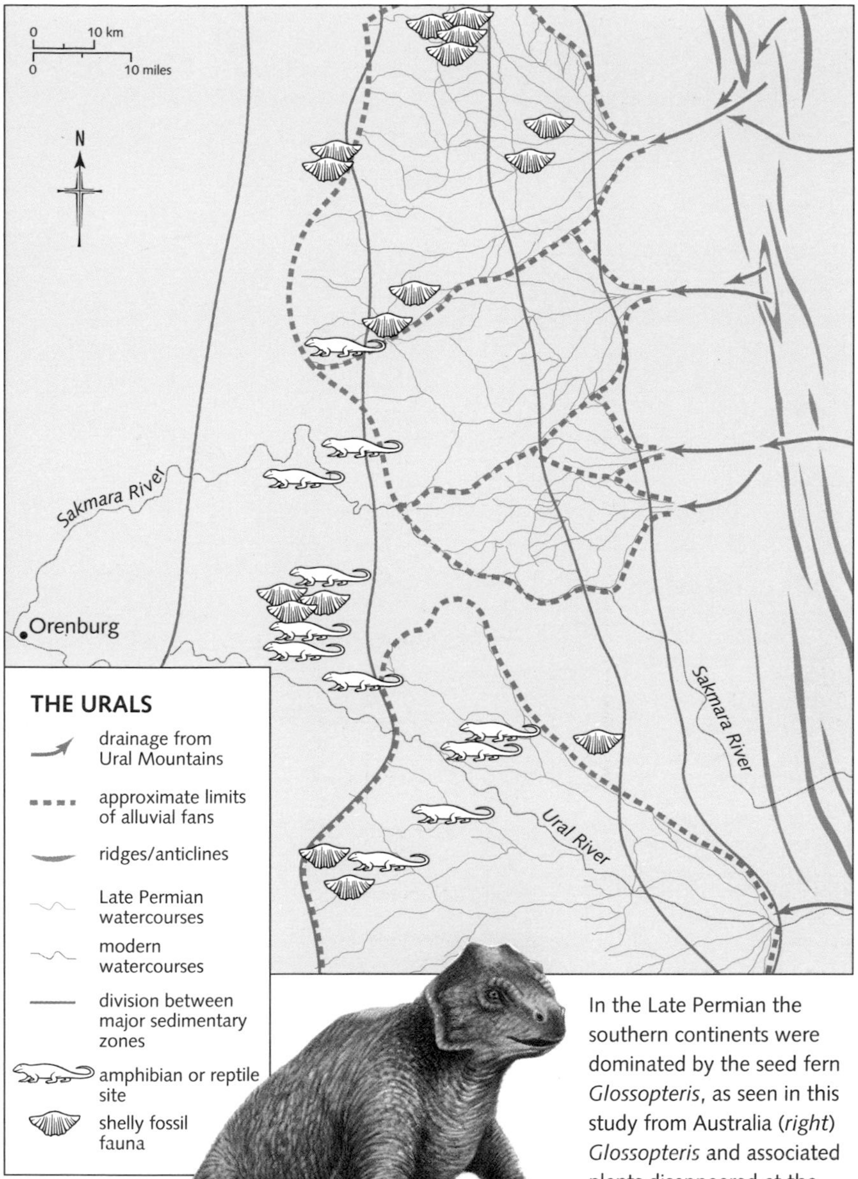

Another possible explanation for the extinction is that major earth movements associated with the fusion of Pangea caused vast volumes of basalt lava to be erupted in Siberia. The volcanic eruptions poured sulphur dioxide into the atmosphere, causing dramatic cooling and poisoning plants and animals. Also, the Ural Mountains reached their maximum height after tens of millions of years of uplift. Massive alluvial fans of debris formed across the low-lying plains, and the dramatic extinctions can be seen among the famous reptile faunas of central Russia.

After the extinction event, there was a barren phase in the rocks, marked by a lack of oxygen and scarcity of life. Only a few species were left behind, on land and in the sea. It took at least 2 million years for communities to rebuild themselves and 40 million years in all for life to become as diverse as it had been before the event.

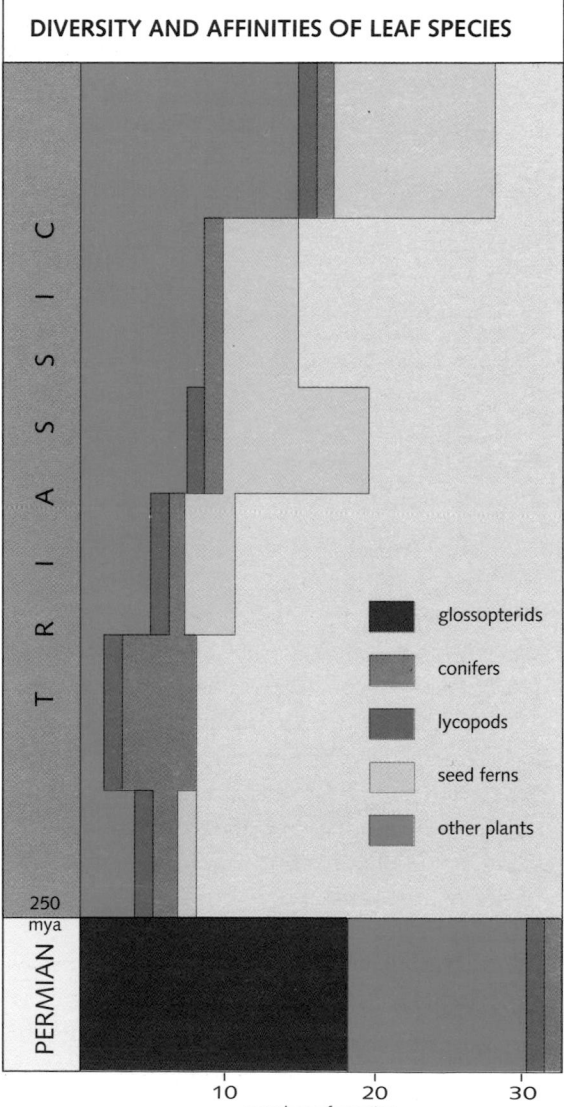

DIVERSITY AND AFFINITIES OF LEAF SPECIES

glossopterids

conifers

lycopods

seed ferns

other plants

TRIASSIC

PERMIAN

250 mya

10   20   30
number of species

THE URALS

↗ drainage from Ural Mountains

- - - approximate limits of alluvial fans

― ridges/anticlines

∿ Late Permian watercourses

∿ modern watercourses

— division between major sedimentary zones

🦎 amphibian or reptile site

🐚 shelly fossil fauna

0   10 km
0   10 miles

N

Sakmara River

Orenburg

Sakmara River

Ural River

In the Late Permian the southern continents were dominated by the seed fern *Glossopteris*, as seen in this study from Australia (*right*) *Glossopteris* and associated plants disappeared at the end of the Permian. Floral diversity recovered over the first 8 million years of the Triassic, but the plant groups were now different.

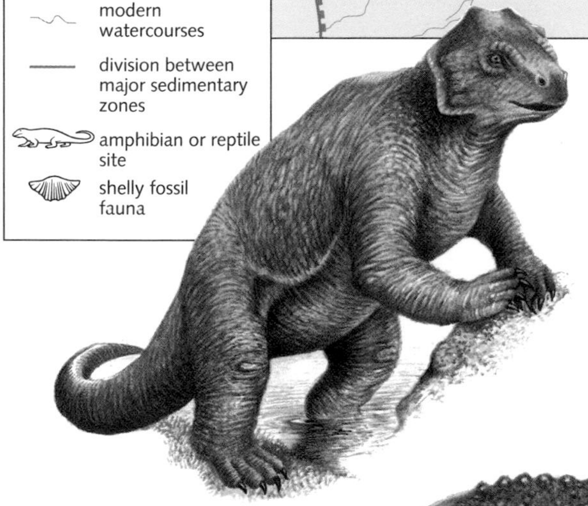

Pareiasaurus (*above*) from South Africa and *Scutosaurus* (*right*) from Russia are both pareiasaurs – herbivorous reptiles from the Late Permian. These shared reptile groups show how similar life was worldwide during the Permian. The pareiasaurs disappeared during the extinction event.

The KT Event

A massive impact that may have killed the dinosaurs

The mass extinction at the Cretaceous-Tertiary (KT) boundary 65 million years ago is the most famous of all extinctions, since it saw the end of the dinosaurs. A much smaller catastrophe than the Permian-Triassic event (*see pages 70–71*), it has fascinated scientists and the general public over the years. This has led to an interesting situation: too many theories, and still no resolution of what precisely happened 65 million years ago.

More than 100 theories for the extinction of the dinosaurs have been published, many of them wholly ridiculous (that the dinosaurs were too stupid, or too undersexed, to survive; that caterpillars ate all the dinosaurs' plant food; or that dinosaurs had become racially senile). Most of these theories are very hard to test, and most refer only to the dinosaurs. Of course, many other plant and animal groups also died out 65 million years ago: the large marine reptiles, such as plesiosaurs and mosasaurs; the flying pterosaurs; and marine groups such as ammonites and belemnites and many of the microscopic floating foraminifera.

Since 1980, a great deal of impressive research has gone into the KT event. This has been an interdisciplinary effort, calling on the skills of geologists, paleontologists, geochemists, atmospheric scientists, ecologists and even astronomers. The event has been studied in exquisite detail, made possible by the superb preservation of many hundreds of sections, in sediments deposited in all kinds of environments, around the world. It is now clear that many marine groups, such as foraminifera and ammonites, died out instantly, but that others, like dinosaurs and pterosaurs, may have declined more slowly.

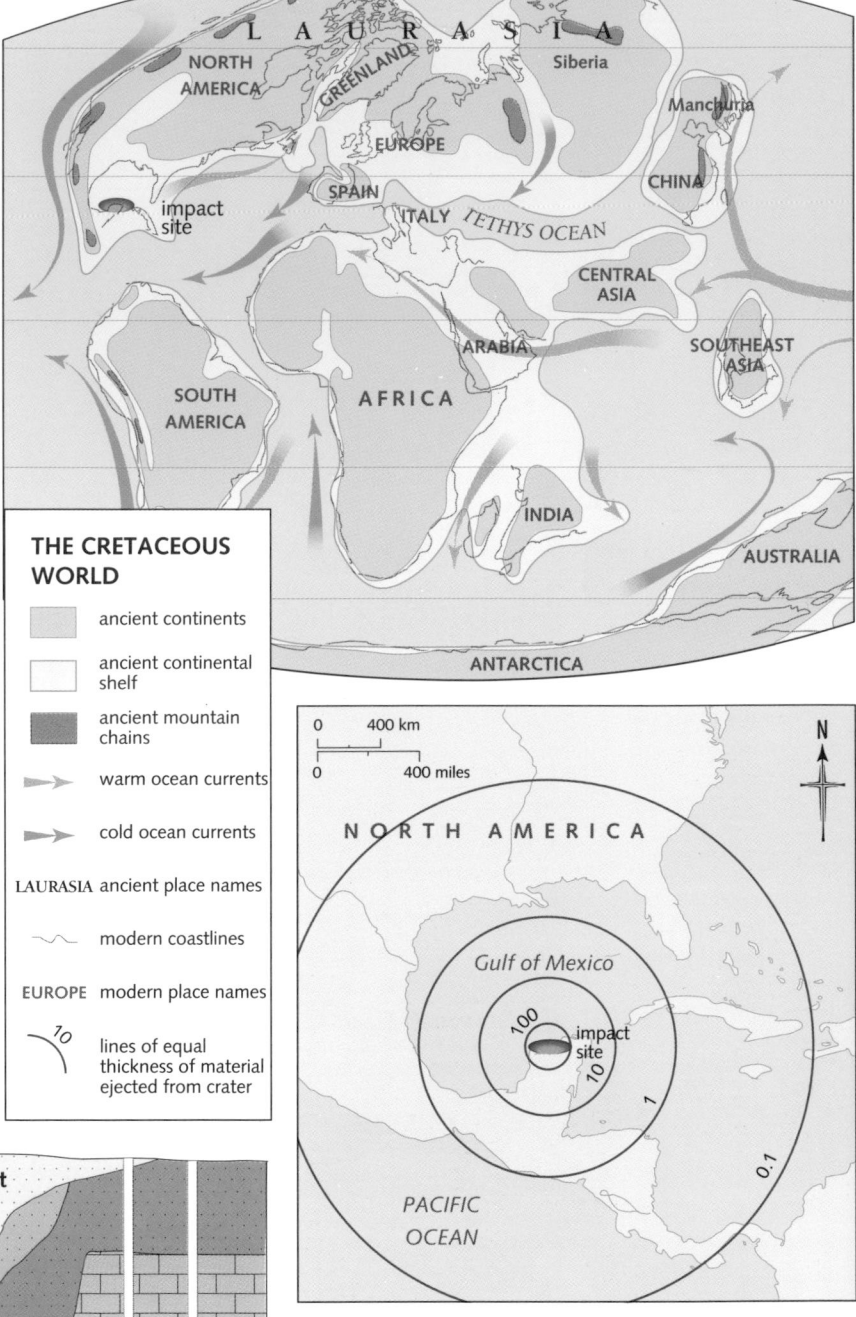

THE CRETACEOUS WORLD

- ancient continents
- ancient continental shelf
- ancient mountain chains
- warm ocean currents
- cold ocean currents
- LAURASIA ancient place names
- modern coastlines
- EUROPE modern place names
- *10* lines of equal thickness of material ejected from crater

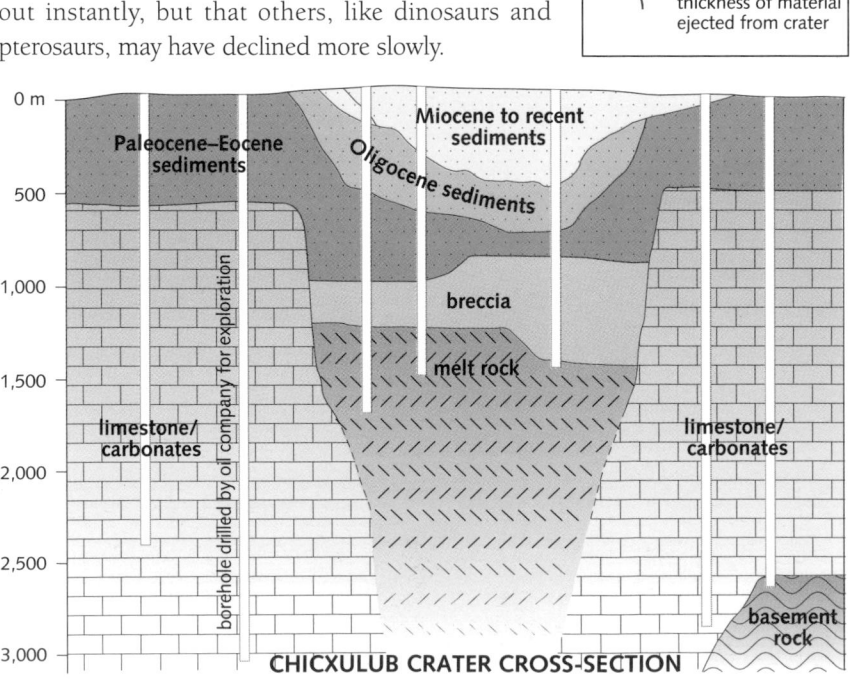

CHICXULUB CRATER CROSS-SECTION

The Chicxulub crater, located in southeastern Mexico, is almost certainly the site where a vast asteroid hit the Earth 65 million years ago. The impact created a deep crater which has since been covered by younger sediments and which has been identified from borehole records (*left*). Studies of the whole Caribbean area confirm the location. As geologists sample KT boundary rocks around the site, they find additional evidence for impact up to 1,000 km away (*above*). This evidence consists of glassy spherules melted and thrown out of the crater, shocked quartz grains produced at high pressure, and disruption of coastlines where a vast tidal wave, or tsunami, hit the shore after the impact.

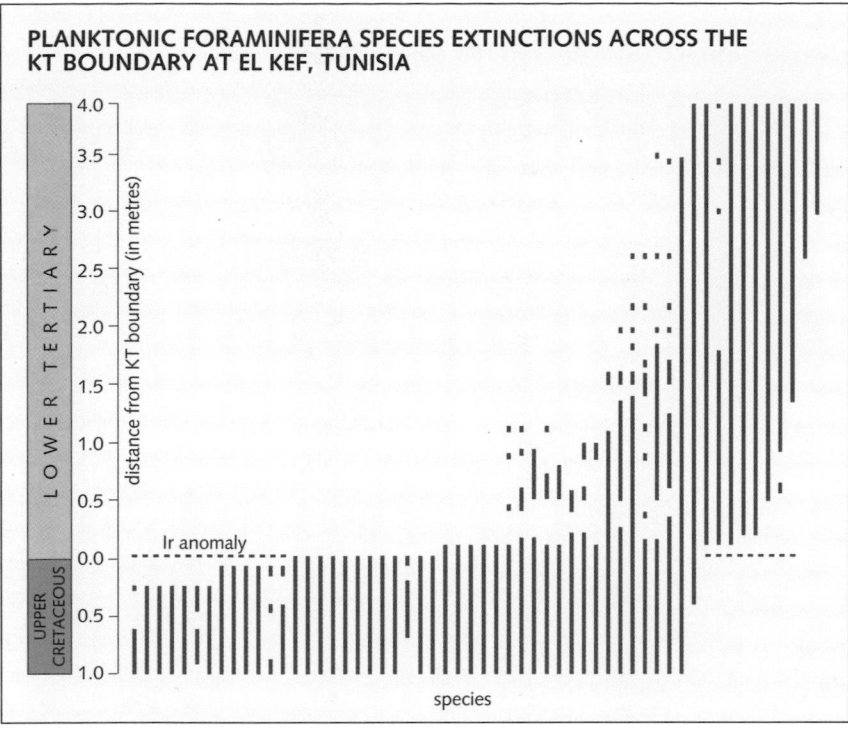

PLANKTONIC FORAMINIFERA SPECIES EXTINCTIONS ACROSS THE KT BOUNDARY AT EL KEF, TUNISIA

The study of the event was revolutionized by the proposal, in 1980, that a giant meteorite, or asteroid, had hit the Earth 65 million years ago, throwing up a vast dust cloud as it drove into the ground. The evidence for an impact consists of high levels of iridium (an element found at high concentration only in meteorites and other extraterrestrial objects) at the boundary, as well as glassy spherules, indicating melting, and shocked quartz, indicating high pressure impact. Opponents of the hypothesis have argued that these could all have been produced by huge volcanic eruptions happening in India at the time, but there are problems in explaining the geochemistry of the glassy spherules and the abundance of shocked quartz by purely volcanic means.

A weakness of the theory was that the crater had not been found. However, a convincing crater, 150 kilometres or more in diameter and surrounded by evidence of fallout, was identified in 1990 at Chicxulub on the Yucatán Peninsula in Mexico.

The dinosaurs were not the only group to die out at the end of the Cretaceous period. Some of the most dramatic extinctions occurred in the sea. The planktonic foraminifera (*above*) were virtually wiped out. This range chart from Tunisia shows that more than 30 species died out instantaneously. The coiled ammonites (*below*) entirely disappeared. Ammonites had existed since the Devonian, and had come close to extinction several times (*right*), but had recovered and radiated each time. The KT event finally wiped them out. Only their distant relatives, the squid, octopus and nautilus (*below*), survive.

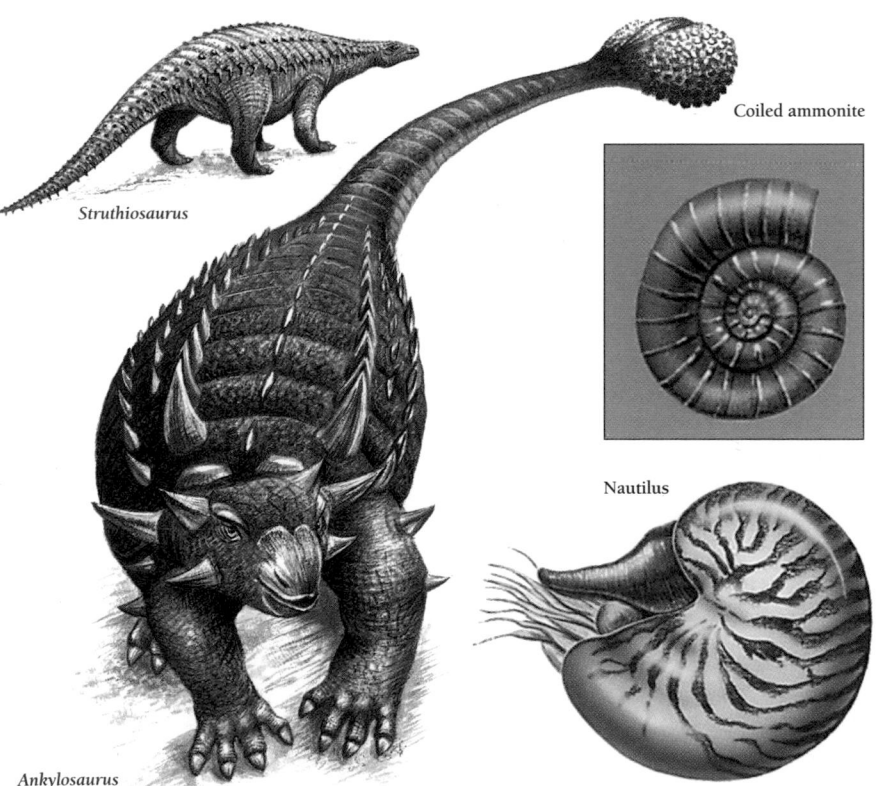

Struthiosaurus

Ankylosaurus

Coiled ammonite

Nautilus

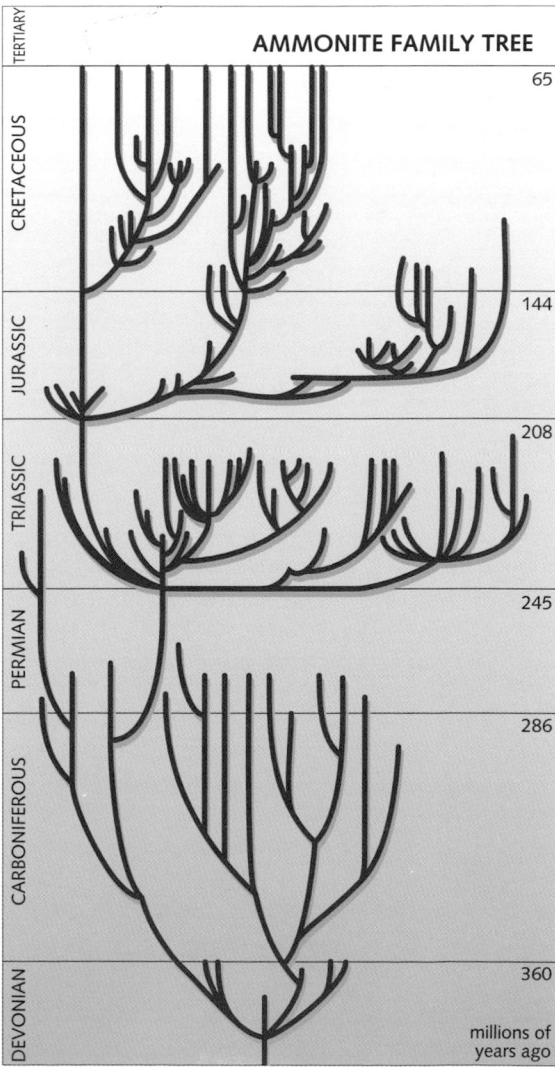

AMMONITE FAMILY TREE

TERTIARY

CRETACEOUS — 65

JURASSIC — 144

TRIASSIC — 208

PERMIAN — 245

CARBONIFEROUS — 286

DEVONIAN — 360

millions of years ago

Recovery and Radiation
The success of the Paleocene mammals

If extinction marks the end of a line, radiation marks its start. Life has diversified hugely since its origin 3,500 million years ago, but that increase in diversity has happened in fits and starts. Extinction events have, from time to time, cut diversity back. At other times, radiation events build diversity up again, often to levels never before achieved.

Radiation can be the result of two main situations. Sometimes a group radiates (diversifies) following the appearance of some adaptation that enables the group to exploit a new way of life. An example of this might be the origin of flowers and pollination by insects. When this happened, perhaps 120 million years ago, the new flowering plants began to diversify rapidly because they could breed much more effectively than other plant types.

The other kind of radiation occurs when a mass extinction has cleared out ecospace for survivors to radiate into. One of the best examples of this is seen among the Paleocene mammals, which arose 220 million years ago, at about the same time as the dinosaurs. Mammals existed throughout the entire age of the dinosaurs, but mainly as small nocturnal animals. They were warm-blooded even then, and many may have produced live young rather than laying eggs. They also had larger brains than the reptilian dinosaurs. Despite these supposed advantages, the mammals failed to replace the dinosaurs for 165 million years. Only when the dinosaurs had been wiped out, perhaps by catastrophic environmental stresses following an impact (*see pages 72–73*), were the mammals able to exploit their adaptations.

During the Paleocene and the Early Eocene, mammals diversified spectacularly, evolving from rat- and cat-sized generalists to whales, bats, horses, monkeys and rabbits. In the first 15 million years after the extinction of the dinosaurs, mammals essentially achieved their modern diversity. Indeed, they may have been rather more diverse, since some major Paleocene and Eocene mammal groups have since died out. This shows the vast potential of evolution when the brakes are taken off.

The Paleocene was a time of rapid origination of new mammal species. It was also a time of biogeographic ferment. New mammal groups were evolving in different parts of the world and moving between the continents. Elements of the midwestern North American Paleocene mammal fauna may have arrived there by a number of routes, some passing across open ocean from South America, others following routes through Africa, Europe and Asia.

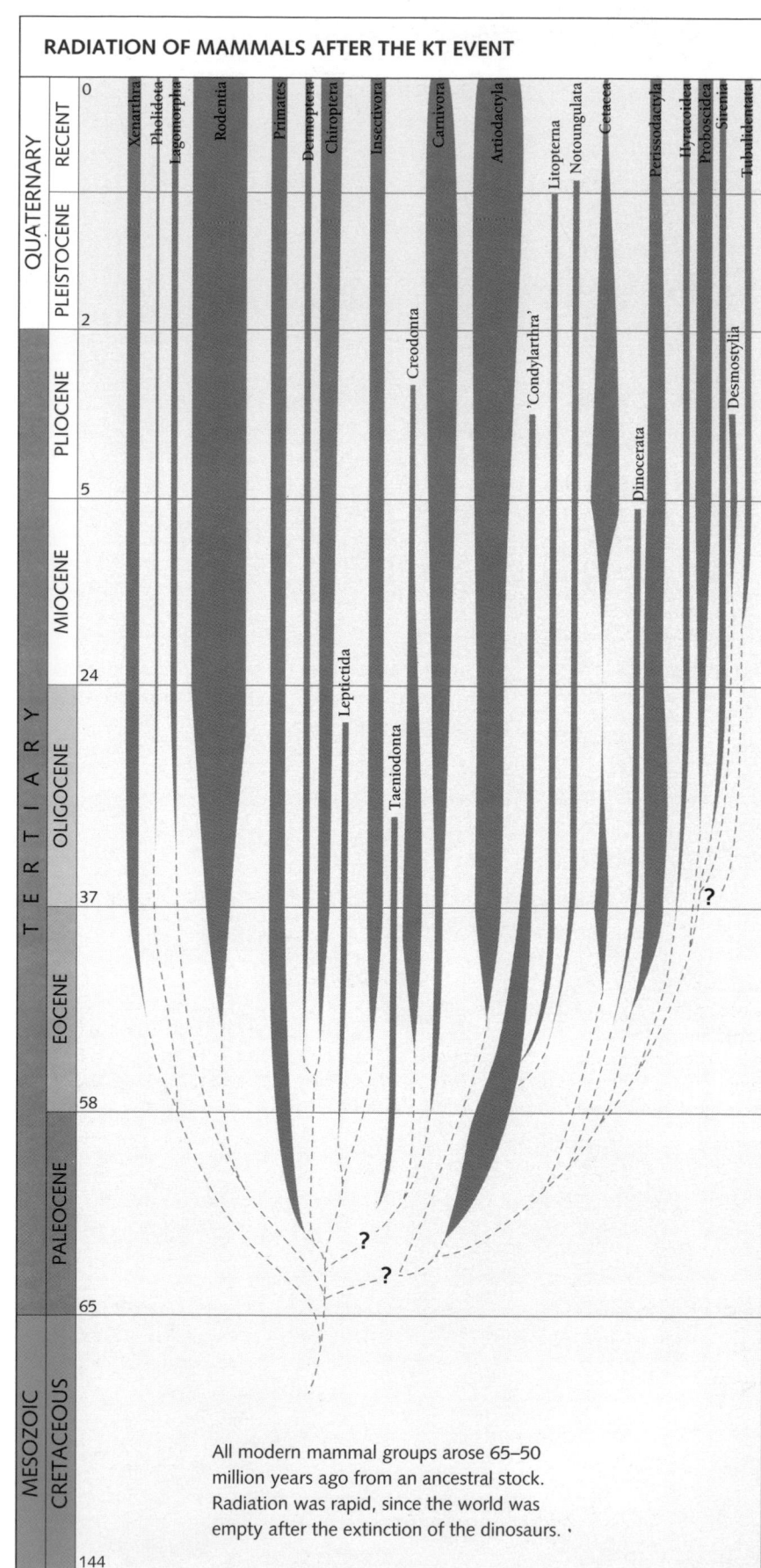

RADIATION OF MAMMALS AFTER THE KT EVENT

All modern mammal groups arose 65–50 million years ago from an ancestral stock. Radiation was rapid, since the world was empty after the extinction of the dinosaurs. ·

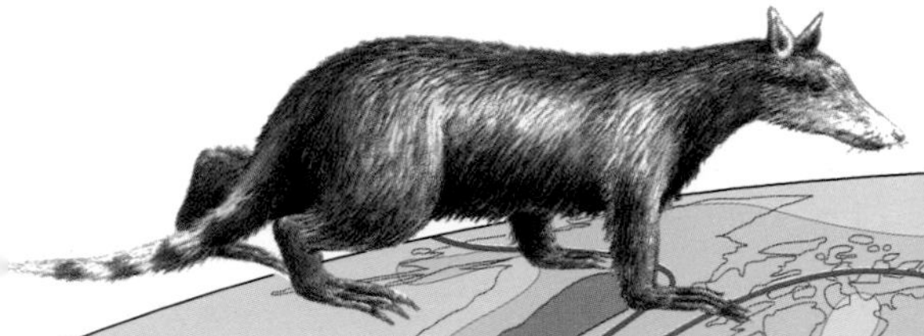

Metacheiromys

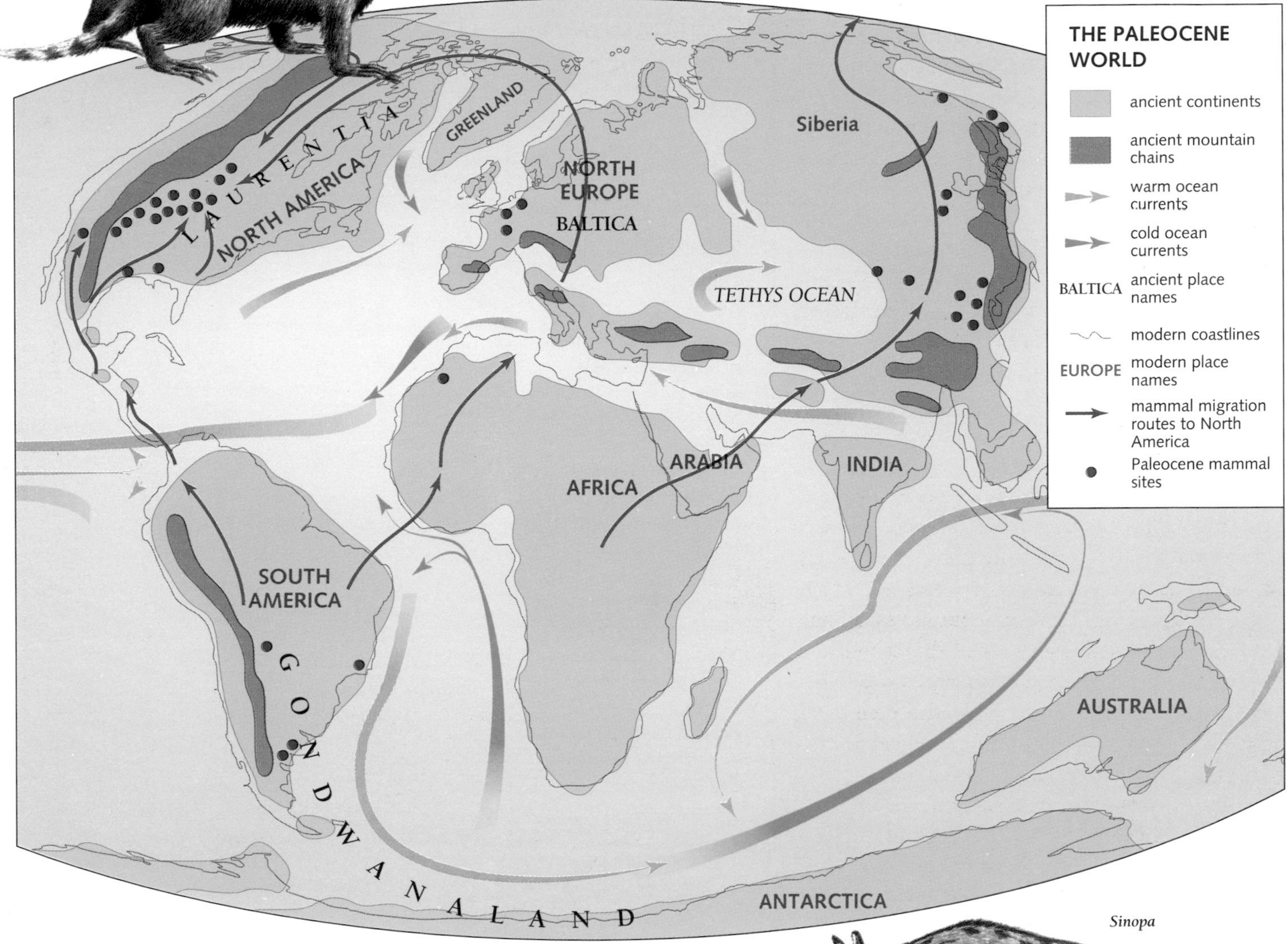

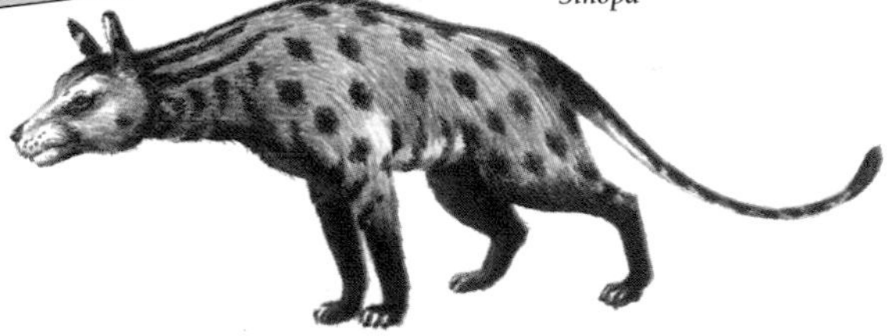

THE PALEOCENE WORLD

- ancient continents
- ancient mountain chains
- warm ocean currents
- cold ocean currents
- BALTICA ancient place names
- modern coastlines
- EUROPE modern place names
- mammal migration routes to North America
- Paleocene mammal sites

LAURENTIA
GREENLAND
NORTH AMERICA
NORTH EUROPE
BALTICA
Siberia
TETHYS OCEAN
AFRICA
ARABIA
INDIA
SOUTH AMERICA
GONDWANALAND
ANTARCTICA
AUSTRALIA

Sinopa

North American mammals
Paleocene mammals in North America arrived from other parts of the world by a variety of different routes (*above*). Mammal groups that had originated in North America also migrated to all other continents.

India as a ferry
India acted as slow-moving ferry (*below*), taking Paleocene mammals from Africa, then drifting northwards towards Asia. These mammals evolved, and later passed into Asia.

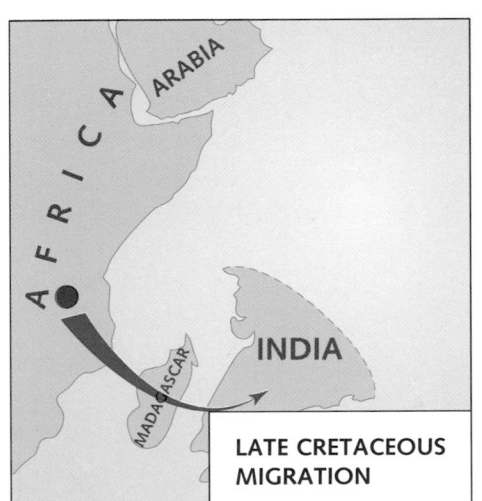

LATE CRETACEOUS MIGRATION

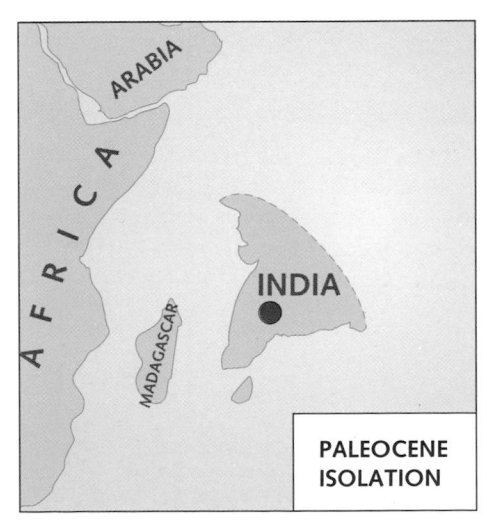

PALEOCENE ISOLATION

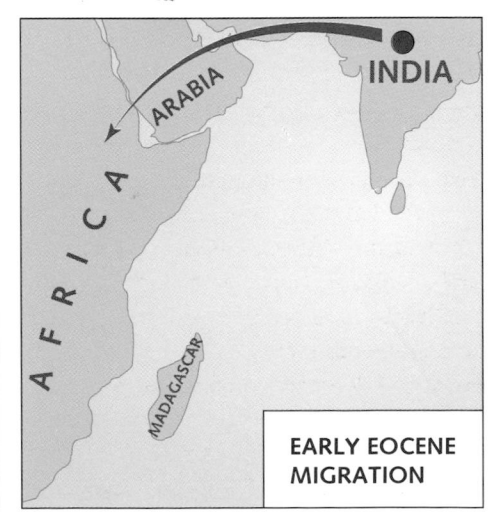

EARLY EOCENE MIGRATION

Punctuation and Gradualism
Speciation in snails and bryozoans

For years, paleontologists assumed that life evolved continuously, sometimes changing slowly and sometimes more quickly. Clearly, it was difficult to dissect the precise details of geologically rapid processes, like small-scale evolution and speciation, from the fossil record. This would require organisms that were abundant as fossils, occurring in a complete sequence of rocks without any major gaps. The paleontologist would also have to be sure that the fossil species could be precisely identified. This can be difficult when there are only skeletons or shells to examine.

Detailed studies of excellently preserved fossil sequences have shown that a common feature of evolution is stasis. Most species do not evolve continuously, but seem to show minimal change, often for millions of years. This observation led two young American paleontologists, Niles Eldredge and Stephen Jay Gould, to propose in 1972 that stasis was a normal part of evolution, and that change happened in a short burst of rapid evolution at the time of speciation.

This model of evolution by punctuated equilibrium looks rectangular. When evolution is plotted against time on the vertical axis, stasis is shown by straight vertical lines, indicating no net change, and

LAKE TURKANA

Lake Turkana
The Earth's crust is opening up along the East African rift valley. This process has produced great north-south fissures and active volcanoes over the past 10 million years. Lakes in these fault-valleys, such as Lake Turkana in Tanzania (*right*), have been occupied by snails and freshwater bivalves during this time, and huge numbers of their shells may be collected in the well-dated sediments. Millions of specimens were collected through the past 3 million years and measured in huge detail.

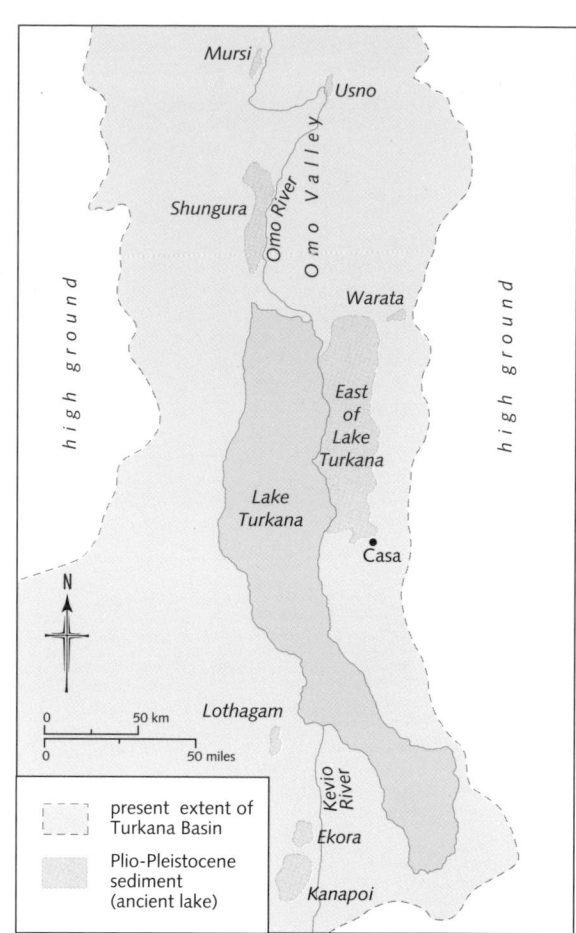

Evolution or short-term change? (*right*)
The freshwater snails and bivalves from Lake Turkana provide a detailed record of evolution over the past 3 million years. Many millions of specimens from 11 different species were collected, centimetre by centimetre, through the ancient lake sediments. The rocks are dated precisely by periodic volcanic ash beds, and lake level can be assessed by evidence from the sediments. Most of the species lineages showed bursts of evolutionary activity, when dramatic changes occurred in the shape of their shells. This happened primarily at times of sudden changes in lake level. Could it have been due to evolution, or had it merely been a temporary change of shape induced by environmental stresses?

SEDIMENTS OF LAKE TURKANA

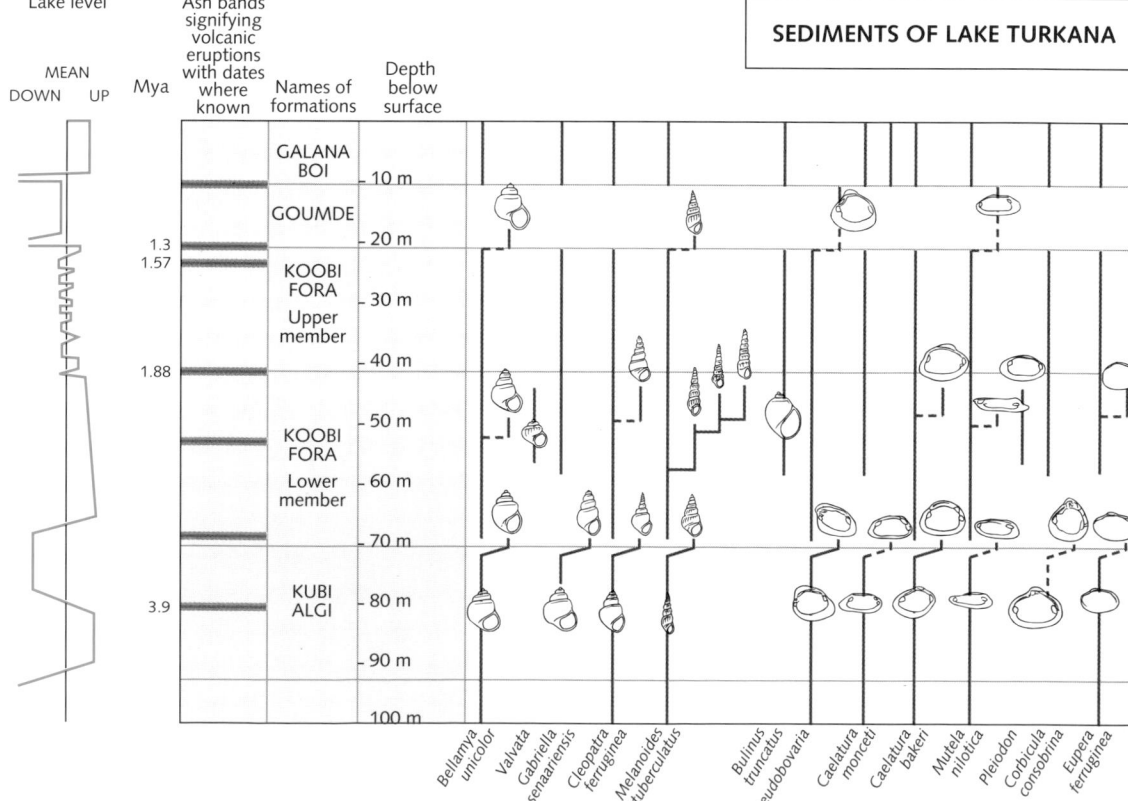

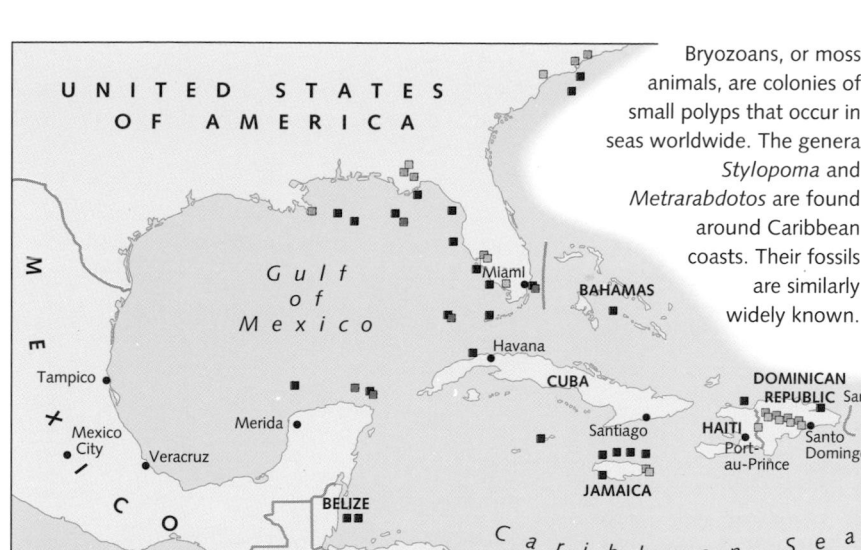

Bryozoans, or moss animals, are colonies of small polyps that occur in seas worldwide. The genera *Stylopoma* and *Metrarabdotos* are found around Caribbean coasts. Their fossils are similarly widely known.

In a more recent study of bryozoans (small colonial animals) in the Caribbean, a variety of methods confirmed that two living genera had evolved according to a punctuated pattern. Generally, the species changed very little, but at particular times some change in the environment induced bursts of rapid speciation, before a further 'normal' phase of minimal change. Whether this is a typical pattern of species evolution is still unclear.

BRYOZOANS IN THE CARIBBEAN

Metrarabdotos
- ▪ living specimen
- ▫ fossil specimen

Stylopoma
- ▪ living specimen
- ▫ fossil specimen

speciation by straight horizontal lines, indicating geologically instantaneous change. The classic model of evolution, phyletic gradualism, consists of gently sloping lines, as species evolve at different rates. Species originate from time to time, but there is no dramatic burst of evolution. Since 1972, paleontologists have sought examples of punctuated equilibrium and phyletic gradualism, but the search was not as straightforward as some people had assumed.

One of the most detailed studies undertaken so far was of snails and freshwater bivalves in Lake Turkana in the East African rift valley. The lineages show stasis, with a few dramatic sideways shifts. This was interpreted at first as an example of punctuated equilibrium, but critics pointed out that the sideways jumps, representing dramatic changes in shell shape, coincide with rapid changes in lake level. This provides some evidence that the changes were not truly evolutionary: they were short-term changes in the external appearance of the molluscs, but were not genetically coded.

The evolution of bryozoans
(diagrams, *right*)
The evolution of the Caribbean bryozoan genera *Stylopoma* and *Metrarabdotos* has been reconstructed using a variety of techniques. The phylogeny of *Stylopoma* (*top right*) is a cladogram (*see page 26*) based on external characteristics. The phylogeny of *Metrarabdotos* (*bottom right*) is based on the order of occurrence of fossils. Both of these patterns show a strongly punctuational pattern.

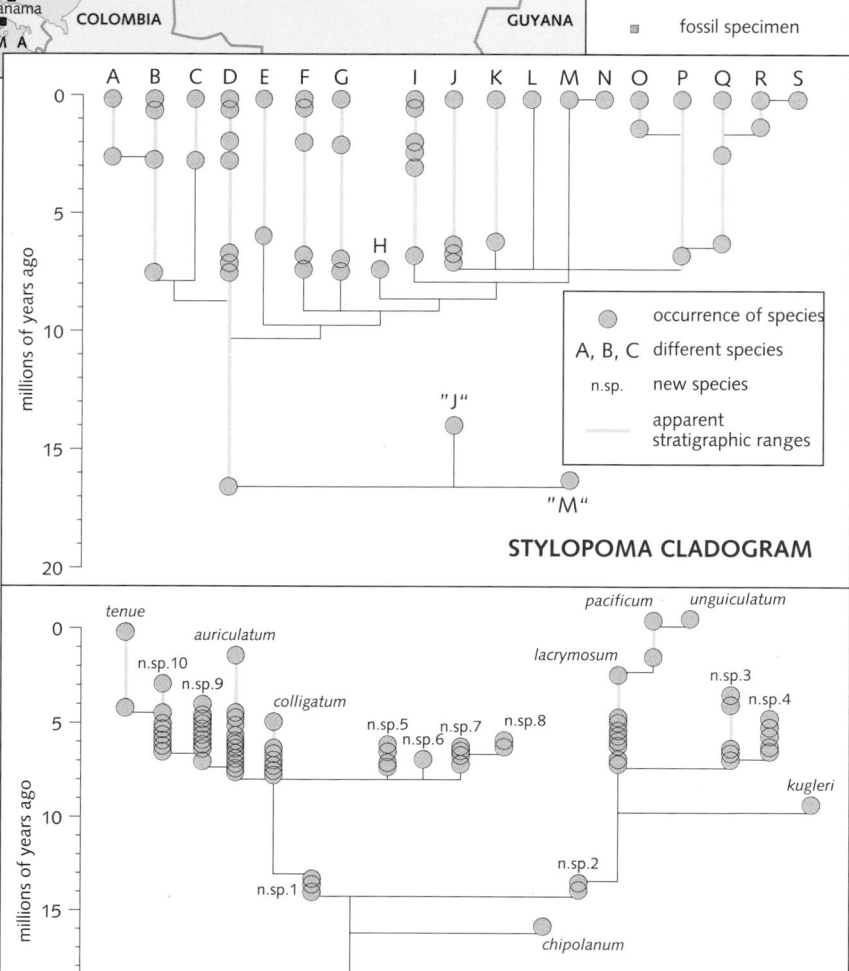

occurrence of species
A, B, C different species
n.sp. new species
apparent stratigraphic ranges

STYLOPOMA CLADOGRAM

METRARABDOTOS PHYLOGENY

The Horse Family
The evolution of horses in North America

The evolution of horses is one of the best-studied cases of long-term evolution, and is presented in textbooks as an excellent example of evolution. The first horse fossils were found around 1800 in Europe, and Eocene specimens found about 40 years later showed that horses had once been small, many-toed animals. The horse family tree became a classic later in the 19th century, when dozens of fossil specimens had been unearthed in the Tertiary rocks of North America.

Edward Cope and Othniel Marsh, better known for their protracted rivalry over new dinosaur fossils, also had teams of collectors exploring the new American states in the Midwest and seeking mammal specimens. They found abundant complete horse skeletons in all Tertiary rocks back to the Eocene. In general, the horses became smaller and had more toes the further back in time they went. When Thomas Henry Huxley, Darwin's strong supporter, visited North America in the 1870s, he was shown the remarkable progression of horse specimens, but the story was still misunderstood at that time.

The classic horse evolution story tells of simple one-way progress from the terrier-sized, four-toed Eocene *Hyracotherium*, through pony-sized, three-toed Oligocene and Miocene forms, to the large one-toed Pliocene and modern horses. Their teeth were seen to have become deeper. *Hyracotherium* browsed on soft leaves and lived a secretive life in the forests. Later horses moved out on to the developing grasslands, and required greater height for fast running and deep-rooted teeth to grind the silica-rich grasses.

This basic story is still regarded as superficially correct, but it is now recognized that the pattern was not a single line of advance. Since 1870, many thousands of horse fossils have been collected worldwide. These show how horses evolved largely in North America and repeatedly sent out emigrants to other parts of the world. In addition, the line of evolution in North America was not simple, and there were many sideways excursions. Much of the history of horses was mediated by geographic changes in the distribution of grasslands.

The oldest horses, *Hyracotherium* and related forms, arose in North America in the Eocene and migrated by a variety of routes to Europe and Asia. It was probably possible for them to reach western Europe via Greenland at that time, since the North Atlantic was still narrow enough and Greenland was not too cold. Other hyracotheres entered eastern Asia across the Bering Strait. There may also have been

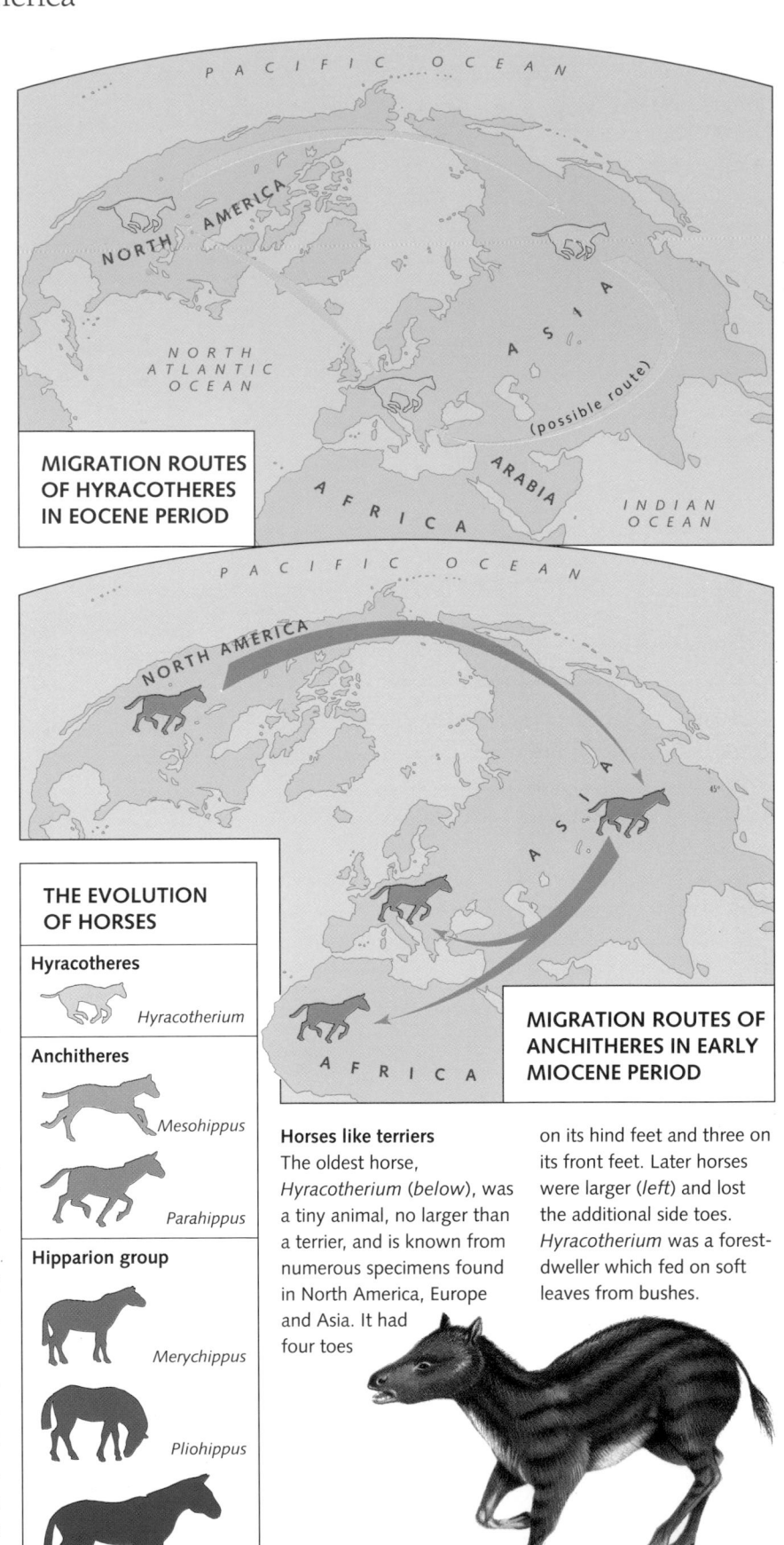

MIGRATION ROUTES OF HYRACOTHERES IN EOCENE PERIOD

MIGRATION ROUTES OF ANCHITHERES IN EARLY MIOCENE PERIOD

THE EVOLUTION OF HORSES

Hyracotheres
Hyracotherium

Anchitheres
Mesohippus
Parahippus

Hipparion group
Merychippus
Pliohippus
Equus

Horses like terriers
The oldest horse, *Hyracotherium* (below), was a tiny animal, no larger than a terrier, and is known from numerous specimens found in North America, Europe and Asia. It had four toes on its hind feet and three on its front feet. Later horses were larger (*left*) and lost the additional side toes. *Hyracotherium* was a forest-dweller which fed on soft leaves from bushes.

some migration from Asia to Europe. By Early Miocene times, the North Atlantic route was no longer passable by land mammals, and the next major migration phase could only use the Alaska–Siberia route. The anchitheres, a group of pony-sized horses, evolved in North America and moved across Asia to Europe and Africa. These were the last of the forest-dwelling short-toothed horses.

Two further waves of evolution were marked by the origin of *Merychippus* and the *Hipparion* group in the Miocene, and of modern horses in the Pliocene. Both groups seem to have originated in North America, and they conquered the world successfully after travelling across the Bering Strait to Asia.

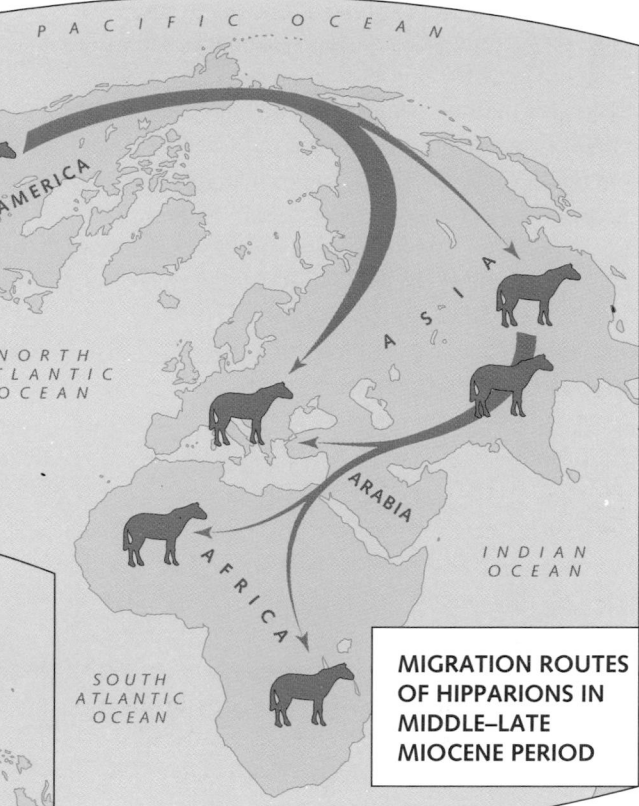

MIGRATION ROUTES OF HIPPARIONS IN MIDDLE–LATE MIOCENE PERIOD

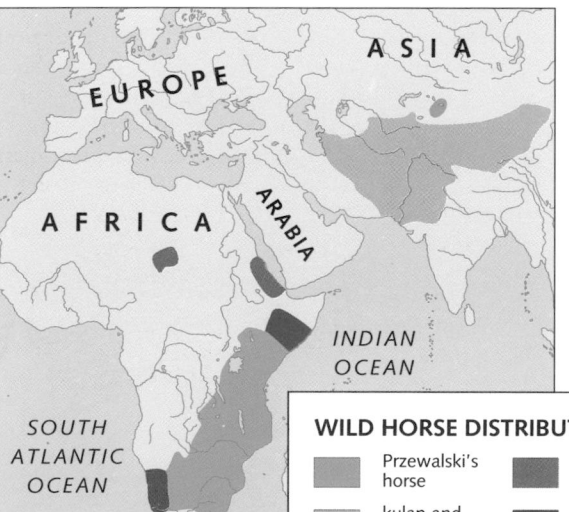

MIGRATION ROUTES OF EQUUS, HIPPIDION AND ONOHIPPIDIUM IN PLIOCENE–PLEISTOCENE PERIODS

Modern horses
Modern wild horse distributions (*below left*) are much reduced in comparison with the earlier record of the group. Horses and zebras, which are classified into several species of *Equus*, today occur naturally only in Asia and Africa. *Equus* was killed off in the Americas at the end of the last ice age, perhaps by hunting. Aboriginal horse populations in Europe and northern Asia disappeared later, also as a result of human activity. *Below*: Przewalski's horse, which is native to a small area in central Asia, is close to the ancestral stock of all living horses.

In the Middle and Late Miocene, the hipparions (*above right*) advanced across the Bering Strait and through Asia to Europe and Africa. They colonized Europe, Asia and Africa. This was followed by the radiation of modern horses (*above*). These reached the places formerly occupied by the hipparions, as well as South America.

WILD HORSE DISTRIBUTION (PRESENT-DAY)

- Przewalski's horse
- kulan and onager
- African ass
- Grevy's zebra
- plains zebra
- mountain zebra

Living Fossils
Lungfish and their distant ancestors

There are many plants and animals today that are called 'living fossils' – the coelacanth, the tuatara, the bowfin fish, lungfish, the mollusc *Neopilina*, the horseshoe crab and the brachiopod *Lingula*. They are called living fossils because they look very similar to creatures of long ago. Some of them, such as the coelacanth and *Neopilina*, were found only quite recently and, in each case, they caused a great shock to biologists because they looked just like creatures known only from fossils that are hundreds of millions of years old. But is there more to living fossils than this emotional impact?

These living fossils have been

Their close relatives the coelacanths have a similar history. They were thought to have died out by Jurassic times, but a specimen fished up in the Indian Ocean in 1938 caused a sensation. Fisheries scientists were unable to identify it until one of them consulted a book about fossil fish. Since 1938 several more specimens of the new coelacanth *Latimeria* have been obtained and studied, but that extraordinary gap of 150 million years in the fossil record is still unfilled.

The living coelacanth, *Latimeria* (below), is a large fish, some 2 metres long. It looks exactly like its smaller fossil relatives, but the youngest fossil coelacanth, such as *Diplurus* (below), is 150 million years old.

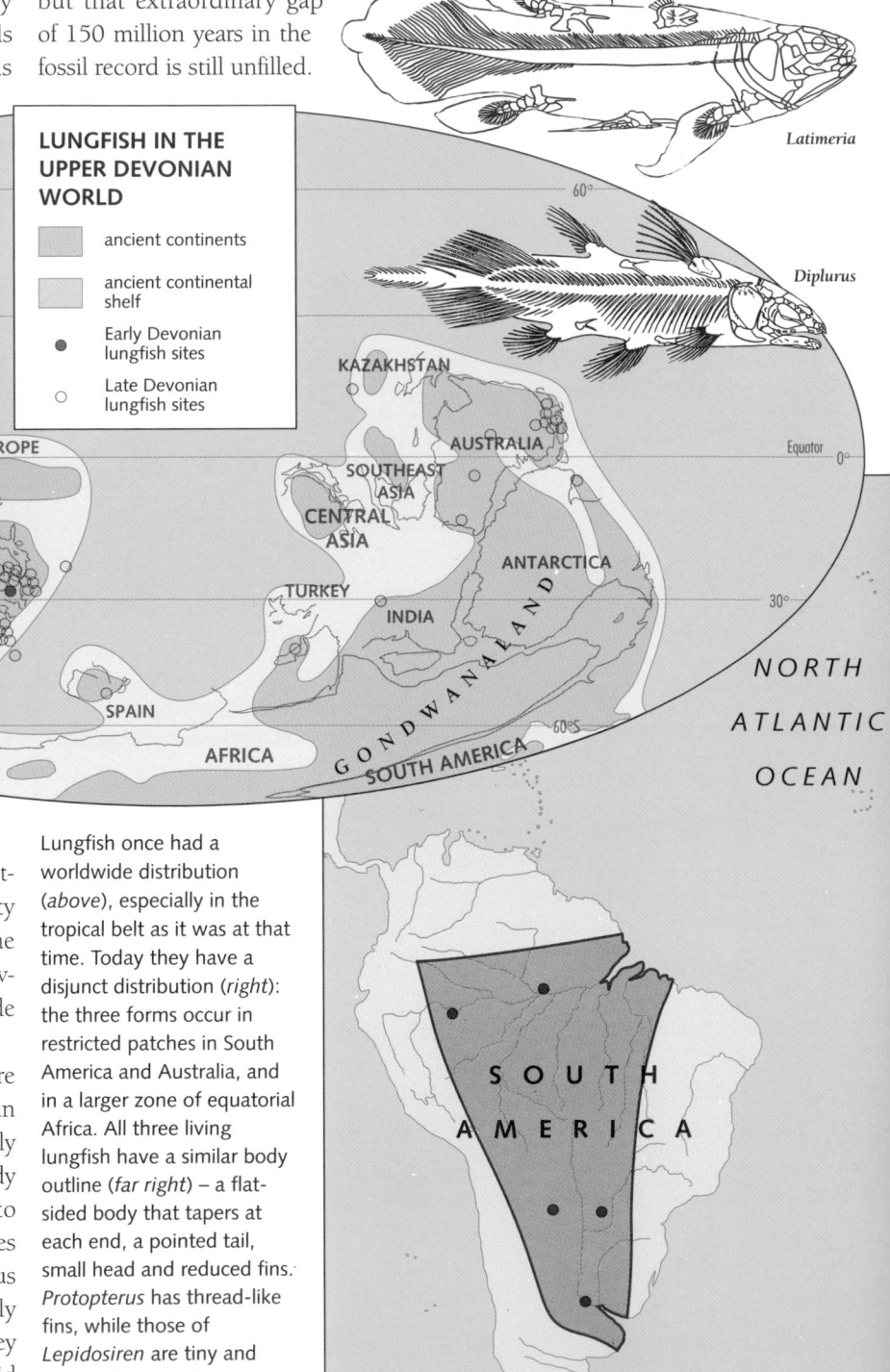

Latimeria

Diplurus

LUNGFISH IN THE UPPER DEVONIAN WORLD

- ancient continents
- ancient continental shelf
- ● Early Devonian lungfish sites
- ○ Late Devonian lungfish sites

explained in many ways: as slowly evolving specialists that escaped competition by having unique narrow preferences; as long-lived generalists that escaped competition by being adaptable to a wide range of niches; or as low-diversity groups that have existed for a very long time. On the other hand, there may be nothing special about living fossils; perhaps they just lie at one end of a wide spectrum of evolution rates.

One well-studied example of living fossils are the lungfish, known from 20–30 species at a time in Devonian rocks worldwide. Lungfish evolved quickly in the Devonian, almost achieving their modern body shapes within 50 million years, but then seem to have stagnated. There were only four or five species worldwide at any time, and very little further obvious change took place. Now lungfish are known only from Australia, South America and Africa, where they live a specialized life, estivating in burrows to avoid the dry season when other fish die.

Lungfish once had a worldwide distribution (*above*), especially in the tropical belt as it was at that time. Today they have a disjunct distribution (*right*): the three forms occur in restricted patches in South America and Australia, and in a larger zone of equatorial Africa. All three living lungfish have a similar body outline (*far right*) – a flat-sided body that tapers at each end, a pointed tail, small head and reduced fins. *Protopterus* has thread-like fins, while those of *Lepidosiren* are tiny and those of *Neoceratodus* are rather broader.

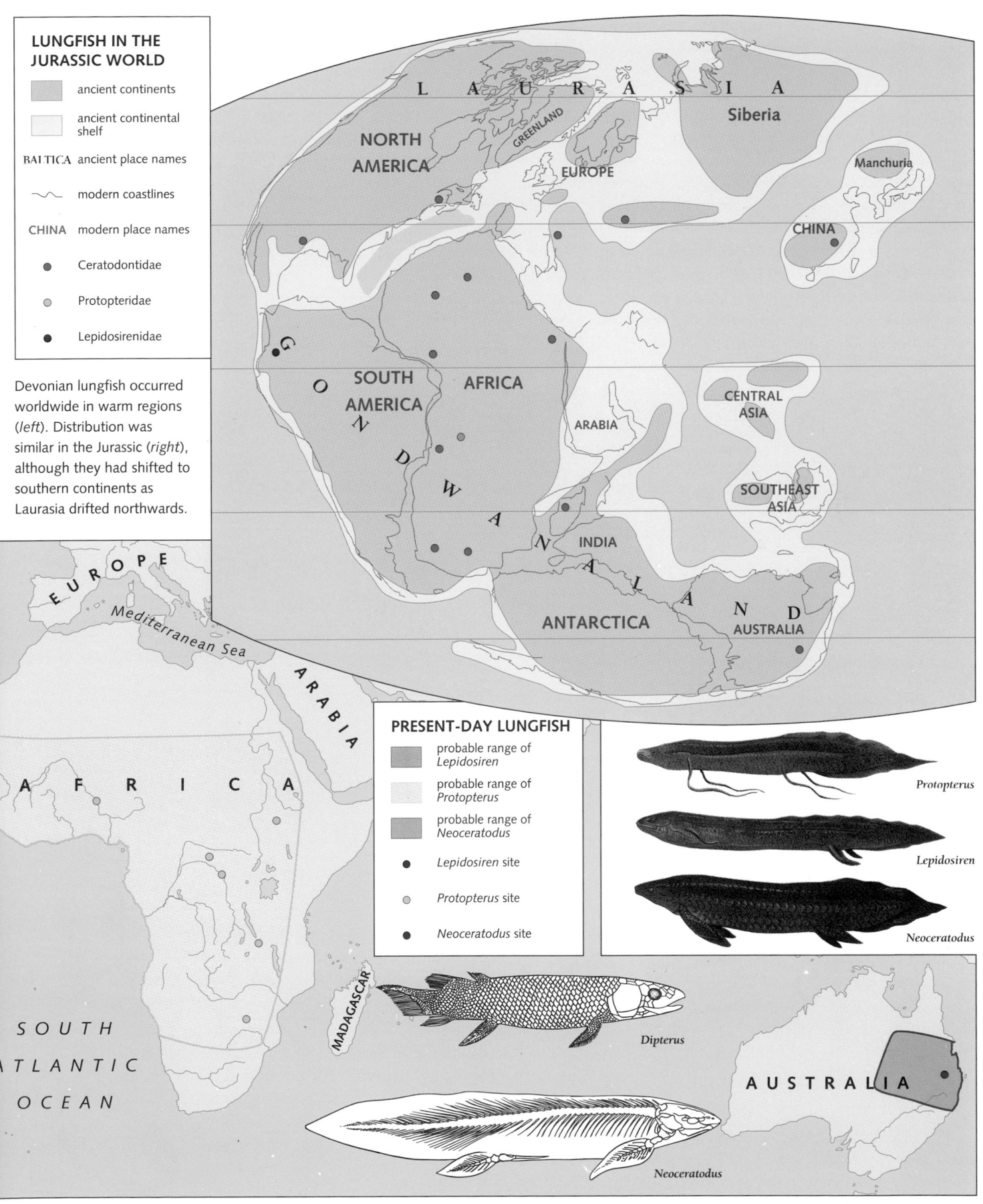

LUNGFISH IN THE JURASSIC WORLD

- ancient continents
- ancient continental shelf

BALTICA   ancient place names

~~~   modern coastlines

CHINA   modern place names

- ● Ceratodontidae
- ● Protopteridae
- ● Lepidosirenidae

Devonian lungfish occurred worldwide in warm regions (*left*). Distribution was similar in the Jurassic (*right*), although they had shifted to southern continents as Laurasia drifted northwards.

PRESENT-DAY LUNGFISH

- probable range of *Lepidosiren*
- probable range of *Protopterus*
- probable range of *Neoceratodus*
- ● *Lepidosiren* site
- ● *Protopterus* site
- ● *Neoceratodus* site

Protopterus

Lepidosiren

Neoceratodus

Dipterus

Neoceratodus

The Tethys Ocean
Evolution in an ancient seaway

The Tethys Ocean ran beneath the present Mediterranean Sea and eastwards across the Middle East and much of Asia. To the west, because the Atlantic had not opened up, it also extended over the area of the Caribbean. The Tethys Ocean existed for over 300 million years, girdling the equatorial realm for much of that time. This major seaway has now disappeared entirely, but it has left its mark on the biogeographic distributions of many fossil groups, and on some living groups too.

Tethys may have originated in Devonian times, some 390 million years ago, a time of major continental collision to the north and associated mountain building. The ocean had certainly begun to open by the beginning of the Carboniferous 350 million years ago. It extended as a narrow arm of sea from the Salt Range of Pakistan in the southwest, and one in Southeast Asia. Evidence for these paleobiogeographic zones comes from studies of the fossils of the ancient shallow shelf sea that ran along the southern margin of the Tethys Ocean, from the present-day Mediterranean area to modern Japan and Southeast Asia. Microscopic planktonic organisms, such as fusulinids, and larger shellfish give clear evidence for the former existence of this vast ocean.

The Tethyan margins remained similar throughout the Triassic and Jurassic, even during the initial opening up of the Atlantic. In the Jurassic, there were two major provinces of marine life at the western end where the ocean reached western Europe. A northern or boreal province extended over Britain, Germany, northern France and much of eastern Europe. The

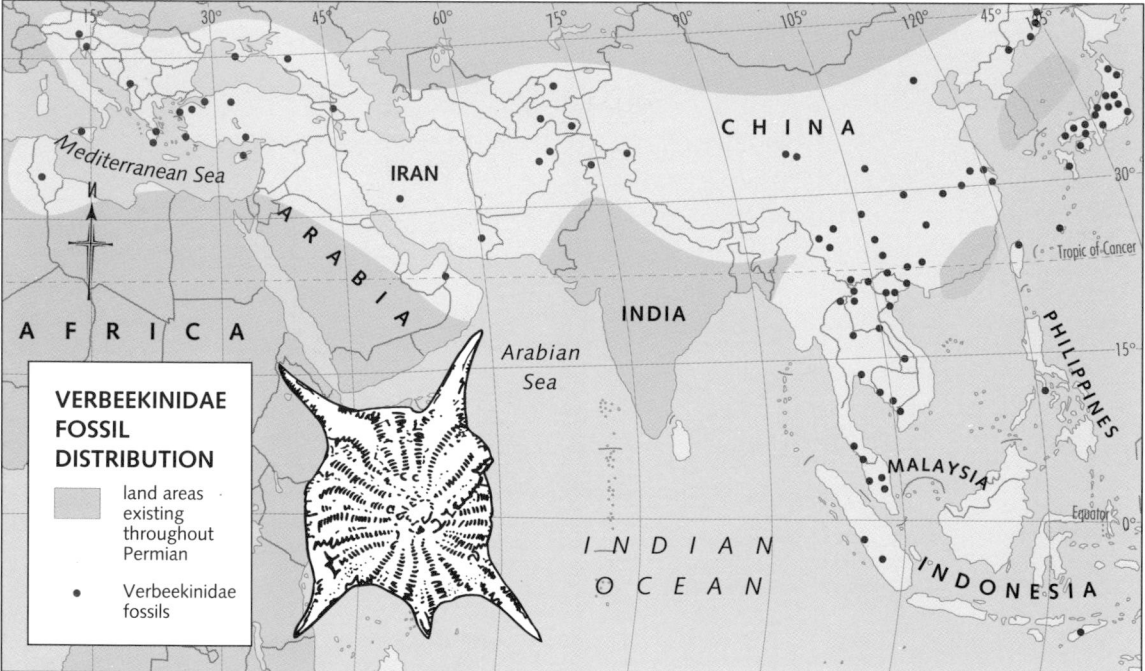

VERBEEKINIDAE FOSSIL DISTRIBUTION

land areas existing throughout Permian

• Verbeekinidae fossils

Verbeekinid foraminifera (members of the plankton) are found throughout the broad strip of rocks extending across the shallow seas of eastern Europe and Asia from the Permian and earlier (*left*). This belt led to the concept of a Tethyan ocean. When Tethys is opened up (*top right*), the distribution of these foraminifers makes sense. Molluscs from the Miocene (*above* and *below*) show that Tethys survived until 5–10 million years ago.

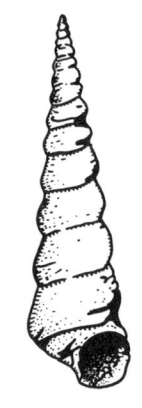

the west, separating the supercontinents Laurasia and Gondwanaland.

The nature of Tethys changed dramatically in the Permian and Triassic periods, when Laurasia and Gondwanaland coalesced to form a single supercontinent, Pangea. The narrow Tethys Sea of the Carboniferous closed off but a huge ocean opened to the east, its coastline extending across central Asia, much of central Europe (then the salty Zechstein Sea), northern Africa, Arabia, north India, north Australia and Southeast Asia. The ocean margins supported very different faunas of bivalves, brachiopods and corals, and major provinces are distinguished: one in the north on the southern shores of Angaraland (now Siberia), one in the Himalayas and

boreal faunas contained very different shellfish, particularly ammonites from the southern or Mediterranean/Tethyan province.

The Tethys Ocean survived the great mass extinction 65 million years ago, and came to an end about 20 million years ago as a result of several events. First, India crossed from the southern margin of Tethys, breaking away from Gondwanaland 100 million years ago. It drifted north and finally collided with Asia. Africa and Arabia were also moving north, and contacted southern Europe about 20 million years ago. As Africa drove north, it raised the Alps. Tethys split at its eastern end, one branch stretching from the Alps to the Caspian Sea and the other lying south of a new land mass over Turkey. The last gasp of Tethys was an enclosed ocean basin in the northern belt, over the Black Sea and the Caspian Sea. This finally emptied of water 5 million years ago. The modern Mediterranean followed about 4 million years ago, when the waters of the Atlantic broke through the Straits of Gibraltar.

The history of Tethys has been reconstructed from studies of the geology of large tracts of Europe and Asia. The concept of Tethys had been proposed long before continental drift became an acceptable model for the history of the Earth's surface (*see pages 86–87*), but early paleontologists were puzzled by the close proximity of different marine faunas. When Tethys is unzipped, it is obvious that the faunas of Southeast Asia and India once lay far away from those of China, Siberia and southern Europe.

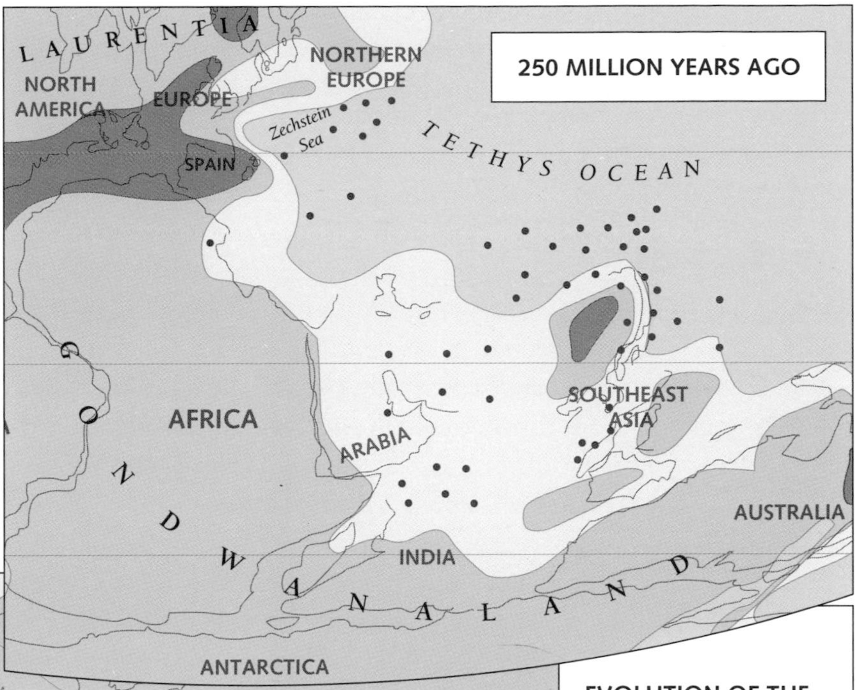

250 MILLION YEARS AGO

LAURENTIA
NORTH AMERICA
EUROPE
NORTHERN EUROPE
Zechstein Sea
SPAIN
TETHYS OCEAN
GONDWANALAND
AFRICA
ARABIA
SOUTHEAST ASIA
INDIA
AUSTRALIA
ANTARCTICA

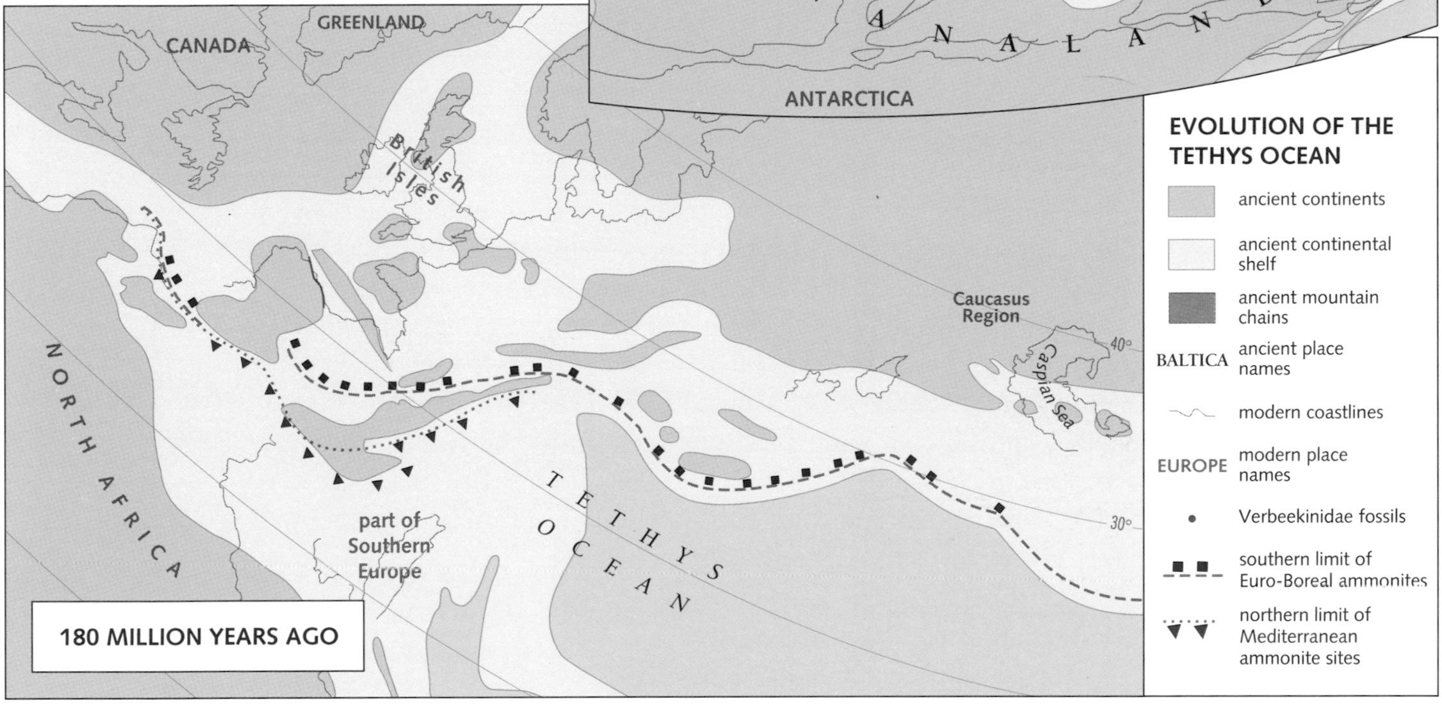

180 MILLION YEARS AGO

GREENLAND
CANADA
British Isles
Caucasus Region
BALTICA
Caspian Sea
40°
NORTH AFRICA
part of Southern Europe
TETHYS OCEAN
30°

EVOLUTION OF THE TETHYS OCEAN

	ancient continents
	ancient continental shelf
	ancient mountain chains
BALTICA	ancient place names
	modern coastlines
EUROPE	modern place names
•	Verbeekinidae fossils
■■ ■	southern limit of Euro-Boreal ammonites
▼▼	northern limit of Mediterranean ammonite sites

Paleobiogeography

The Tethys Ocean was at its height from Permian (*top right*) to Jurassic (*above*) times. At its eastern end, the ocean was so wide that the marine faunas within it were quite isolated. At its western end, the ocean was narrower. During the Jurassic, ammonites were common swimming marine animals. A distinct Euro–Boreal fauna is found on the northern shore, and a Mediterranean fauna is found in the south.

The end of Tethys

Tethys closed up rapidly between 20 and 10 million years ago. India moved from the south to meet Asia, closing the eastern end. Africa moved towards southern Europe, and Tethys became pinched off. Land rose in the region of Turkey, splitting Tethys in two (*right*). By 5 million years ago, the southern strip had become land, and Tethys had become a restricted sea over the Black Sea and the Caspian Sea.

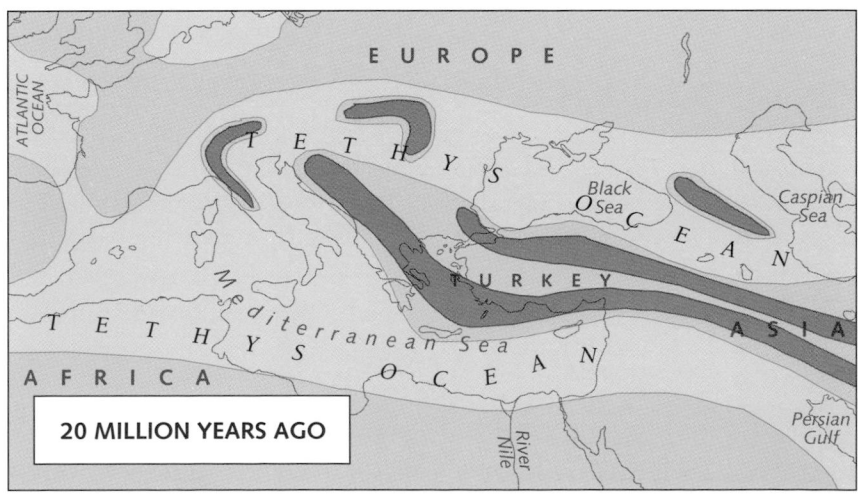

EUROPE
ATLANTIC OCEAN
TETHYS OCEAN
Black Sea
Caspian Sea
TURKEY
ASIA
Mediterranean Sea
AFRICA
TETHYS OCEAN
River Nile
Persian Gulf

20 MILLION YEARS AGO

Section V: Continental Movement

Continental movement has had profound effects on the pattern of life on Earth. Continents and ocean basins are in a constant state of flux as a result of plate tectonics, and this has split formerly continuous species distributions, and brought others together, often with dramatic evolutionary results.

Until the 1960s, most biologists and geologists viewed the Earth as static. The layout of the continents and oceans had apparently not changed since the origin of the Earth. But seemingly inexplicable distributions were found. Why do marsupials occur essentially only in Australia and South America, for example? Fossils were known from North America and Europe, so the ancestral forms must have hiked overland from Europe, across Asia and a chain of islands, to Australia. Other distributions were even more difficult to explain. Why should the Permian reptile *Mesosaurus* be found only in a small area on Brazil's east coast and a small area in west Africa? Why do modern arum lilies occur in patches in South America and south-eastern Asia, or India and South Africa? Why should *Nothofagus*, the southern beech, occur only in patches around the Antarctic, in Chile, New Zealand, Australia and New Guinea, with fossils on Antarctica?

The solution favoured by most biologists and paleontologists was a network of land bridges in former times. These routes of dry land between the continents had to be postulated to explain some of the scattered, or disjunct, distributions. Hence, a land bridge between South America and Africa was built across the narrowest part of the South Atlantic. Another was constructed over Greenland to link North America and Europe, and others were built across the Pacific. These were used to explain the distributions despite the absence of any geological evidence, and despite the impossibility of imagining such structures in waters that are today kilometres deep.

Continental drift

The solution had been hinted at in the 1860s, when it was noted how the coasts of Africa and South America matched closely. The idea was developed in 1912 by Alfred Wegener, who argued that anomalous distributions could be explained by the slow and continuous drift of the continents. He reconstructed the former positions of the continents by looking at ancient plant and animal distributions, and realized that during the Permian and Triassic there had been a single supercontinent, Pangea, formed by the amalgamation of older continental masses, essentially Gondwanaland (South America, Africa, India, Australia and Antarctica) and Laurasia (North America, Europe and parts of Asia). After the Triassic, Pangea had been broken up by the opening of the Atlantic Ocean and the breakup of Gondwanaland, when India, Australia and Antarctica drifted away from Africa. This idea was taken up by a few enthusiasts in the 1920s and 1930s, but most geologists were violently opposed to the disturbing idea that the Earth was in constant upheaval. Even more dismissive were geophysicists, who asserted that there was no known mechanism that could drive millions of tonnes of rock around over millions of years.

Alex Du Toit in South Africa extended Wegener's evidence for a southern glaciation in the Carboniferous and Permian. Glacier movement was clearly shown in rocks of that age in South America, southern Africa, India and Australia. Fixed-earth geologists had to assume that the ice sheet had extended over much of the present southern hemisphere. But the fact that it extended into India, which is now in the northern hemisphere, was inexplicable, as the ice sheet would have had to cross the Equator. When Gondwanaland is restored, there is no problem. But du Toit and the other supporters of continental drift were ignored or ridiculed in the 1930s and 1940s.

Plate tectonics

The dénouement came in 1963, when British geophysicists Fred Vine and Drummond Matthews announced evidence for changes in the Earth's magnetization in rocks in the Indian Ocean. The Earth's magnetic field had flipped over many times, giving phases of 'normal' magnetization and 'reverse' magnetization (when the magnetic poles reverse). Magnetic stripes reported on the Atlantic floor were seen to form a mirror image pattern on either side of the Mid-Atlantic Ridge. The Ridge is a site of volcanic activity, and geophysicists proposed that it was a source of new oceanic crust for the entire length of the Atlantic. The mirror image stripes indicate an identical history of production of new crust on either side of the Ridge, with the rocks becoming older the further they are from it. They document a history of some 200 million years, confirming Wegener's proposal that the Atlantic began to open 200 million years ago. The mechanism was becoming clear: Europe and Africa on one side and the Americas on the other are being forced apart a few millimetres a year by the constant production of new ocean floor.

As new crust is produced down the middle of the Atlantic, and along another ridge system that runs from California across the Pacific, south of New Zealand, and over the southern Indian Ocean to Africa, the continents move to accommodate it. The Earth's crust is divided into five major plates (Eurasia, Africa, Australia, South America and North America) and numerous smaller ones, and there is constant movement. The driving force comes from the molten rocks (magma) of the Earth's mantle, the zone beneath the solid crust. Magma rises at the mid-ocean ridges. Some emerges and solidifies as new crust, but most meets the cold base of the crust, and moves sideways and then downwards, forming great rotating convection cells. It is these convection cells in the mantle that drive continental drift. This is the plate tectonic motor that physicists did not predict in the 1940s.

Glossopteris and Gondwana
Ancient plants and continental drift

One of the classic pieces of evidence for continental drift is the distribution of the seed fern *Glossopteris* in Permian rocks. In Victorian times, this plant was identified from such far-flung parts of the southern hemisphere as South Africa, Argentina, India, Australia and Antarctica. The distribution was hard to explain on modern geography, not least because of the huge spans of ocean to be crossed by seeds. Even more bizarre was the climatic span of the genus: how could a single plant type be found from the South Pole to the Equator, and, in the case of India, north of the Equator too?

In the 19th century, geographers noted the close match between the west coast of Africa and the east coast of South America and speculated that the two continents might once have lain side by side. In 1912, German climatologist Alfred Wegener gathered evidence suggesting that there had been a southern supercontinent (Gondwanaland) consisting of Africa, South America, Antarctica, India and Australia, and that the continents had since drifted apart. Most astonishing

was the idea that India had fitted between Africa and Australia when it was part of Gondwanaland and must then have drifted northwards before fusing with Asia.

Wegener called this idea continental drift. It was based on a variety of evidence, including the obvious fit of the continents and paleontological evidence such as the distribution of *Glossopteris*. The latter made sense when it was plotted on the ancient supercontinent. Wegener also noted that some reptiles of similar age confirmed the fit. A small freshwater reptile, *Mesosaurus*, was found in small patches of Lower Permian rocks in eastern Brazil and western Africa, and these patches would have fitted neatly together when the continents were joined.

In addition, the geology of the time confirmed the pattern. In the Late Carboniferous and Early Permian, glaciations affected Gondwanaland. The striations (which show the direction of glacier movements) make sense in a reconstructed map, since

The supercontinent known as Gondwanaland lay in the southern hemisphere in Permian and Triassic times (*below*) and was still fused with the northern supercontinent Laurasia at this time, forming Pangea. Fossil plants and animals prove this arrangement of the continents, since the distributions are inexplicable on a modern world geography. The unusual fish-eating reptile *Mesosaurus* occurs in a small basin in Brazil and west Africa. *Glossopteris* and associated plants are found from Antarctica to India – not so far apart in the Permian world, but now very distant from each other.

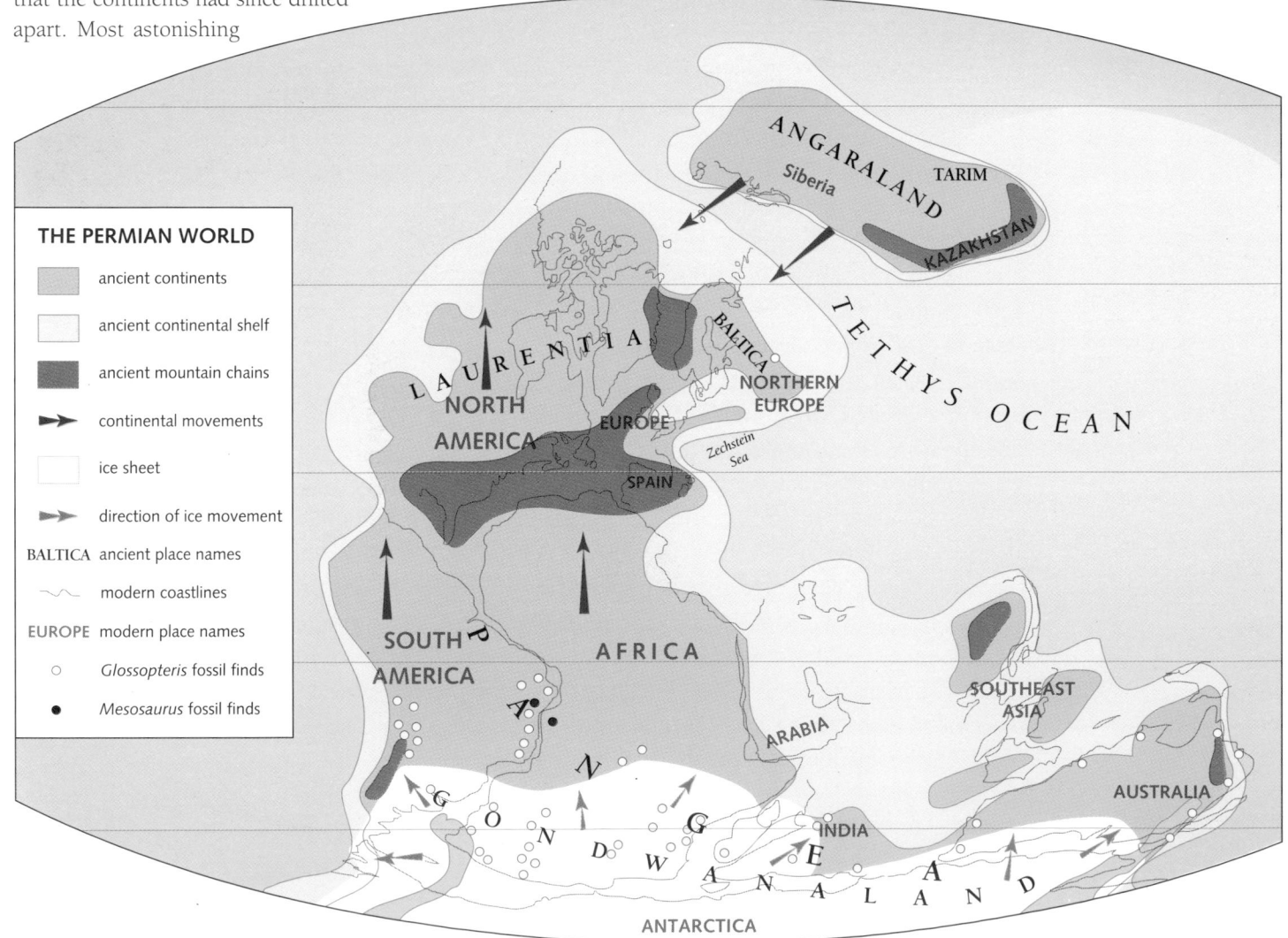

THE PERMIAN WORLD

- ancient continents
- ancient continental shelf
- ancient mountain chains
- continental movements
- ice sheet
- direction of ice movement
- BALTICA ancient place names
- modern coastlines
- EUROPE modern place names
- ○ *Glossopteris* fossil finds
- ● *Mesosaurus* fossil finds

Mesosaurus (*below*) was a small reptile less than a metre in length. It had a long narrow snout lined with dozens of needle-like teeth and was almost certainly a fish-eater, seizing fish in its mouth and holding them fast with its long teeth. A freshwater dweller, it could not have crossed the sea between South America and Africa where fossil remains have been found and thus backs up the theory of continental drift.

Mesosaurus

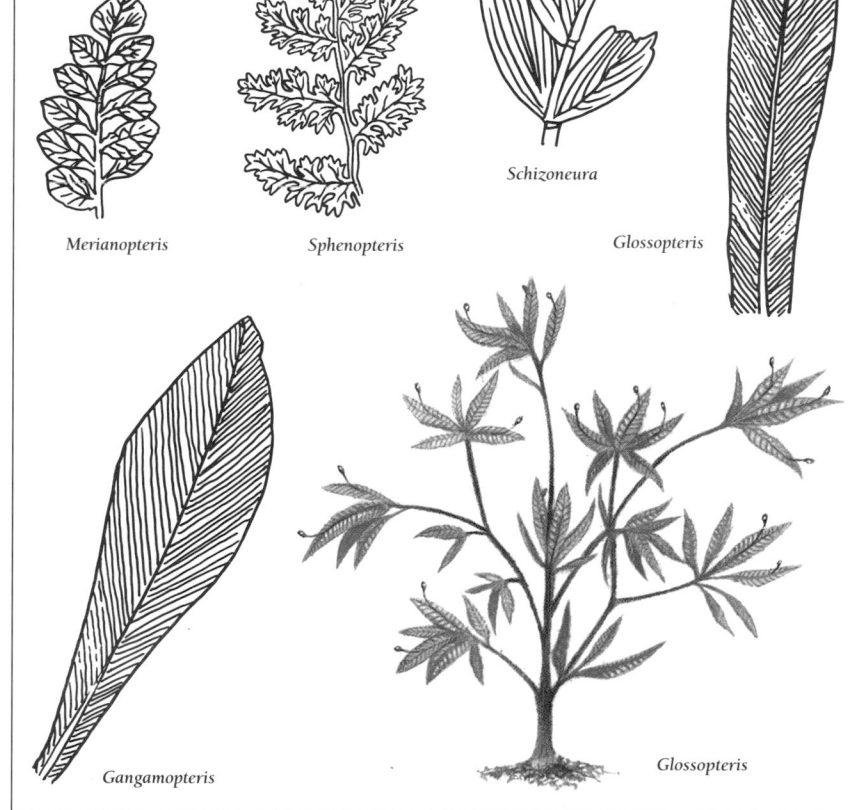

THE PRESENT-DAY DISTRIBUTION OF LATE PALEOZOIC LAND FLORAS

- northern floras
- southern floras
- Asiatic floras

Below: Typical *Glossopteris* plants, currently found from the South Pole to north of the Equator. This distribution makes no sense climatically, while the Late Permian distribution makes no sense on the modern map (*above*).

Merianopteris   *Sphenopteris*   *Schizoneura*   *Glossopteris*

Gangamopteris   *Glossopteris*

they point outwards from a glacial centre somewhere over southern Africa. On a modern world map they make no sense at all. Those in India, for instance, suggest that the ice moved over the Equator.

When Wegener published his ideas, most geologists ignored or ridiculed them. In particular, geophysicists declared that the movement of whole continents around the globe, like lubricated jigsaw pieces, was impossible. There was no known process within the molten rock of the Earth's interior to drive the continents. However, new evidence in the 1950s and 1960s showed that Wegener had been right. Geologists found evidence from geophysics showing that the magnetization of ancient rocks could only be understood if continents were moved out of their present places. When certain iron-bearing rocks are formed, they take up a magnetic signal in line with the poles. Today, after hundreds of millions of years of drift, those magnetized rocks are no longer in alignment. Studies of the molten rocks in the Earth's mantle showed that there are great circulating plumes that rise and spread beneath the solid crust. These natural convection cells power the drift of the continents, by a process known as plate tectonics.

Carabid Beetles
Dispersal, pulses and continental drift

Certain groups of plants and animals are particularly useful in the study of the distribution of life forms on Earth. Carabid beetles have been studied by biogeographers since the last century. There are over 40,000 known and described species, and the drawings below and opposite how diverse these are in shape and size. They come in winged forms and in non-winged forms. They are found in almost every region of the world except Antarctica, and members of the family occupy most land habitats and altitudes from sea level to over 5,000 metres.

The key to the present diversity of carabids lies in their history. Early proto-carabid fossils have been found in rocks of Permian age (300–250 million years old) in Russia. At that time the continents of the Earth were combined in one continuous land mass known as Pangea (literally, 'all Earth'). The first splitting of carabid beetles into groups or tribes is thought to have started in the Permian period.

In common with other groups, carabids are thought to have diversified and developed new species and groups by allopatric speciation: that is, by the geographical separation of a part of the population, followed by its separate development. One of the preconditions for this process is that there should be no movement back and forth between the populations. This would enable mating to continue and would maintain a flow of genes between the two populations, preventing the development of a separate species. For allopatric speciation to work the two populations must remain isolated – by physical or ecological barriers.

But this in itself presents a problem for biogeographers, for if one group of beetles has managed to get across a barrier – a river, a mountain chain or a glacier – then why should other groups not be able to follow? The answer must be that there is some change in conditions that enforces the isolation of the pioneering group. This change could be anything from the diversion of a river channel to an avalanche or a lava flow.

This idea of using physical changes to explain the distribution of species is known as vicariance biogeography (*see pages 92–93*). It is a theory that received its biggest boost when knowledge of the greatest physical changes that have happened on the Earth's surface – the movement of the continents (*see pages 86–87*) – was gained.

The maps on these pages give the history of three groups of carabid beetles as proposed by one

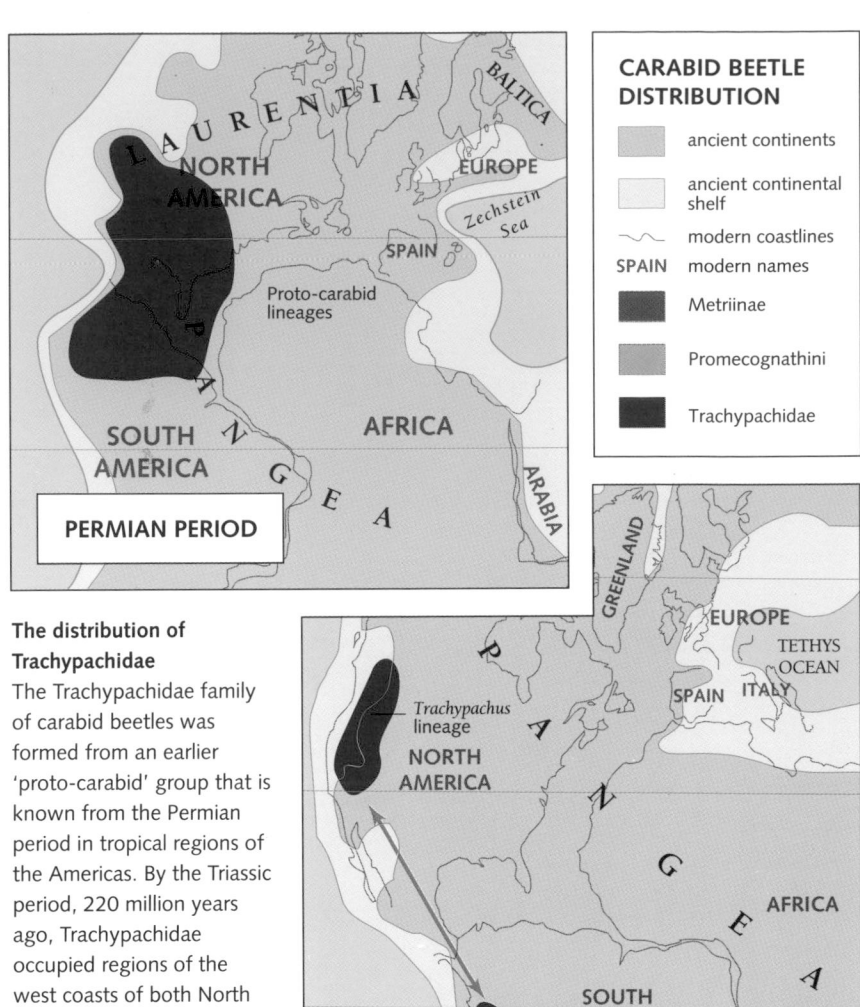

CARABID BEETLE DISTRIBUTION

- ancient continents
- ancient continental shelf
- modern coastlines
- SPAIN modern names
- Metriinae
- Promecognathini
- Trachypachidae

PERMIAN PERIOD

TRIASSIC PERIOD

The distribution of Trachypachidae

The Trachypachidae family of carabid beetles was formed from an earlier 'proto-carabid' group that is known from the Permian period in tropical regions of the Americas. By the Triassic period, 220 million years ago, Trachypachidae occupied regions of the west coasts of both North America and South America outside the tropics. Further movements away from the tropics brought Trachypachidae to northern Europe by the Eocene period, approximately 50 million years ago.

Pherosophus aequinoctialis

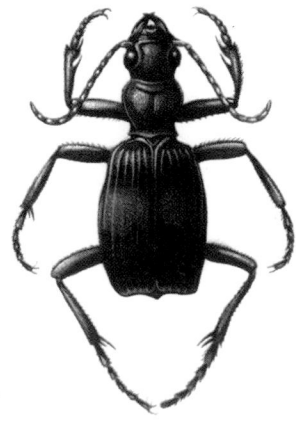

biogeographer. The evolutionary history of such a vast family over a timespan of 300 million years is extraordinarily complex, and it is this complexity that makes the carabid beetles such a fruitful subject. The development of new groups, species and subspecies happens for a variety of reasons. Each major and minor event of the past 300 million years will have had a different effect at a different level on the carabid beetles. At one level continental drift is thought to be the most important event, at another the development of land bridges in the ice ages, and at another the history of forest refuges and cyclical flooding in the Amazon basin (*see pages 112–113*). The carabids serve as a reminder that complex patterns of life cannot be contained by simple explanations.

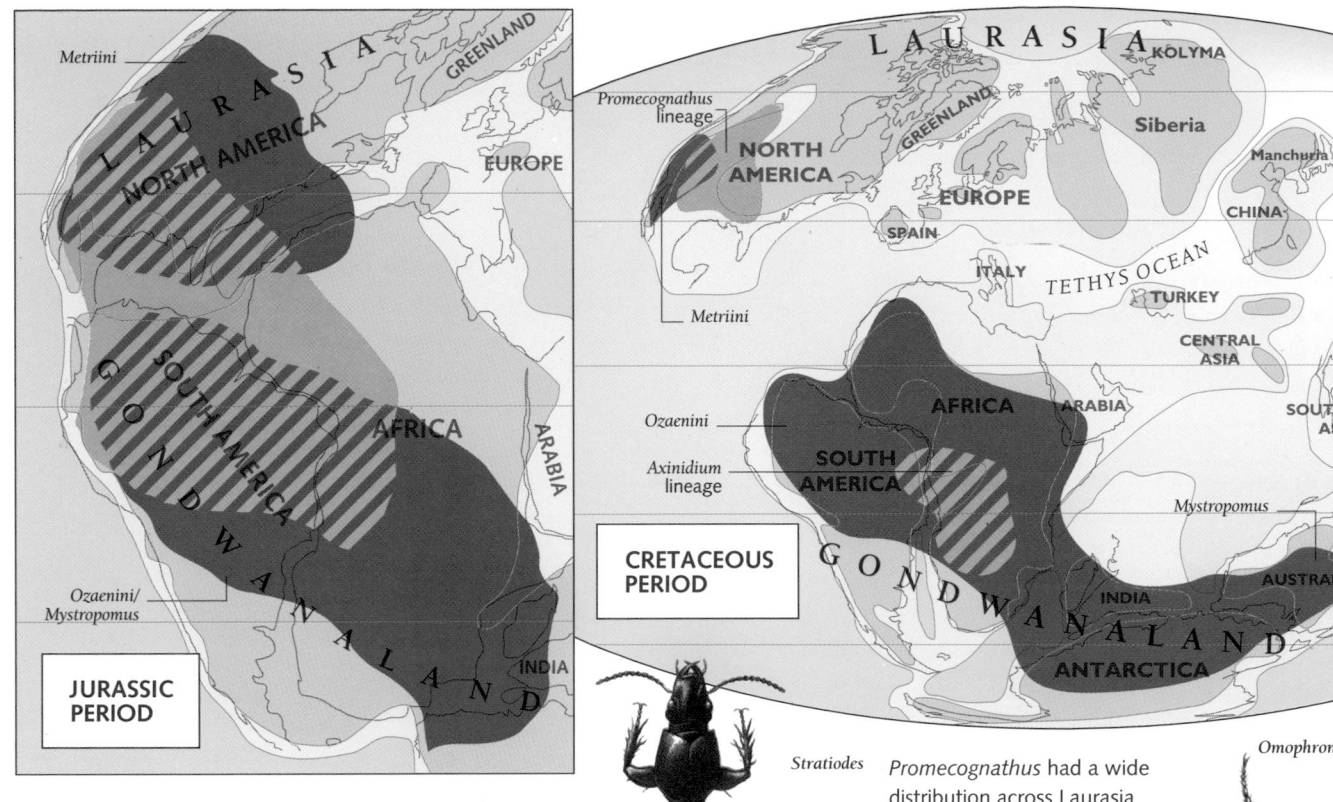

JURASSIC PERIOD

CRETACEOUS PERIOD

Metriinae distribution

The Metriinae spread out of the tropics during the Jurassic period when the continents were still linked. Separate lineages developed in North America and in Gondwanaland as the continents moved apart during the Cretaceous period 100 million years ago. Further continental drifting then isolated the *Mystropomus* lineage in Australia, and separated the

Pasimachus cordicollis

Ozaenini into South American, African, Indian and Southeast Asian populations. The latter three were later reunited genetically through dispersal between them.

Stratiodes

Promecognathus had a wide distribution across Laurasia and Gondwana in the Jurassic, but this had separated and shrunk by the Cretaceous. The map below shows all three families in the Tertiary period.

Omophron gratum

Notiophilus specularis

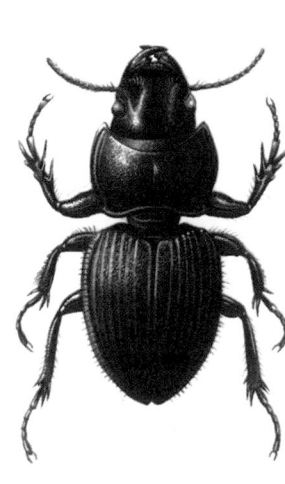

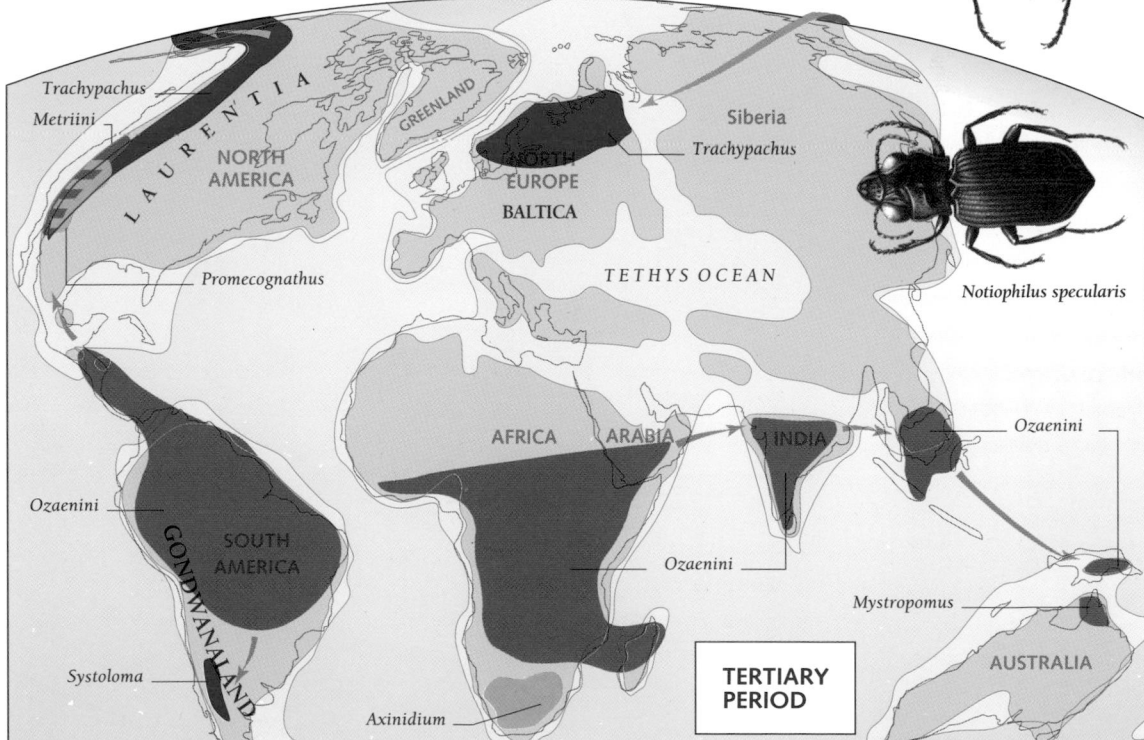

TERTIARY PERIOD

Dinosaurs of North America and Africa

Shared distributions in a united world

Spectacular dinosaur fossils have been found all over the North American continent. The most abundant remains have come from the Morrison Formation, dated as Late Jurassic (about 150 million years old). The limestone, sandstone and shale of the Morrison beds cover an area that extends from northern New Mexico to the border between the United States and Canada.

Morrison fossils have been known since the 1870s, and many of the most famous dinosaurs, such as *Apatosaurus* (*Brontosaurus*), *Allosaurus* and *Diplodocus* were found in this rock unit. Along with fossil plants, shellfish, fish, lizards, turtles, flying pterosaurs and mammals, these fossils, and the

rocks, give a picture of a subtropical lowland area with broad rivers and pools. Palm trees, seed ferns and other exotic trees stood in groups around the watercourses, and the dinosaurs lived in abundance. The most striking thing is that the dinosaur scene in the Late Jurassic was the same all over the world. There were no local faunas on each continent.

In the Late Jurassic, Pangea was still largely intact, although the Atlantic had begun to open. North America lay just north of the Equator, and Africa had begun to separate from the southeast margin of Europe. However, the extraordinary dinosaurs of East Africa prove that migration routes between all the continents were still open. Dinosaur faunas from

During the Late Jurassic, mountain building took place in the western states of North America (*below*). Volcanoes sent ash long distances, and the new Atlantic Ocean extended a branch northwards from the Caribbean area to Texas. Dinosaurs lived on the plains east of the mountains. Rapid erosion of the mountains built up great thicknesses of sediment (the Morrison Formation) and this entombed hundreds of dinosaur skeletons.

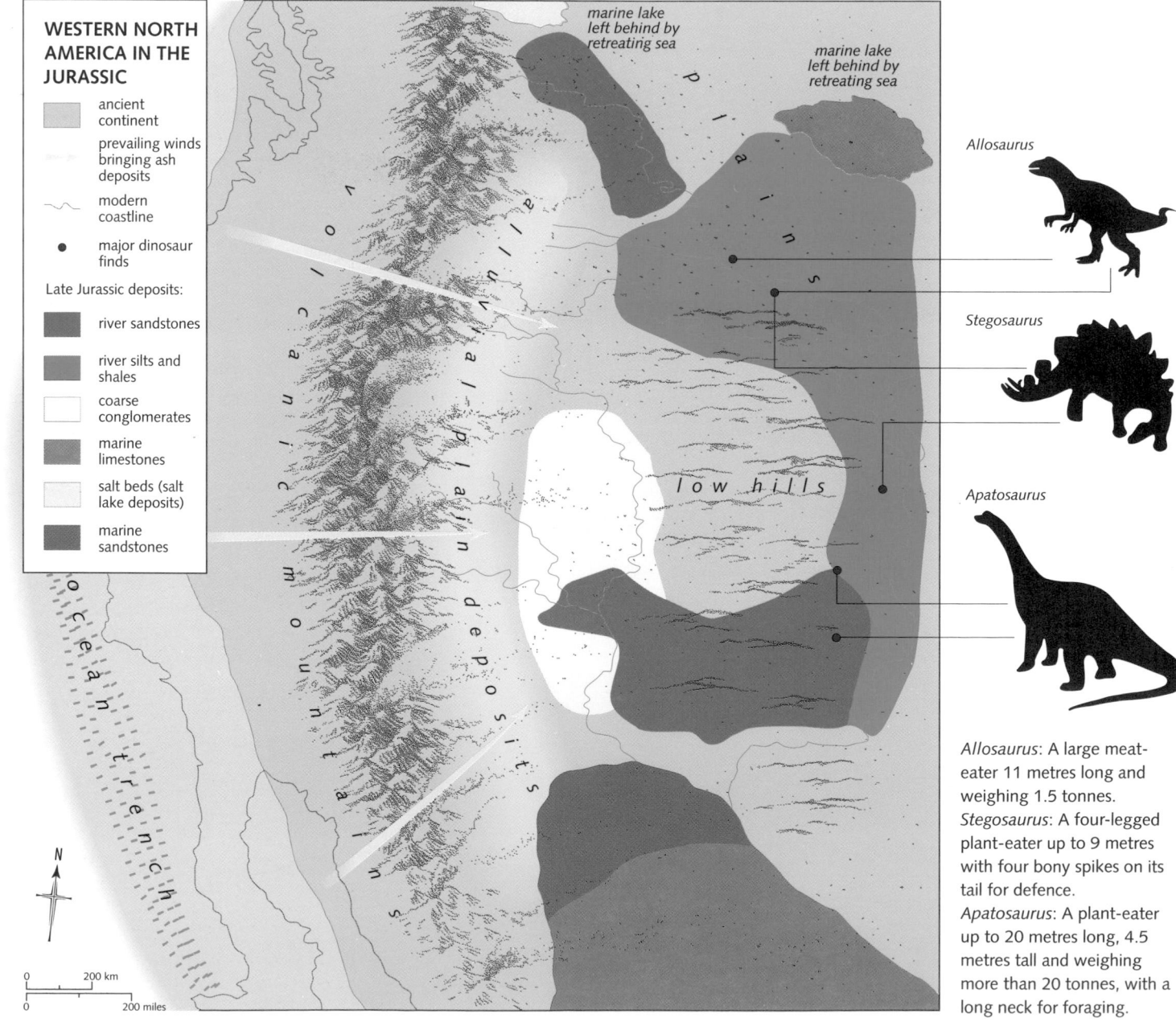

WESTERN NORTH AMERICA IN THE JURASSIC

- ancient continent
- prevailing winds bringing ash deposits
- modern coastline
- major dinosaur finds

Late Jurassic deposits:
- river sandstones
- river silts and shales
- coarse conglomerates
- marine limestones
- salt beds (salt lake deposits)
- marine sandstones

Allosaurus

Stegosaurus

Apatosaurus

Allosaurus: A large meat-eater 11 metres long and weighing 1.5 tonnes.
Stegosaurus: A four-legged plant-eater up to 9 metres with four bony spikes on its tail for defence.
Apatosaurus: A plant-eater up to 20 metres long, 4.5 metres tall and weighing more than 20 tonnes, with a long neck for foraging.

marine lake left behind by retreating sea

marine lake left behind by retreating sea

0   200 km
0   200 miles

JURASSIC DINOSAUR SITES

ancient continents

ancient continental shelf

ancient mountain chains

BALTICA ancient place names

∿ modern coastlines

CHINA modern place names

• dinosaur sites

Tendaguru in Tanzania are nearly identical to those of the Morrison Formation. In particular, the regions share the giant sauropods *Brachiosaurus* and *Barosaurus*. Other dinosaurs, such as the stegosaurs *Stegosaurus* (North America) and *Kentrosaurus* (Africa) and the meat-eating theropods, are very similar. The ornithopod *Dryosaurus* is found in both Africa and North America.

Dinosaur families seem to have had worldwide distributions until the end of the Jurassic, when more endemism set in.

The Jurassic world (*top*) was dominated by the supercontinent Pangea. The Atlantic Ocean began to unzip between North Africa, southern Europe and eastern North America just before the beginning of the Jurassic. By Late Jurassic times, much of the Atlantic had opened up, but dinosaurs could migrate over northern Europe to North America.

Dinosaurs were usually preserved by river deposits. The carcass (*top*) was covered with sand or mud and the bones were infiltrated with minerals. Over time, the rocks are buried and may be tilted by earth movements (*above*). Erosion may later expose part of the skeleton (*left*), and it may be discovered.

By Cretaceous times (*below*), the Atlantic had opened up more fully, although the northern part was still a dry land route. The southern continents separated increasingly throughout the Cretaceous. At the beginning of the period, however, dinosaurs could island-hop from Europe to Africa, and from South America to Antarctica and Australia.

Each continent had its own forms but some interchange could still take place. Some groups died out in one part of the world but lived on in others. The giant sauropods that dominated the Late Jurassic scene, for example, became largely extinct, except in South America.

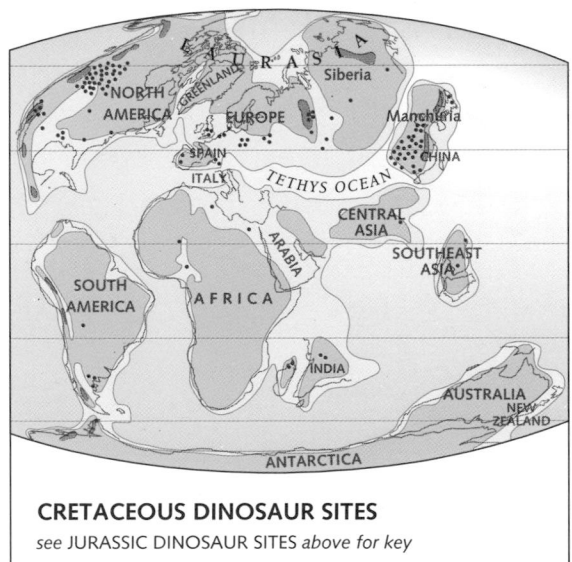

CRETACEOUS DINOSAUR SITES
see JURASSIC DINOSAUR SITES above for key

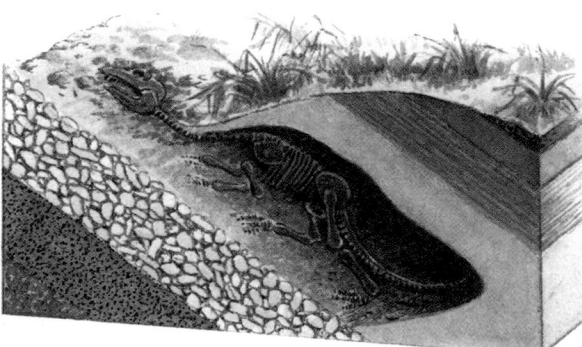

Gondwanaland and the Southern Beech
Present-day flora and continental movement

Single species that occur in areas that are great distances apart are said to have disjunct distributions. These are both a puzzle and a potential source of historical information, and are common in both living organisms and the fossil record. There is particular interest in species which have both fossil and present-day distributions, especially where the historical record differs from current evidence.

Unlike other members of the Fagaceae family of oaks, beeches and sweet chestnuts, the southern beech (*Nothofagus*) is found only in the southern hemisphere. The two main areas in which it occurs, Australasia and South America, are some 7,000 kilometres apart. The conflicting theories as to how this came about demonstrate two different views of how plant and animal distribution should be interpreted.

Dispersal biogeography, which is built on the notion that present distribution is the result of expansion from a historical 'heartland', holds that *Nothofagus* seeds must have dispersed from South America to Australasia, or vice versa. Vicariance biogeography takes the opposite view – that disjunct distributions are due to the dispersal of once contiguous continents. This is supported in the case of *Nothofagus* by fossil specimens found far outside the present range. Combined with the discovery of continental drift and the past linking of the southern continents, this suggests that *Nothofagus* previously had a continuous distribution across all the southern continents, which is now much reduced.

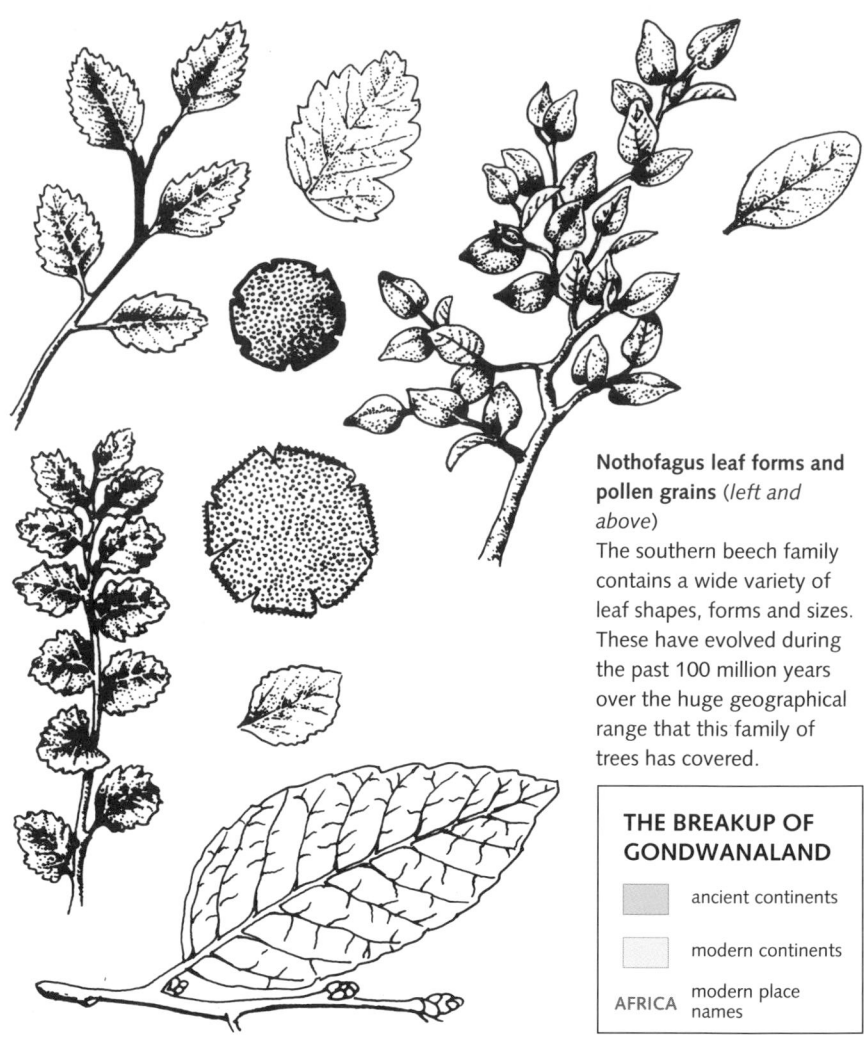

Nothofagus leaf forms and pollen grains (*left and above*)
The southern beech family contains a wide variety of leaf shapes, forms and sizes. These have evolved during the past 100 million years over the huge geographical range that this family of trees has covered.

THE BREAKUP OF GONDWANALAND

	ancient continents
	modern continents
AFRICA	modern place names

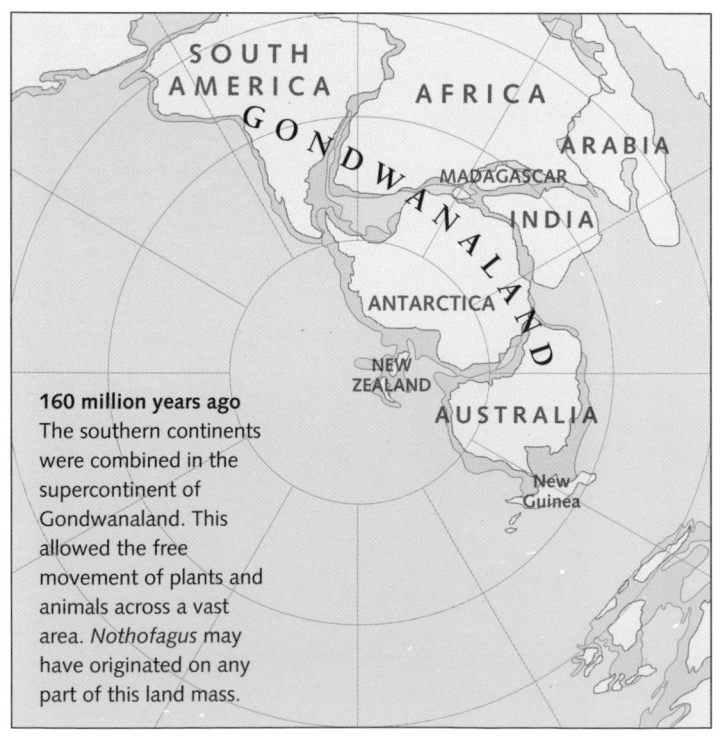

160 million years ago
The southern continents were combined in the supercontinent of Gondwanaland. This allowed the free movement of plants and animals across a vast area. *Nothofagus* may have originated on any part of this land mass.

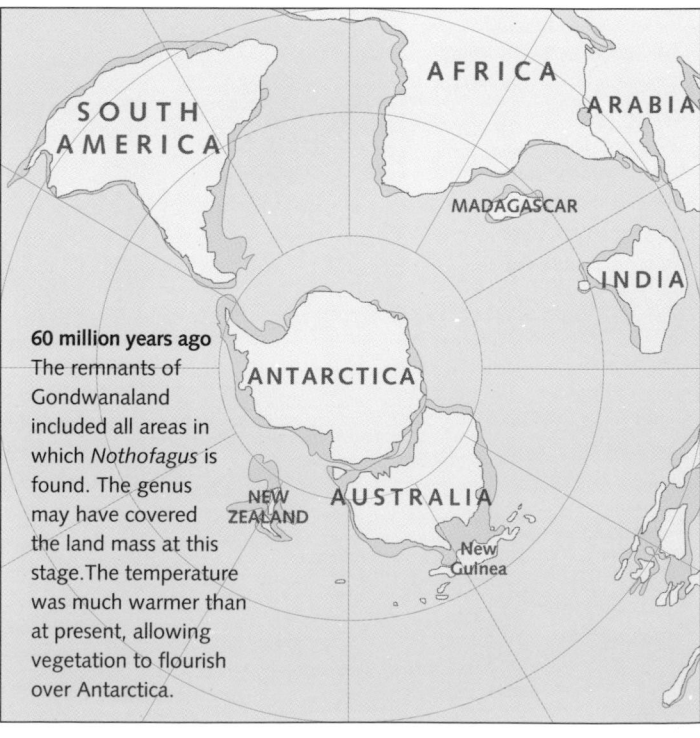

60 million years ago
The remnants of Gondwanaland included all areas in which *Nothofagus* is found. The genus may have covered the land mass at this stage. The temperature was much warmer than at present, allowing vegetation to flourish over Antarctica.

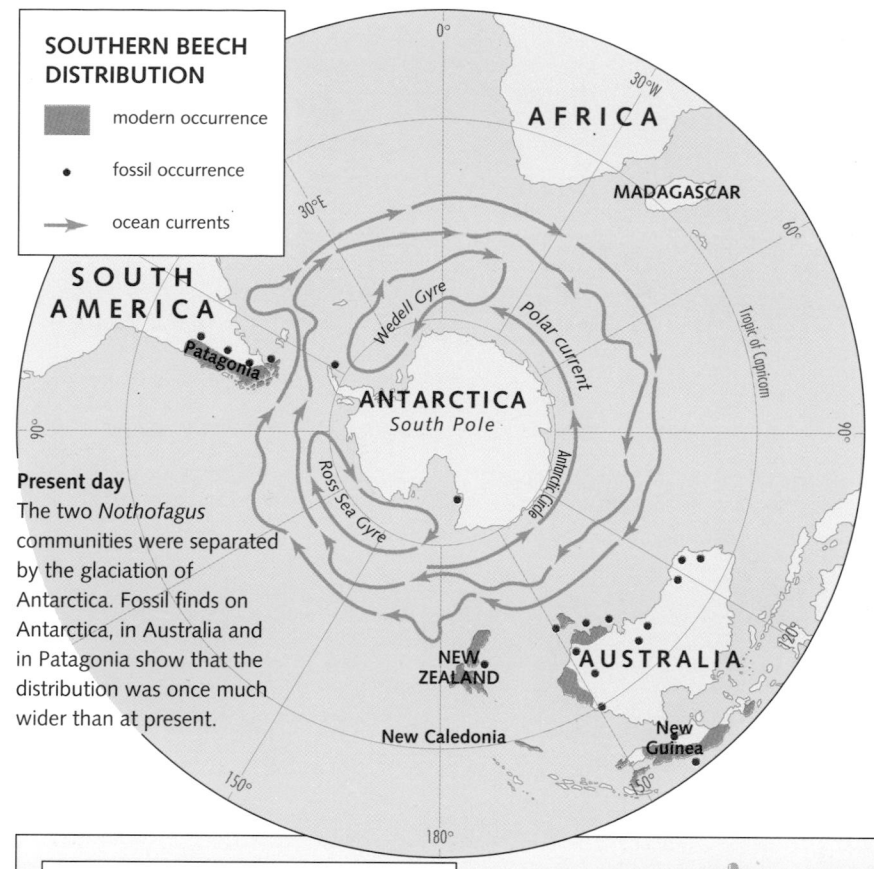

SOUTHERN BEECH DISTRIBUTION

- modern occurrence
- • fossil occurrence
- → ocean currents

Present day
The two *Nothofagus* communities were separated by the glaciation of Antarctica. Fossil finds on Antarctica, in Australia and in Patagonia show that the distribution was once much wider than at present.

The separation of the continents

Our knowledge of continental movement does not rule out the dispersal of *Nothofagus* from one site to a number of isolated locations, but it does suggest a more likely explanation. The actual route of the spread of the genus across the continents and its centre of origin are not known, and probably never will be, because we do not have enough fossil information. In any case, as one researcher in this field has pointed out, our aim should be to understand the flora and fauna of a region, not to trace the exact route of a single species.

A map of the distribution of *Nothofagus*, combined with a cladogram (a family tree based on observed characteristics), gives an insight into how the development of new characteristics and species is related to geographical separation. For example, tracheids – elongated cells found in certain woods – are present only in the *Nothofagus* groups of New Guinea and New Caledonia. The fact that they are not found in southern beeches in other places suggests that this group became isolated from the others at an early stage. The same thinking can be applied to other characteristics, and allows us to build up a picture of the pattern of geographical separation, which has obvious parallels with the breakup of Gondwanaland.

This picture can become confused, as characteristics mayarise independently in separate locations. This is known as convergent evolution.

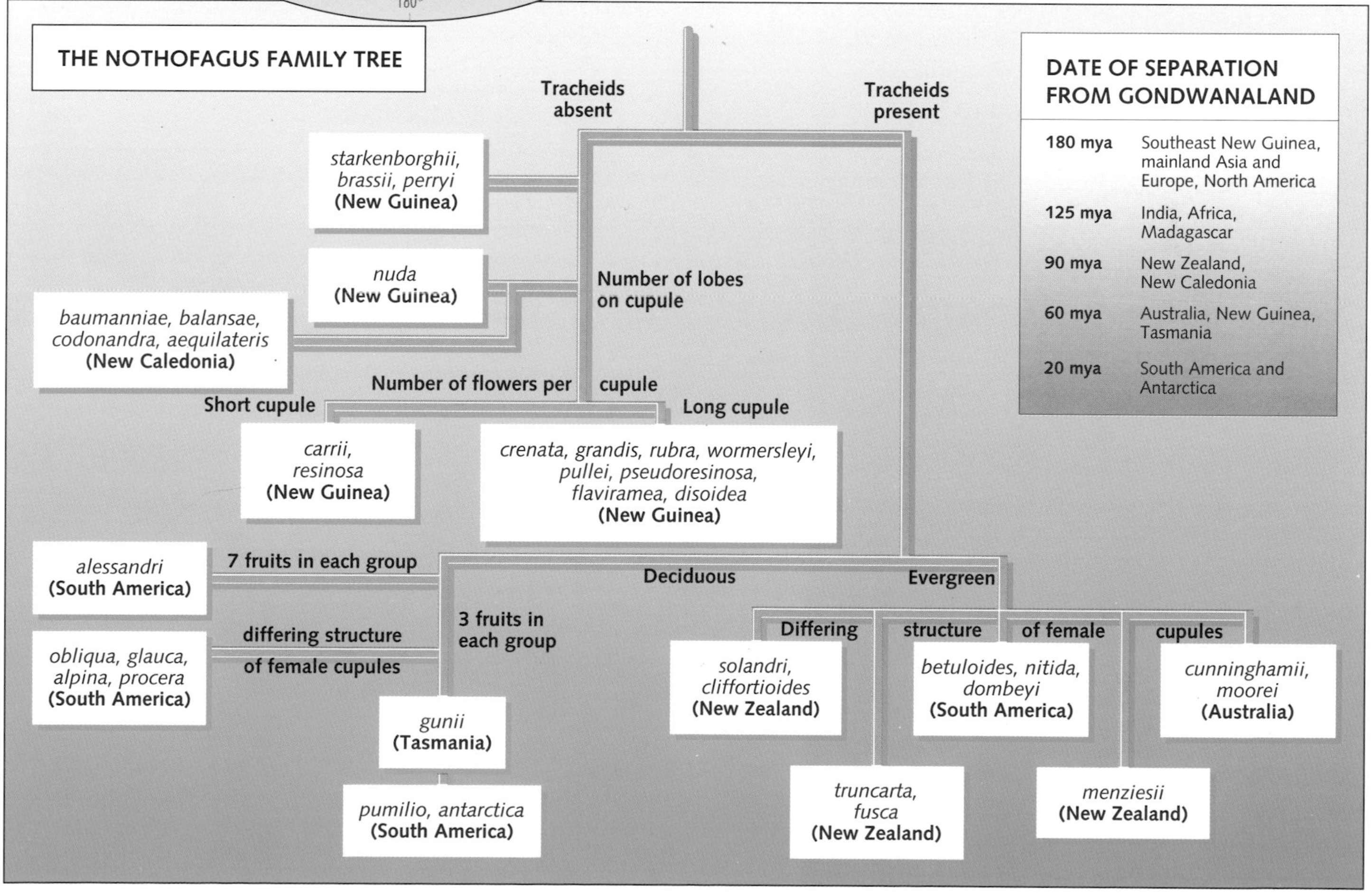

THE NOTHOFAGUS FAMILY TREE

Tracheids absent

Tracheids present

starkenborghii, brassii, perryi (New Guinea)

nuda (New Guinea)

baumanniae, balansae, codonandra, aequilateris (New Caledonia)

Number of lobes on cupule

Number of flowers per cupule

Short cupule

Long cupule

carrii, resinosa (New Guinea)

crenata, grandis, rubra, wormersleyi, pullei, pseudoresinosa, flaviramea, disoidea (New Guinea)

alessandri (South America)

7 fruits in each group

Deciduous

Evergreen

obliqua, glauca, alpina, procera (South America)

differing structure of female cupules

3 fruits in each group

Differing

structure

of female

cupules

solandri, cliffortioides (New Zealand)

betuloides, nitida, dombeyi (South America)

cunninghamii, moorei (Australia)

gunii (Tasmania)

truncarta, fusca (New Zealand)

menziesii (New Zealand)

pumilio, antarctica (South America)

DATE OF SEPARATION FROM GONDWANALAND

180 mya	Southeast New Guinea, mainland Asia and Europe, North America
125 mya	India, Africa, Madagascar
90 mya	New Zealand, New Caledonia
60 mya	Australia, New Guinea, Tasmania
20 mya	South America and Antarctica

Arum Lilies

Ancient geography and modern distribution

Many modern plants and animals have distributions that are extraordinary when viewed on a modern world map. For decades, zoologists and botanists were bamboozled by plants that are today found only in Africa and South America, for instance. They could understand that the plants were showing a preference for tropical climatic conditions, but how on Earth had they got from South America to Africa?

Early this century, there was a trend among biogeographers to reconstruct land bridges wholesale. The end result of this was a globe covered almost entirely by land. The northern route from Siberia to Alaska (*see pages 116–117*) is easy to accept, but how could the broad tropical zone of the Pacific be bridged? Continental drift (*see pages 86–87*) resolved the dilemma, although it took from 1912 to the mid 1960s for the idea of a worldwide network of land bridges to die. Of greatest interest for understanding the modern world is the last 200 million years of continental drift and the breakup of Pangea.

Angiosperms, the flowering plants, dominate the landscape today. The group arose more than 120

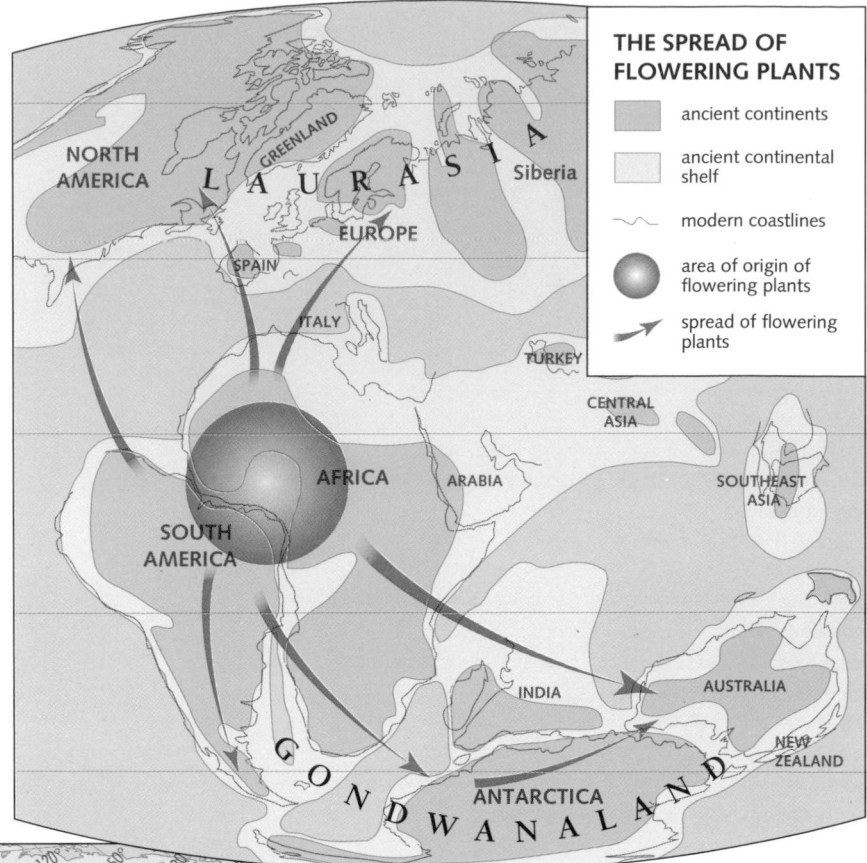

THE SPREAD OF FLOWERING PLANTS
- ancient continents
- ancient continental shelf
- modern coastlines
- area of origin of flowering plants
- spread of flowering plants

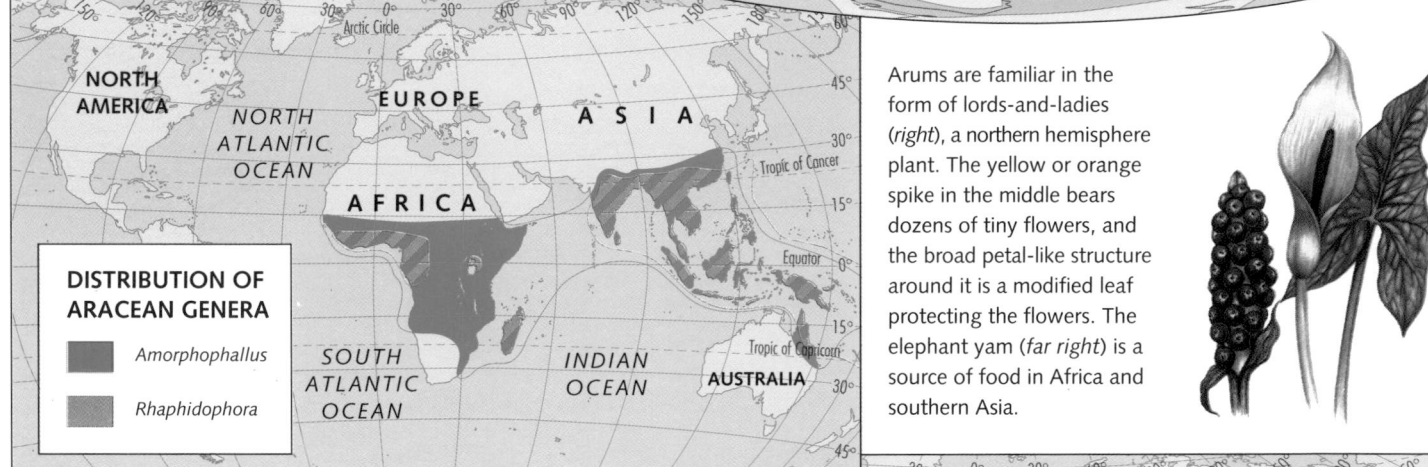

DISTRIBUTION OF ARACEAN GENERA
- *Amorphophallus*
- *Rhaphidophora*

Arums are familiar in the form of lords-and-ladies (*right*), a northern hemisphere plant. The yellow or orange spike in the middle bears dozens of tiny flowers, and the broad petal-like structure around it is a modified leaf protecting the flowers. The elephant yam (*far right*) is a source of food in Africa and southern Asia.

million years ago, perhaps in the region of Africa and South America, and spread rapidly worldwide. Today there are dozens of orders of angiosperms, from magnolias to oaks, from daffodils to grasses. The Aracean genera (arum lilies and relatives) occurs worldwide, but particularly in tropical and subtropical realms. Many of the arums have unusual flowers consisting of a fleshy spadix, or spike, in the centre, made from dozens of minute flowers. This flowering spike is surrounded by a leaf-like bract, which is often mistaken for the flower. Araceans are familiar in the wild in the northern hemisphere as lords-and-ladies and the

Monstereae   Lasieae

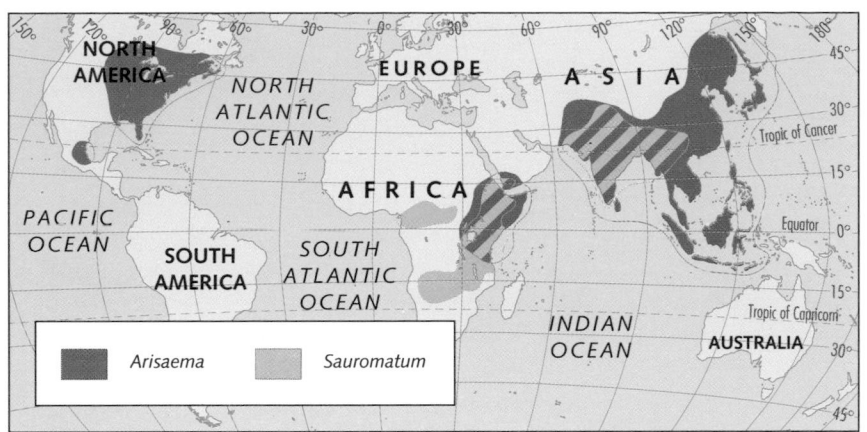

arum lily, and are also familiar as house plants such as the Swiss cheese plant and the leopard lily.

Tropical forms show disjunct distributions split by all the major oceans – the Indian, the Pacific and the Atlantic. These all relate to the origin of the group in the Cretaceous, and to the subsequent drifting apart of what was once a continuous tropical belt of land. However, there are relatively few transatlantic distributions among arums – most distributions span the Pacific. This suggests that there may have been a climatic barrier across the South Atlantic Ocean as it opened up.

Tropical distribution
Some arums, such as *Amorphophallus*, *Rhaphidophora*, Monstereae and Lasieae (*far left* and *below left*) show a disjunct distribution across the Indian Ocean. Perhaps the Asiatic elements were transported by India when it drifted away from Africa. *Arisaema* and *Sauromatum* (*above*) have a similar distribution, but occur further north.

Many arums, such as *Homalomena*, *Schismatoglottis* and *Spathiphyllum* (*right*) show transpacific distributions. The genera do not occur in Africa and their ancestors may have taken a southern route through Antarctica and Australia. Some African arums show relationships with genera in the Americas (*bottom right*), but there are few transatlantic links.

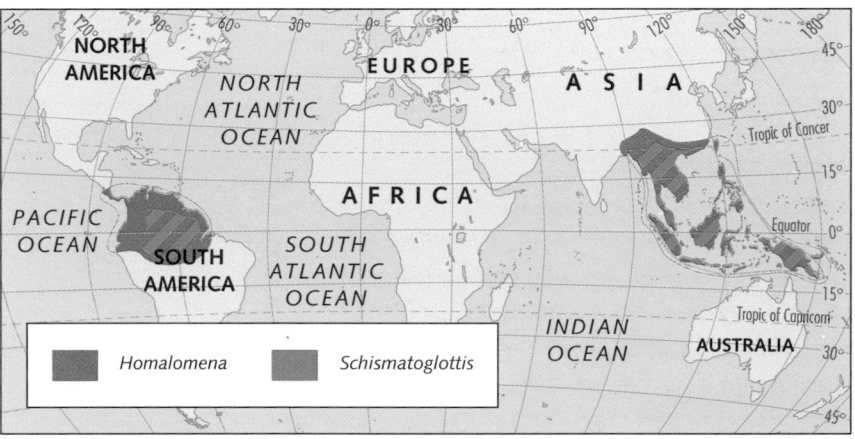

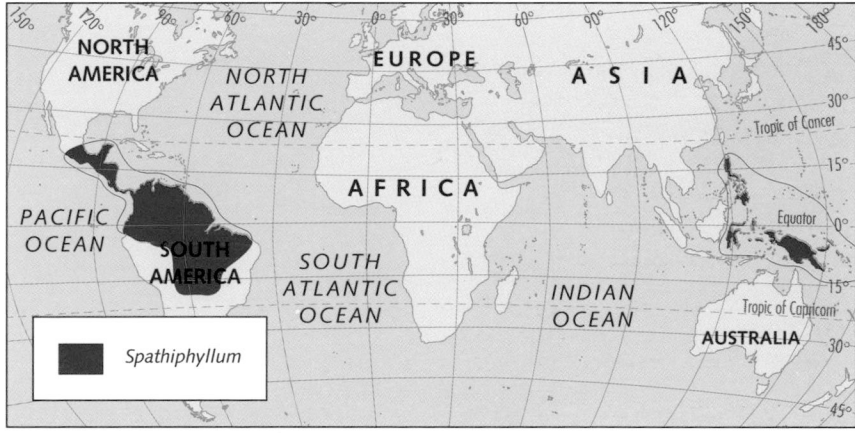

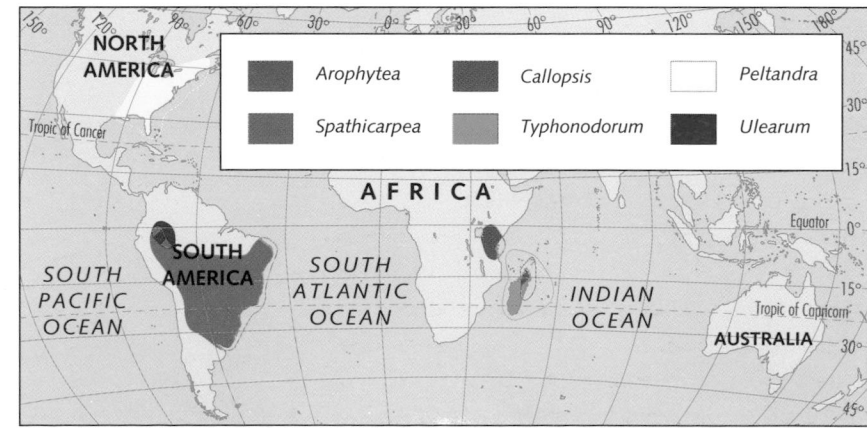

March of the Marsupials
Evolution in isolation

Marsupials – the pouched mammals such as kangaroos, wombats and koalas – are best known from Australia, but it is no surprise to find that they inhabit neighbouring islands in Southeast Asia, such as New Guinea and Borneo, too. What is surprising is that marsupials are also present in South America, especially opossums. These small tree-living insect-eaters have extended their range into Central America and the southeastern United States. There is no question that the marsupials of Australia and South America are related, but where did they originate, and how did they come to have such a dramatically disjunct distribution?

It has been known since the 19th century that marsupials arose in the Americas. The oldest specimens were known from the mid Cretaceous (100 million years ago) of the midwestern states, where they were found side-by-side with dinosaurs. More recently, similarly ancient fossils have been found in South America, and it is evident that marsupials became moderately diverse in the Americas before being almost totally wiped out by the KT event (*see pages 72–73*). In Tertiary times marsupials were relatively common in Europe, but they died out about 40 million years ago. The oldest marsupial fossils to be found in Australia are about 40 million years old,

so it has always been assumed that the group reached that region at a relatively late stage.

The history of marsupials and the reason for their present distribution seemed clear earlier this century. The idea of continental drift was strongly opposed by experts on fossil mammals, such as George Gaylord Simpson, who proposed that the group migrated from the Americas by a northern route over Greenland to Europe, and then across Asia to Australasia. The weak point of this argument was that no fossil specimens of marsupials were yet known from any part of Asia along the proposed route, but Simpson expected that such specimens would come to light.

The view that a northern route had existed was rapidly overthrown by the acceptance of the theory of continental drift in the 1960s. An alternative southern route was proposed, with marsupials migrating from the Americas northwards into Europe, and south into Antarctica, and then to Australia. The existence of this route was confirmed in the 1980s by the discovery of marsupial fossils in the Tertiary of Antarctica. Marsupial fossils have recently been found in Africa too. The animals had presumably arrived there from Europe, but were shortlived on the African continent.

Ring-tailed wallaby
Petrogale xanthopus

Pouched mammals
Marsupials have a breeding system halfway between egg-laying reptiles and the majority of mammals, which produce advanced young. Marsupials are born at an early embryonic stage. They then creep up to the pouch, where they are nurtured until they achieve independence. Australasian marsupials, such as kangaroos, wallabies, koalas and wombats, are the most familiar to us, but the opossums found in the Americas are marsupials too. This disjunct distribution is difficult to explain.

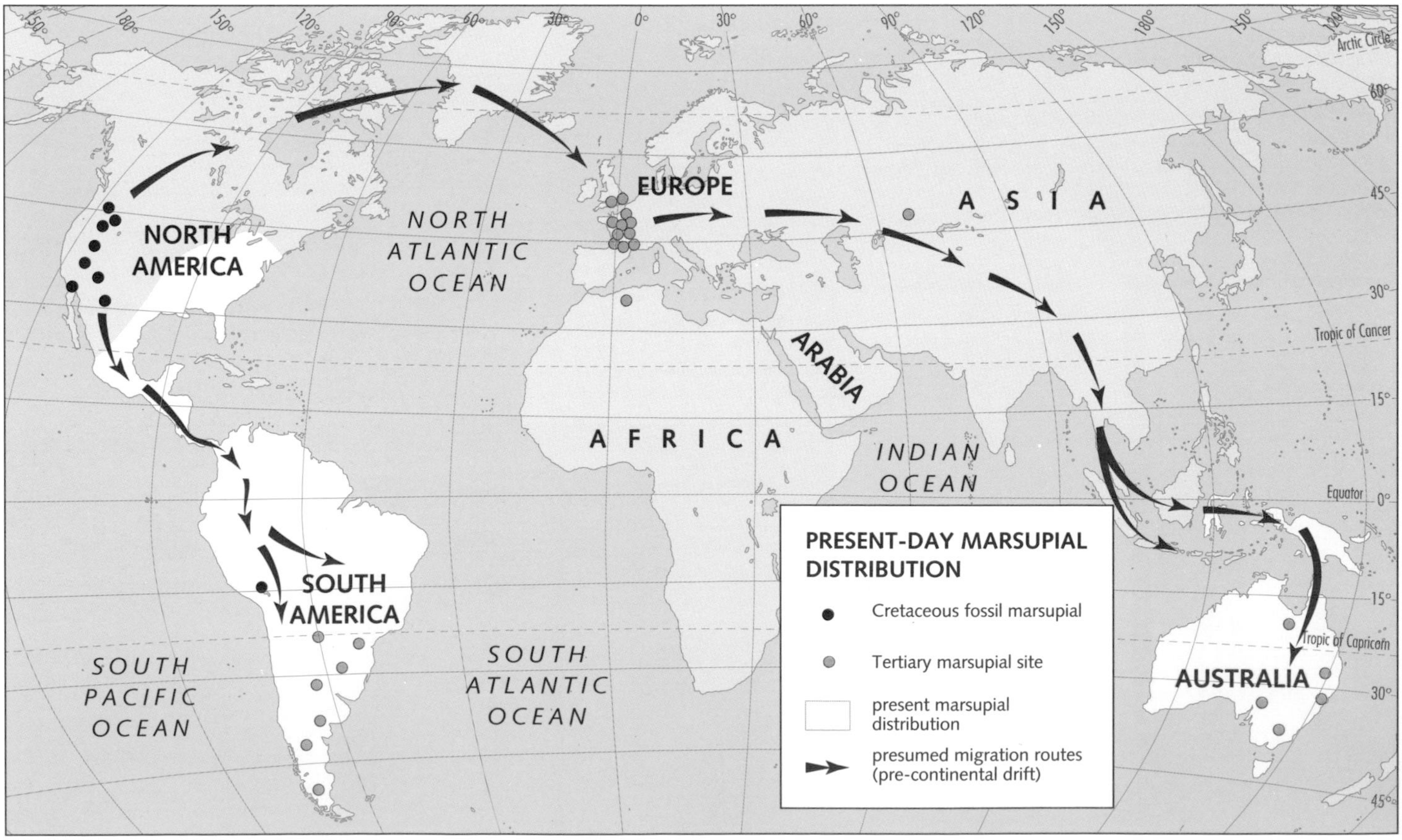

NORTH AMERICA

NORTH ATLANTIC OCEAN

EUROPE

ASIA

ARABIA

AFRICA

INDIAN OCEAN

SOUTH AMERICA

SOUTH PACIFIC OCEAN

SOUTH ATLANTIC OCEAN

AUSTRALIA

Arctic Circle

Tropic of Cancer

Equator

Tropic of Capricorn

PRESENT-DAY MARSUPIAL DISTRIBUTION

- Cretaceous fossil marsupial
- Tertiary marsupial site
- present marsupial distribution
- presumed migration routes (pre-continental drift)

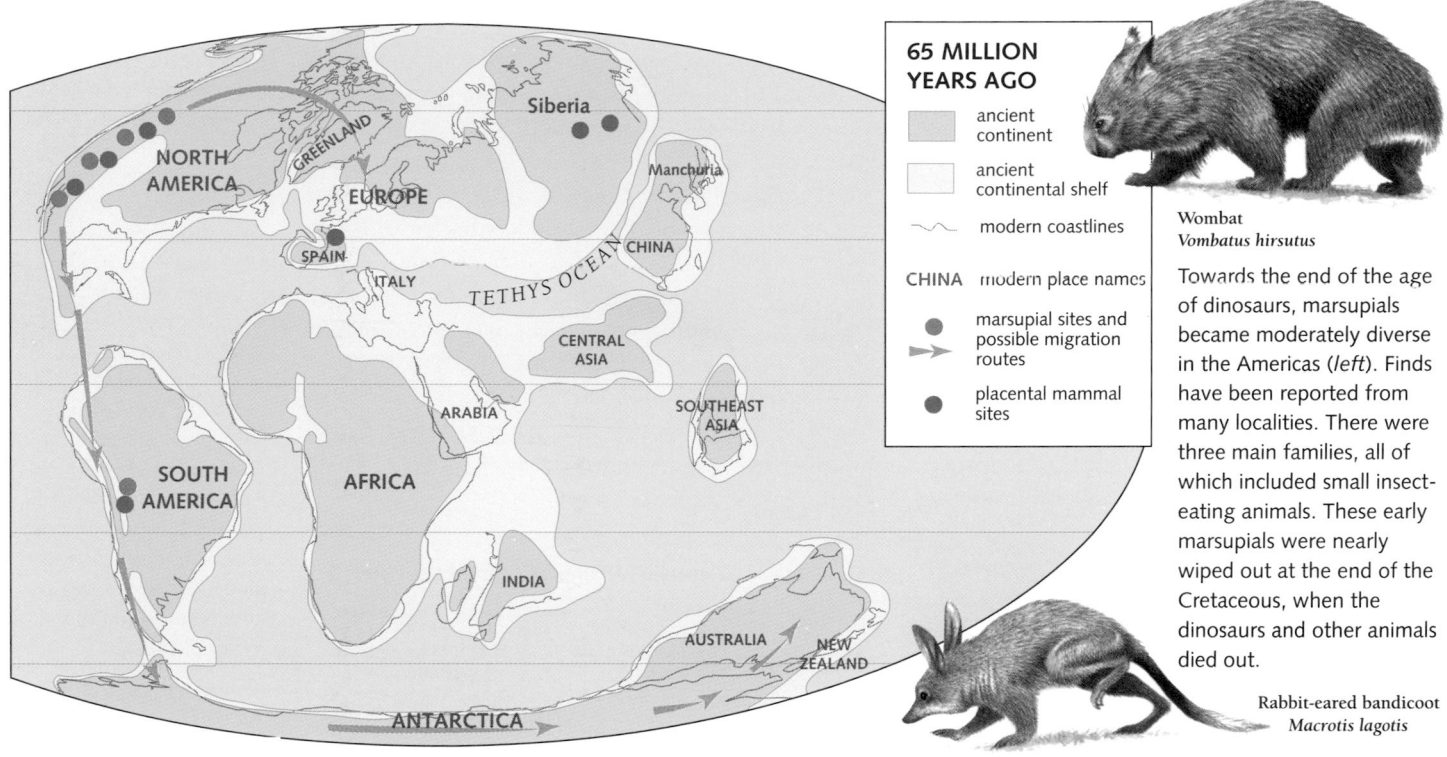

65 MILLION YEARS AGO

	ancient continent
	ancient continental shelf
~~	modern coastlines
CHINA	modern place names
●	marsupial sites and possible migration routes
●	placental mammal sites

Wombat
Vombatus hirsutus

Towards the end of the age of dinosaurs, marsupials became moderately diverse in the Americas (*left*). Finds have been reported from many localities. There were three main families, all of which included small insect-eating animals. These early marsupials were nearly wiped out at the end of the Cretaceous, when the dinosaurs and other animals died out.

Rabbit-eared bandicoot
Macrotis lagotis

Eocene
During the Eocene 50 million years ago, marsupials were abundant in the Americas and Europe. Specimens have also been found in central Asia, Africa and Antarctica.

Left: The molar teeth of marsupials from different parts of the Earth are all similar and opossum-like. From top to bottom: *Amphiperatherium* (Europe), *Alphadon* (North America), *Peratherium* (Kazakhstan) and *Garatherium* (Algeria).

George Gaylord Simpson did not live to see the long-awaited discovery of marsupial teeth in central Asia, which could have vindicated his hypothesis of a northern route of dispersal. In detail, however, the Asiatic marsupial, like the African, is of European type and shows no affinities with Australian marsupials. This was probably another short-lived invasion that led to nothing.

Tasmanian devil
Sarcophilus harrisii

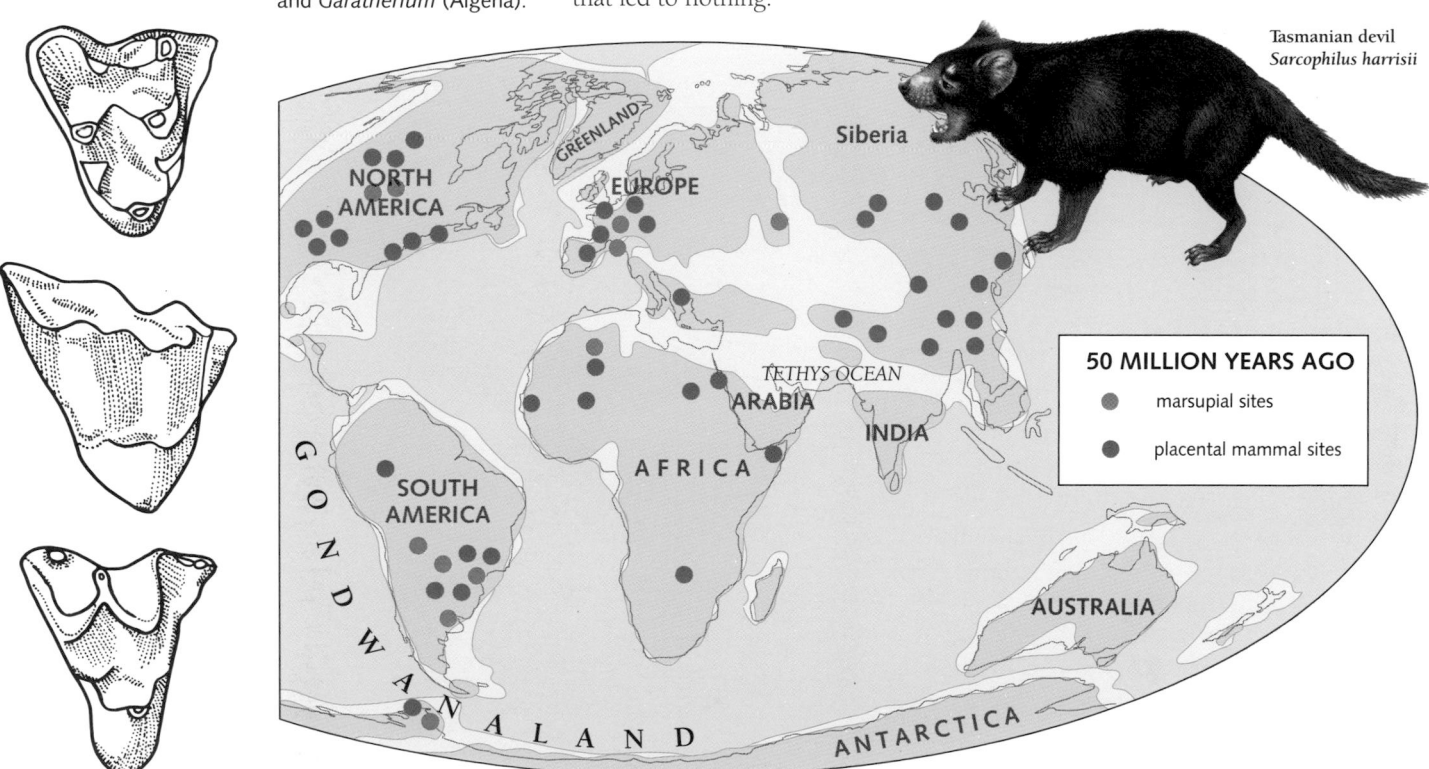

50 MILLION YEARS AGO

| ● | marsupial sites |
| ● | placental mammal sites |

Mastodons, Mammoths and Elephants
The decline of a worldwide group

The proboscideans, a group that includes elephants and their fossil relatives, once had a worldwide distribution. Today the surviving species – the African elephant, *Loxodonta africana*, and the Indian elephant, *Elephas maximus* – are regarded as endangered species. But proboscideans have a history of great adaptability. Modern elephants live in tropical regions, and many fossil forms have been found there too. Some now extinct species occupied Europe and North America, while mammoths even flourished in the tundra and snowfields of the ice ages (*see pages 114–115*).

Proboscideans arose in Egypt 50 million years ago. The best known early form, *Moeritherium*, was a pig-sized animal with a tiny trunk and sharp tusk-like incisor teeth. It probably looked and behaved like a pigmy hippopotamus, living in and near rivers and feeding on soft plant food. Proboscideans are characterised by the trunk (a fleshy extension of the nose region) and the tusks (elongated teeth). The trunk became longer as elephants grew larger. It enabled elephants to pick up food from the ground. Most large mammals solved this problem by having a long neck, but this only works if the head is light. The proboscidean head is heavy because of the weight of

the teeth and tusks, and the neck must be short.

Several proboscidean stocks arose in the Eocene in Africa from an ancestor like *Moeritherium*. The paleomastodonts, *Phiomia* and *Paleomastodon* from the Oligocene of Egypt for instance, had a pair of short tusks above and below. By the Early Miocene, the mastodonts migrated first to India, then to Europe, and finally, by the northern Asiatic route, to North America. A typical early mastodont, such as *Gomphotherium*, was a large animal with four 2-metre-long tusks. It fed by browsing from trees and bushes. Grasslands were spreading worldwide, and

other mammals switched from leaf-eating to grass-eating. Elephants did not make the switch. Mastodonts were particularly successful in the Americas, passing from North America to South America 3 million years ago (*see pages 100–101*) and dying out only with the retreat of the ice in North America and the advance of hunting peoples.

The elephantids, including mammoths and modern elephants, formed the last wave of proboscidean evolution in the Plio-Pleistocene. Mammoths occupied the Old World (northern Europe and Siberia), and some entered North America over the expanded ice sheets. They died out at the end of the ice ages.

Proboscideans evolved solely in Africa from 50 to 35 million years ago, giving rise to a sequence of ever-larger forms. Migration then occurred in three phases, after Africa became linked to the Middle Eastern region (*below*). In the Early Miocene, mastodonts migrated to central Asia. In the Late Miocene they migrated to Europe and North America, and in the Pliocene to South America. Elephants and mammoths migrated in Plio-Pleistocene times from central Asia to Europe, northern Asia, North America and Africa.

EVOLUTION AND MIGRATION OF ELEPHANTS

- original centre of elephant evolution: *Moeritherium* and *Paleomastodon*
- Oligocene migration
- Miocene centre of elephant evolution: *Tetrabelodon*, *Mastodon* and *Elephas*
- Pliocene migrations
- Pleistocene migrations of *Elephas*
- present distribution of Indian elephant (*Elephas maximus*)
- present distribution of African elephant (*Loxodonta africana*)

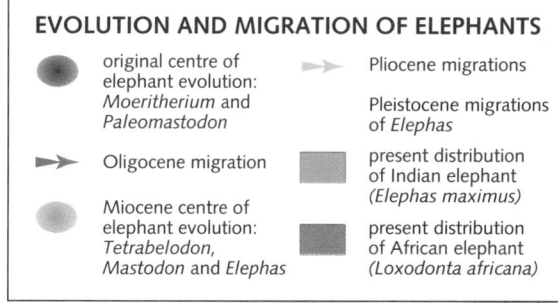

Below: *Ambelodon* from North America had vast shovel-like lower tusks, presumably for scooping up pond plants. It belonged to the large and diverse mastodont group.

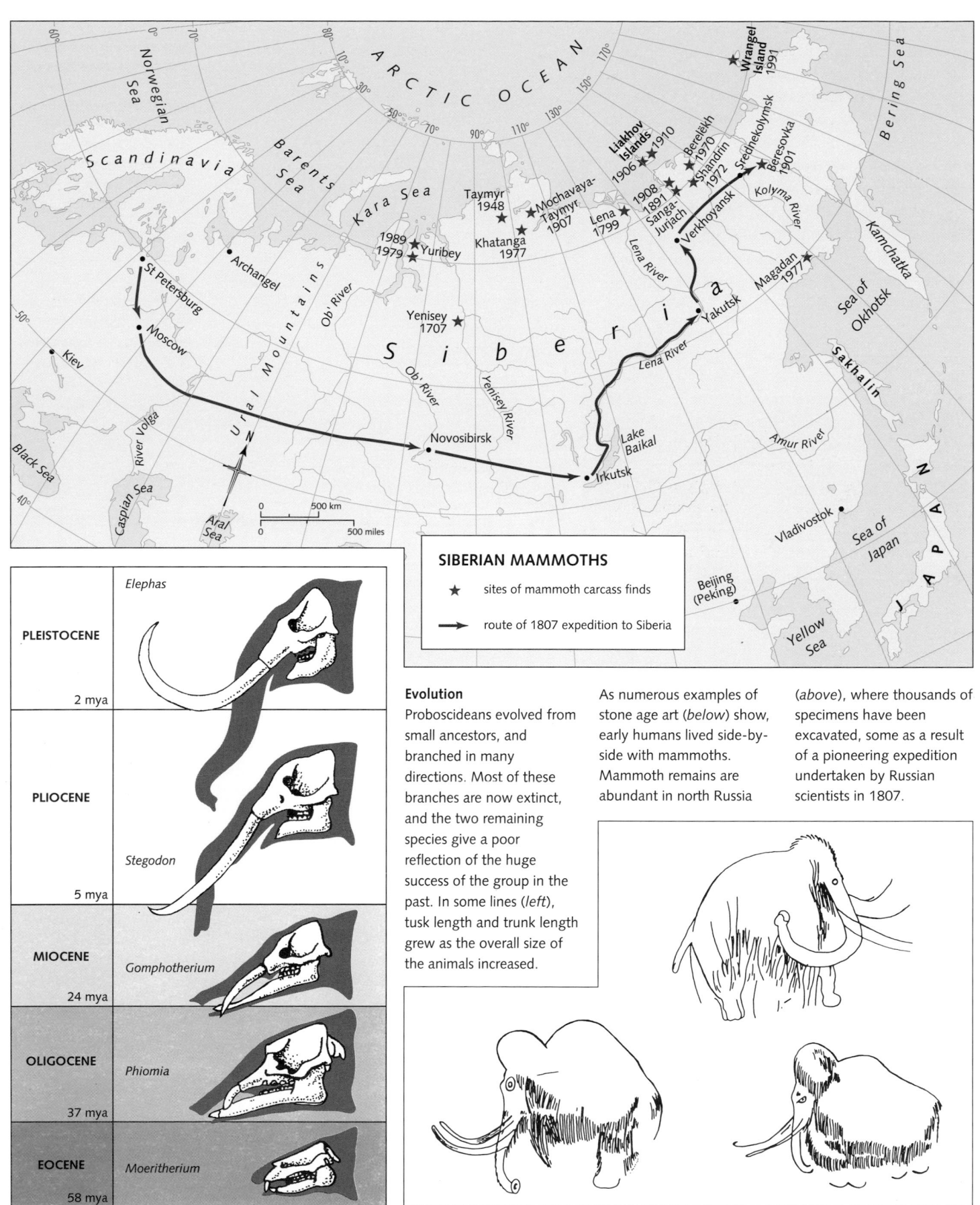

ARCTIC OCEAN

Norwegian Sea

Scandinavia

Barents Sea

Kara Sea

Taymyr 1948

Mochavaya-Taymyr 1907

Liakhov Islands 1906 ★ ★1910

Berelekh 1970
Shandrin 1972

Srednekolymsk

Wrangel Island 1991

Beresovka 1901

Bering Sea

1908
1891
Sanga-Jurjach

Verkhoyansk

Kolyma River

Khatanga 1977

Lena 1799

Lena River

1989 ★
1979 ★ Yuribey

St Petersburg

Archangel

Ural Mountains

Ob' River

Moscow

Kiev

Yenisey 1707

Yenisey River

Ob' River

Siberia

Lena River

Yakutsk

Magadan 1977

Kamchatka

Sea of Okhotsk

Sakhalin

River Volga

Caspian Sea

Black Sea

Novosibirsk

Lake Baikal

Amur River

N

Irkutsk

0   500 km

0   500 miles

Aral Sea

Vladivostok

Sea of Japan

Beijing (Peking)

Yellow Sea

JAPAN

SIBERIAN MAMMOTHS

★ sites of mammoth carcass finds

→ route of 1807 expedition to Siberia

Epoch		
PLEISTOCENE 2 mya	*Elephas*	
PLIOCENE 5 mya	*Stegodon*	
MIOCENE 24 mya	*Gomphotherium*	
OLIGOCENE 37 mya	*Phiomia*	
EOCENE 58 mya	*Moeritherium*	

Evolution

Proboscideans evolved from small ancestors, and branched in many directions. Most of these branches are now extinct, and the two remaining species give a poor reflection of the huge success of the group in the past. In some lines (*left*), tusk length and trunk length grew as the overall size of the animals increased.

As numerous examples of stone age art (*below*) show, early humans lived side-by-side with mammoths. Mammoth remains are abundant in north Russia (*above*), where thousands of specimens have been excavated, some as a result of a pioneering expedition undertaken by Russian scientists in 1807.

The Great American Interchange
A continent-scale experiment in faunal mixing

In the study of evolution, a great deal must be assumed. Modern plants and animals give clues about their history, and fossil evidence provides information about ancient relatives of modern organisms and about ancient geographies, but there are often critical gaps in our knowledge. Biologists can perform small-scale experiments with living plants and animals in an attempt to study the effects of shifts in ecology and geographic changes, but it is difficult to speculate about the effects of major faunal shifts. What, for example, happens when two formerly isolated faunas and floras come into contact? Does one wipe out the other, or do they mix? The Great American Interchange provides such an experiment.

During most of the later Mesozoic and the Tertiary, South America was an island. Its mammal and bird faunas evolved largely in isolation from Africa, Antarctica and North America. By the mid Tertiary, South America was populated by an extraordinary range of animals, many of them unlike anything in other parts of the world. Marsupials (*see pages 96–97*) retained an important role, particularly as carnivores, and there were dog-like, cat-like and even some sabre-toothed forms. The plant-eaters – strange animals such as notoungulates and litopterns, which looked like horses, deer, rhinos and hippos, and edentates (anteaters and sloths), which were quite unlike anything elsewhere – were all endemic mammal groups. The New World monkeys (*see pages 130–131*) are another unique group.

Mammal evolution in North America matched that of the Old World, as there were still land connections that horses and elephants (*see pages 78–79 and 98–99*) could cross. But there was almost no exchange of animals between North and South America until 3 million years ago, when the Isthmus of Panama formed. The classic tale told is of a mass invasion of South America by superior northern mammals, which rapidly decimated the Latin breeds. New work, however, shows that while North American mastodonts, horses, jaguars, sabre-toothed cats, wolves and deer spread through South America, many South American animals headed north (monkeys into Central America; anteaters, ground sloths and opossums into North America). The new animals found suitable niches on both sides and caused little extinction in either direction.

A combination of various factors brought about this change in our understanding. First, many more mammal sites have been excavated in South America, and the fossil record there is now very

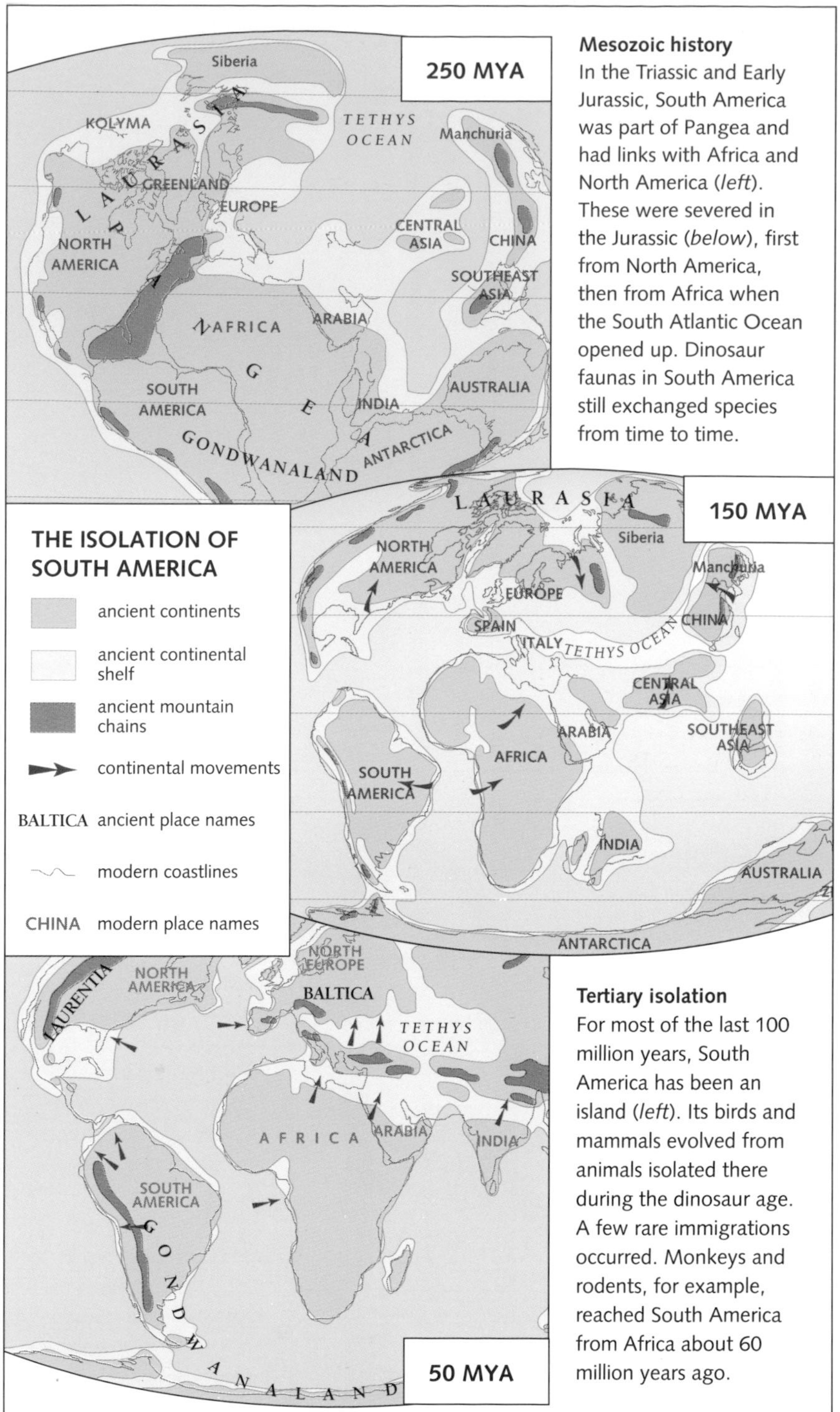

THE ISOLATION OF SOUTH AMERICA

	ancient continents
	ancient continental shelf
	ancient mountain chains
➤	continental movements
BALTICA	ancient place names
∿	modern coastlines
CHINA	modern place names

250 MYA

Mesozoic history
In the Triassic and Early Jurassic, South America was part of Pangea and had links with Africa and North America (*left*). These were severed in the Jurassic (*below*), first from North America, then from Africa when the South Atlantic Ocean opened up. Dinosaur faunas in South America still exchanged species from time to time.

150 MYA

50 MYA

Tertiary isolation
For most of the last 100 million years, South America has been an island (*left*). Its birds and mammals evolved from animals isolated there during the dinosaur age. A few rare immigrations occurred. Monkeys and rodents, for example, reached South America from Africa about 60 million years ago.

Unique mammals
South American mammals mimicked those in the rest of the world. The small marsupial *Argyrolagus* (*left*) had similarities with the rabbit. *Astrapotherium* (*right*) was a large animal that resembled the elephant or rhinoceros.

good. Second, a proper assessment of the interchange has been carried out. Biologists had compared South America with North America, but a true comparison must also include Central America, a zone that was occupied by many species from South America. Third, proper statistical presentations were given, showing an overall increase in diversity both north and south of the Isthmus of Panama. This natural experiment has shown that when faunas meet, extinction is not inevitable. Ecosystems can increase in complexity, and immigrants can often insinuate themselves into previously unoccupied niches.

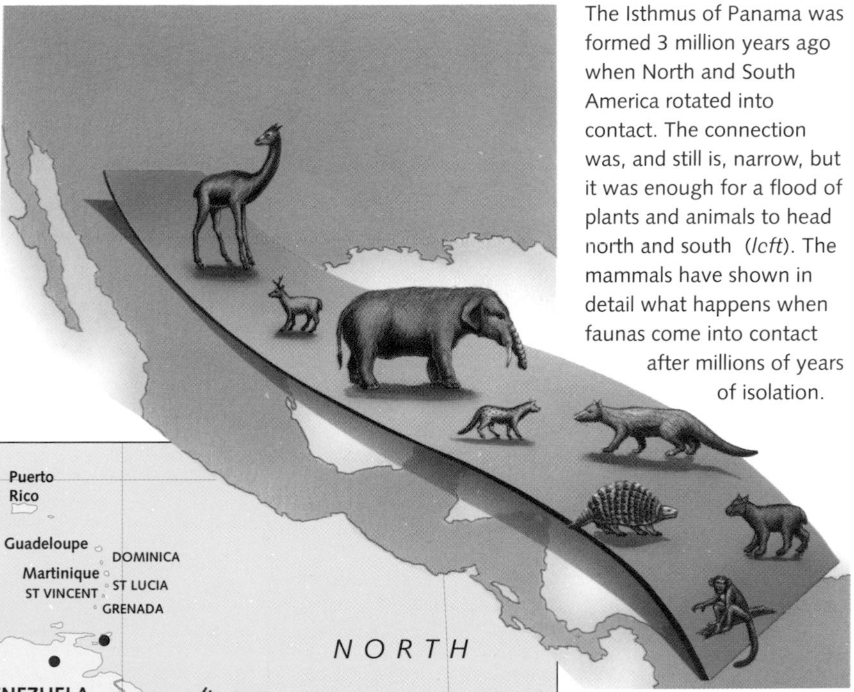

The Isthmus of Panama was formed 3 million years ago when North and South America rotated into contact. The connection was, and still is, narrow, but it was enough for a flood of plants and animals to head north and south (*left*). The mammals have shown in detail what happens when faunas come into contact after millions of years of isolation.

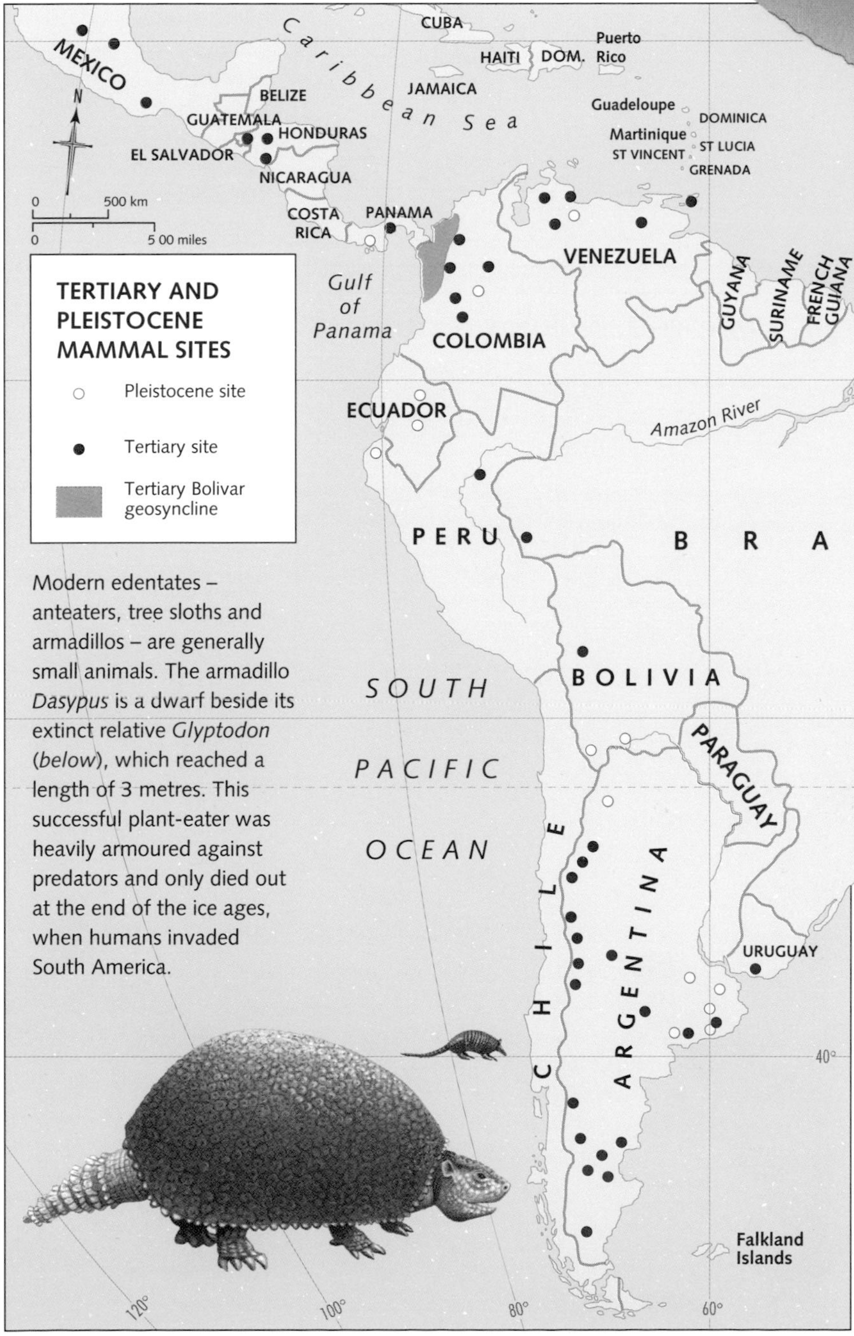

TERTIARY AND PLEISTOCENE MAMMAL SITES

○   Pleistocene site

●   Tertiary site

▨   Tertiary Bolivar geosyncline

Modern edentates – anteaters, tree sloths and armadillos – are generally small animals. The armadillo *Dasypus* is a dwarf beside its extinct relative *Glyptodon* (below), which reached a length of 3 metres. This successful plant-eater was heavily armoured against predators and only died out at the end of the ice ages, when humans invaded South America.

The exchange

At one time it was thought that the exchange of mammals 3 million years ago was unbalanced, with North American species driving South American forms to extinction. Close study of South American faunas shows (*below*) that this was not the case, however. The North American invaders did not compete with established South American forms, and limited extinction took place among the latter groups. Overall diversity increased from 77 to 120 genera.

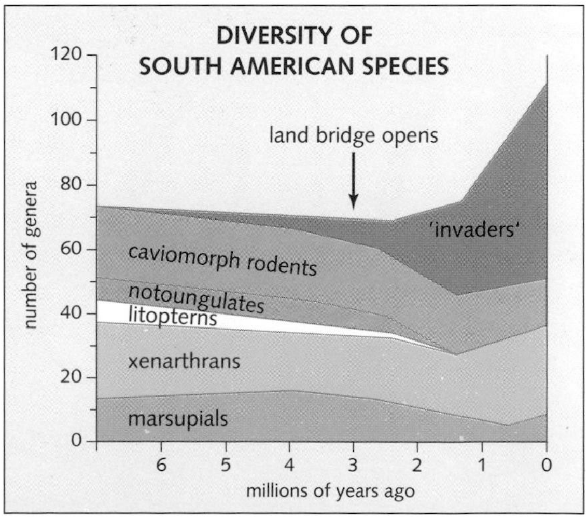

DIVERSITY OF SOUTH AMERICAN SPECIES

The Wallace Line

Continents move apart and come together

The islands between the mainlands of Southeast Asia and Australia provide a living laboratory for the study of evolution. As we would expect, plants and animals have evolved differently on separate islands, while species have migrated between islands and diversified within them. But the islands also carry a more deep-seated history of evolution, which was revealed in one of the most important scientific discoveries of the 19th century.

Alfred Wallace, a biologist working in the region in the 1850s and 1860s, noted that bird populations on the islands belonged to different groups. Those northwest of a particular line (now known as the Wallace Line) had close affinities with Asian groups, while those to the southwest were related to Australian types. The two populations must have evolved in complete isolation from each other.

This was dramatic confirmation of Wallace's ideas about evolution, which contributed to the acceptance of the theory of evolution being developed independently by Darwin at that time. But Wallace did not know that the reason for the separate development of these populations was continental movement – an idea that was not fully accepted for another 100 years. In fact the continents of Australia and Asia, with their associated islands, have moved away from and towards each other in a complex series of movements. The evolution of plant and animal populations was separate for about 50 million years, but that separation is now breaking down, at least for the more mobile species.

1/EARLY CRETACEOUS, 120 MILLION YEARS AGO

SOUTHEAST ASIA

AUSTRALASIA

ANTARCTICA

THE MOVEMENT OF AUSTRALASIA SINCE THE EARLY CRETACEOUS

modern continents

ancient continental shelf

--- plate boundaries

spreading centres

Around 120 million years ago Asia and Australia were separated by a wide stretch of ocean (*above left*). Southeast Asia was made up of plates that had split from Gondwanaland 150 million years ago. Australia was attached to Antarctica until 100 million years ago. Australia's flora and fauna were derived from its link with Antarctica and South America. By 60 million years ago (*below left*) the islands of the two continents were closer together and occasional links developed through the formation of island arcs, over which some species migrated. By 30 million years ago (*above right*) the islands were close enough for increased but intermittent migration to take place. In the last 10 million years (*above far right*), Asia has joined Africa and Europe, introducing other species into its flora and fauna. Movement of Eurasian and Australian species has taken place across the Wallace Line.

2/PALEOCENE, 60 MILLION YEARS AGO

Equator

SOUTHEAST ASIA

AUSTRALASIA

ANTARCTICA

Water buffalo

Kiwi

Duck-billed platypus

Red kangaroo

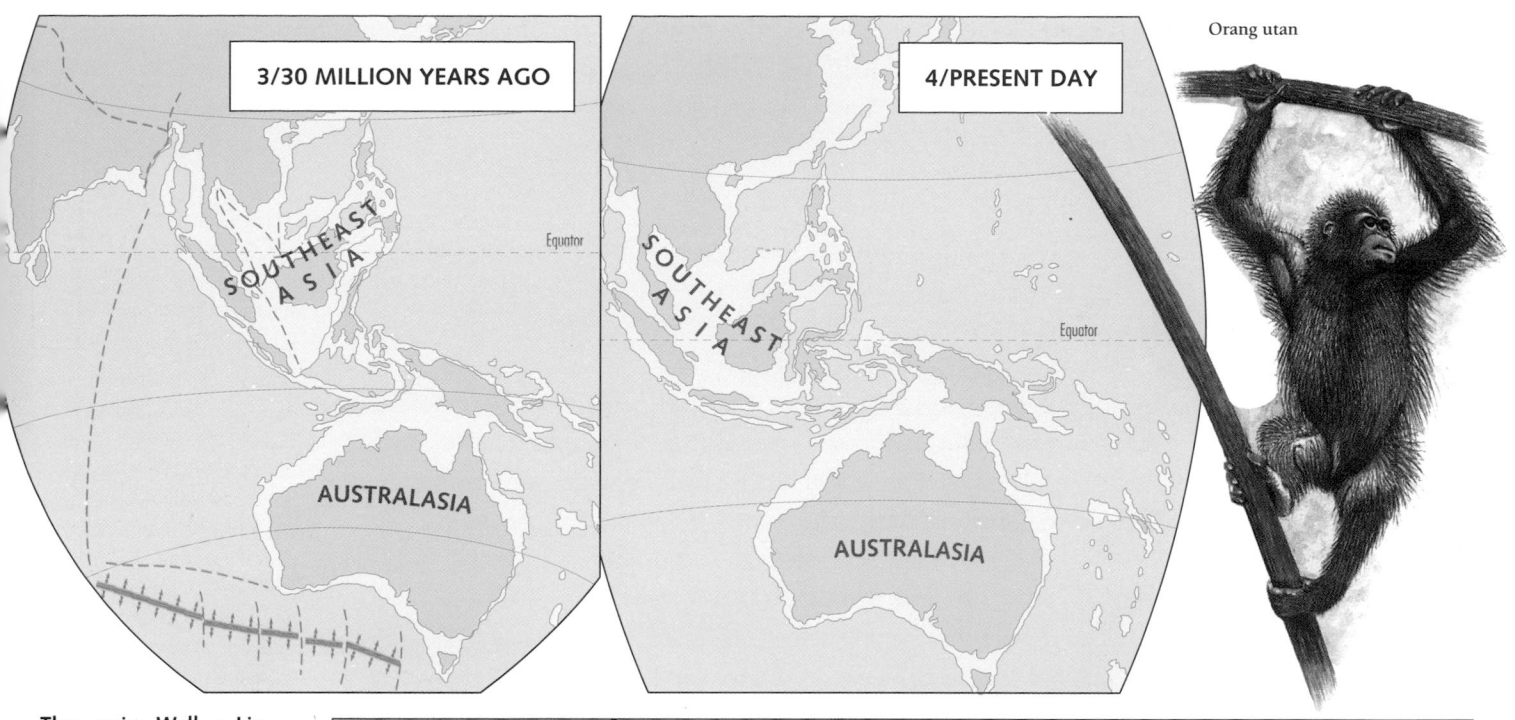

3/30 MILLION YEARS AGO

SOUTHEAST ASIA

Equator

AUSTRALASIA

4/PRESENT DAY

SOUTHEAST ASIA

Equator

AUSTRALASIA

Orang utan

The moving Wallace Line

In the decades following Darwin's publication of *On the Origin of Species*, evolution became the focus of study for many natural scientists and gradually became the reference point for all biological studies. The work of Alfred Wallace was recognized and drew other researchers to Southeast Asia. They found that the lines separating Australian and Asian groups varied in position according to the animal type – a true dividing line between Asia and Australasia does not exist.

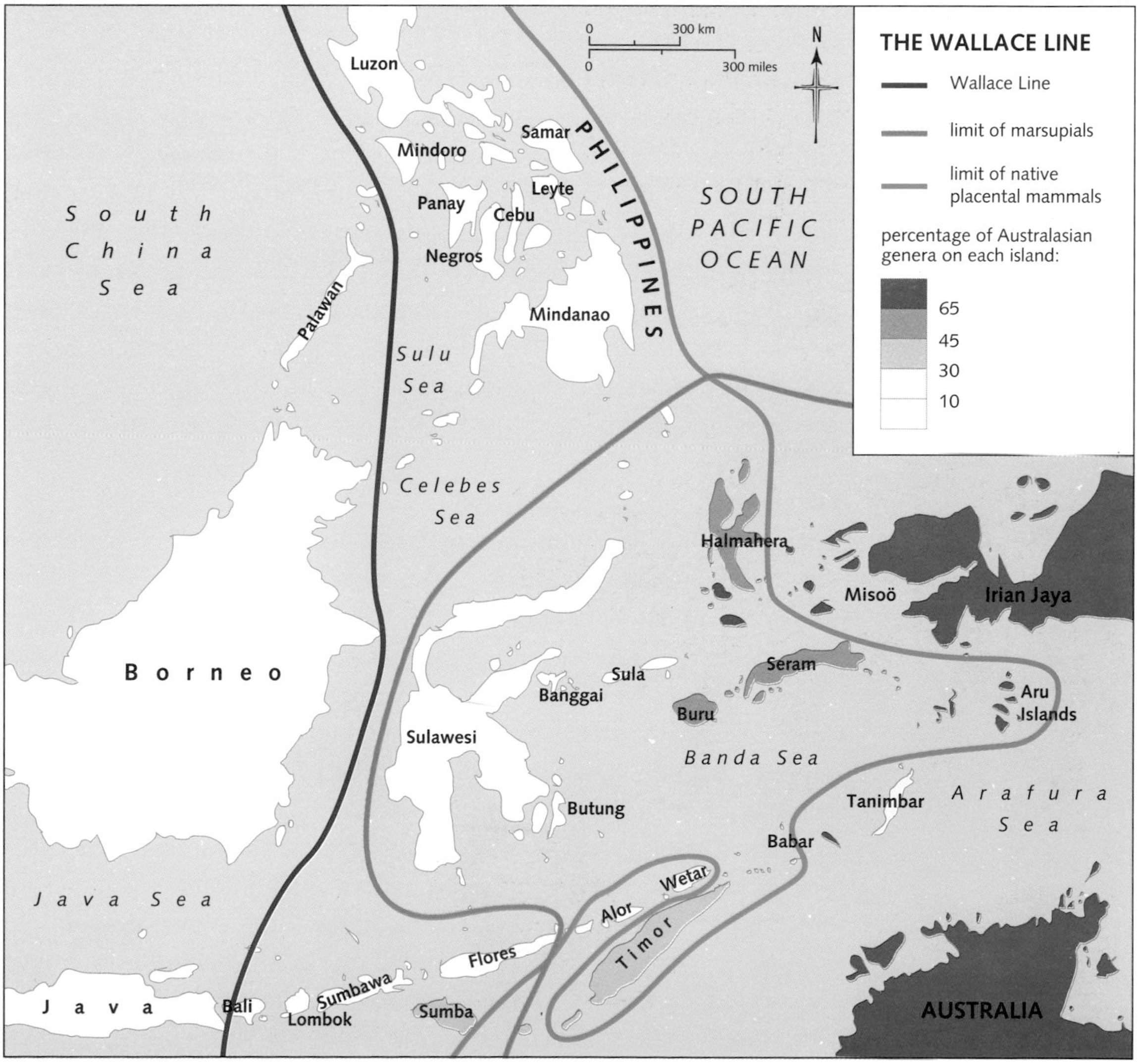

Koala bear

0   300 km
0   300 miles

N

Luzon

Mindoro

Samar

Panay   Leyte
Cebu

Negros

PHILIPPINES

SOUTH PACIFIC OCEAN

Mindanao

South China Sea

Palawan

Sulu Sea

Celebes Sea

Borneo

Sula
Banggai

Buru

Seram

Halmahera

Misoö   Irian Jaya

Aru Islands

Sulawesi

Banda Sea

Butung

Tanimbar   Arafura Sea

Babar

Java Sea

Alor   Wetar

Flores

Timor

AUSTRALIA

Java   Bali   Sumbawa
Lombok   Sumba

THE WALLACE LINE

—— Wallace Line

—— limit of marsupials

—— limit of native placental mammals

percentage of Australasian genera on each island:

65
45
30
10

Section VI: The Legacy of the Ice Ages

The past 15,000 years have witnessed major climatic changes, particularly the retreat of large ice sheets in the northern hemisphere. These geologically recent events have left a profound imprint on the surface of the Earth, and on the distribution of plants and animals.

In Darwin's day, the most recent epochs of the Earth's history, the Pleistocene and Holocene, were the subject of much debate. In Europe and North America, where most geologists worked in the 1830s and 1840s, there are superficial features that indicate some of the major events of the Pleistocene. In particular, geologists noted the dramatic mountain scenery in Scandinavia, Scotland and Canada. Clearly, some huge erosive influences had been at work to carve out the Norwegian fjords and the Scottish lochs and glens, and some extraordinary force had scattered great boulders of granite over the mountainsides, often far from their possible sources.

The debate was between the diluvialists and the glacialists. Many geologists of the 1820s proposed that the Earth had been covered by one or more dramatic floods. Biblical literalists favoured a single flood. Geological sophisticates, such as Georges Cuvier in Paris and Dean William Buckland in Oxford, saw records of major catastrophes further back in geological time, and accepted that the Noachian deluge was only the last in a series of floods. This flood, they argued, explained the granite boulders perched on hillsides in Scotland and Norway. Vast floods had rushed over the continents, carrying great rocks ahead of them. As the floodwaters receded, they carved out the glens and lochs and dumped huge mounds of sand and ground rock on the lower plains. A modification of the deluge idea was that the blocks had been carried in the bottom of icebergs that had floated from arctic waters over northern Europe, and had fallen out as the icebergs melted.

Glaciation

The debate was finally resolved by Louis Agassiz, a Swiss geologist and paleontologist whose studies of the Alps revealed the effects of the modern movements of glaciers and ice sheets. Agassiz also found clear evidence that the Alpine snows had once extended much further, and that the pattern of blocks and dumped sand could be explained only by the advance and retreat of ice sheets in the relatively recent past. He showed that several ice sheets had affected Europe: a main northern one that covered Ireland, Britain as far south as London, Scandinavia, northern Germany and northwestern Russia; and a second substantial Alpine ice sheet that extended from Switzerland some distance into Austria, Germany, France and Italy. Later work has shown that smaller ice sheets were nucleating on other mountain belts, such as the Pyrenees, the Carpathians, the Urals and other ranges in Russia, and in parts of the arctic zones of Siberia, the Himalayas and mountain chains in central Asia and northern China. Agassiz later took up a professorship at Harvard, and observed in the New World all the glacial phenomena he had seen in Europe. At its maximum extent, the great North American ice sheet had covered Greenland, Alaska, Canada and the northern USA, including New England, the Great Lakes, the Rockies and the West Coast mountain belts.

The effects of the glaciations include a broad range of erosive and depositional features. Advancing glaciers move slowly, but they erode because of their vast weight, grinding and pushing rocks ahead of them, and carrying masses of rocks in the base. The slow passage of ice gouges a channel, which typically has a regular U shape, and such valleys are common features of glaciated landscapes. The rocks borne in the base become tools of erosion too, producing deep scratches on the smoothed rock surfaces that they pass over. Such glaciated pavements, covered in numerous parallel scratches, survive abundantly in Canada, Scandinavia and Scotland.

Glaciers also produce characteristic deposits. The base of the ice flow tears great blocks from the landscape, but also grinds rock into a coarse material called glacial till. Blocks may be left in random locations as the glacier's power diminishes. The main mass of smaller boulders, pebbles, sand and mud may be carried much further, and is deposited mainly at the toe of the glacier when it finally comes to a halt. If the glacier stops on land, the till is dumped in the form of a moraine, a ridge marking the maximum advance of the ice sheet. The ice sheet may then melt back, and further till is left in a chaotic deposit. Much of it will be transported in streams of glacial meltwater.

These meltwaters may converge and form lakes ahead of the maximum line of advance of the ice sheets. Fine depositional horizons (varves) form on the lake floor. In summer the glaciers melt back, and coarse sand is carried by the meltwaters into the lakes. In winter the melting slows down, and fine clay and organic matter are deposited. Varves in glacial lakes can provide a detailed record of the seasons over hundreds of years. Other glaciers continue into the sea and break up into icebergs. As these melt, sand and boulders fall into the sediments of the seabed. These exotic blocks (dropstones) stand out around glaciated regions, as they are sharp-edged and cut right through the sedimentary layers because they have fallen some distance.

The Pleistocene, 2 million–10,000 years ago, witnessed five or six major ice advances. The interglacials between these episodes were marked by warmer climates, and we may be living in an interglacial now. The ice sheets had two main effects: they profoundly altered the distributions of climatic belts, and they lowered sea levels worldwide by 100 metres or more at time, since so much water was locked up in the ice. These two phenomena caused huge changes in the distributions of land plants and animals, and the effects may still be seen today, 10,000 years after the retreat of the last major ice sheet.

Ice Advances and Retreats

The decimation of the forests of Europe

One of the central tenets of the theory of evolution is that organisms adapt to suit their environment. As conditions change, plants and animals either change or perish. They do this by 'selecting' certain accidental changes, or mutations, which give them some advantage in the new environment. This selection is made simply by the increased chance that mutated individuals have of surviving to reproduce, and therefore of passing the mutation on to their offspring.

Selection is a time-consuming and extremely messy process. A genetic mutation may be of benefit to an individual, but it may be combined with another mutation that takes this advantage away. A mutation in one generation may not persist to the next, or may remain recessive. In addition, conditions are unlikely to change in a linear fashion. If a region becomes warmer over several thousand years, for example, there will still be occasional years of relative cold, which may place adaptations at risk.

Conditions can change rapidly, but adaptation is a slow process. Many environments are therefore out of equilibrium – organisms are still catching up with changing conditions. In a sense this is true of every environment. The most striking illustration of non-equilibrium, the Earth's 'recovery' from the last ice age, affects almost every part of the world.

At the peak of the last glacial episode, about 18,000 years ago, ice sheets spread across North America and northern Europe and expanded across the southern ocean, altering the underlying landscape and pushing flora and fauna towards the Equator. The ice sheets destroyed plant and animal habitats, and the lowering of temperature over the entire surface of the Earth had effects which are inscribed on the patterns of life today.

In Europe, the study of fossilized pollen grains and preserved plants has enabled paleontologists to build up a picture of the vegetation of the continent before, during and after the last ice age. One thing that has become clear is that plants are still making their way north as the Earth continues to warm up again. In many cases they are finding their way back to latitudes which they occupied before the ice sheets advanced. A good example is the pink rhododendron (*Rhododendron ponticum*), regarded as an invasive newcomer to Ireland, where it is thriving at the expense of 'native' species. Analysis of old peat deposits shows that the plant was actually widespread in Ireland before the last glaciation and is now merely reclaiming its past environment.

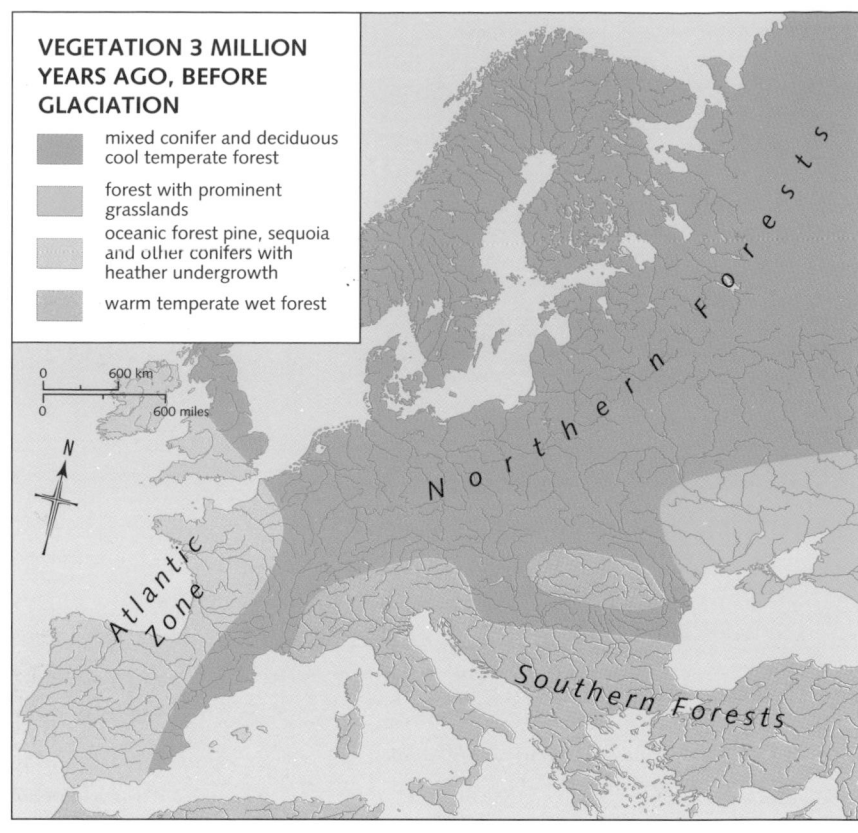

VEGETATION 3 MILLION YEARS AGO, BEFORE GLACIATION

- mixed conifer and deciduous cool temperate forest
- forest with prominent grasslands
- oceanic forest pine, sequoia and other conifers with heather undergrowth
- warm temperate wet forest

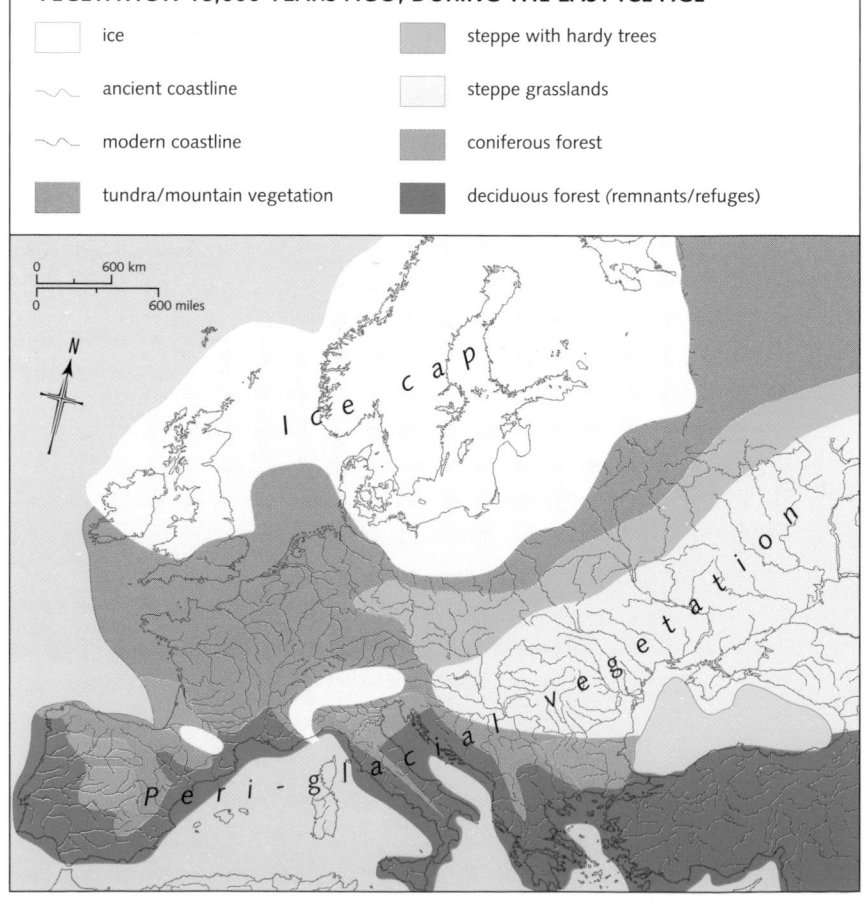

VEGETATION 18,000 YEARS AGO, DURING THE LAST ICE AGE

- ice
- ancient coastline
- modern coastline
- tundra/mountain vegetation
- steppe with hardy trees
- steppe grasslands
- coniferous forest
- deciduous forest (remnants/refuges)

European vegetation through the ice ages
3 million years ago

Before the Pleistocene ice ages, most of Europe was forested. The presence of mixed deciduous forests in the north and warm temperate wet forest in the south indicates a warmer and wetter climate than exists at present. The oceanic forests of the west resembled those of present-day California. Many pre-Pleistocene species have been lost from Europe as a direct result of the ice ages.

18,000 years ago

At their maximum extent the ice sheets extended as far south as southern Britain, Denmark, northern Poland and northwest Russia, with additional glaciation in the Alps. The cooling of the region to the south of the ice sheets pushed the tree line to the southern margins of Europe, covering most of the continent in tundra. Sea levels dropped, linking the British Isles with the mainland of Europe.

Present day

The ice sheets have retreated because Europe, along with the rest of the Earth, has been warming up for about the past 12,000 years. The tree line has moved north again, covering most of Europe in deciduous forest. In cooler regions conifers dominate, while a different type of vegetation has grown up around the Mediterranean, where conditions are dry, with hot summers and mild winters.

Zones of vegetation

There is a rich diversity of vegetation across the continent of Europe, but it still does not match the pre-Pleistocene range of types. Seen here are, from the top downwards, tundra vegetation at Fromkopperan in Norway, coniferous forest at Sunndalen in Norway and Mediterranean vegetation at Monemvasia in Greece.

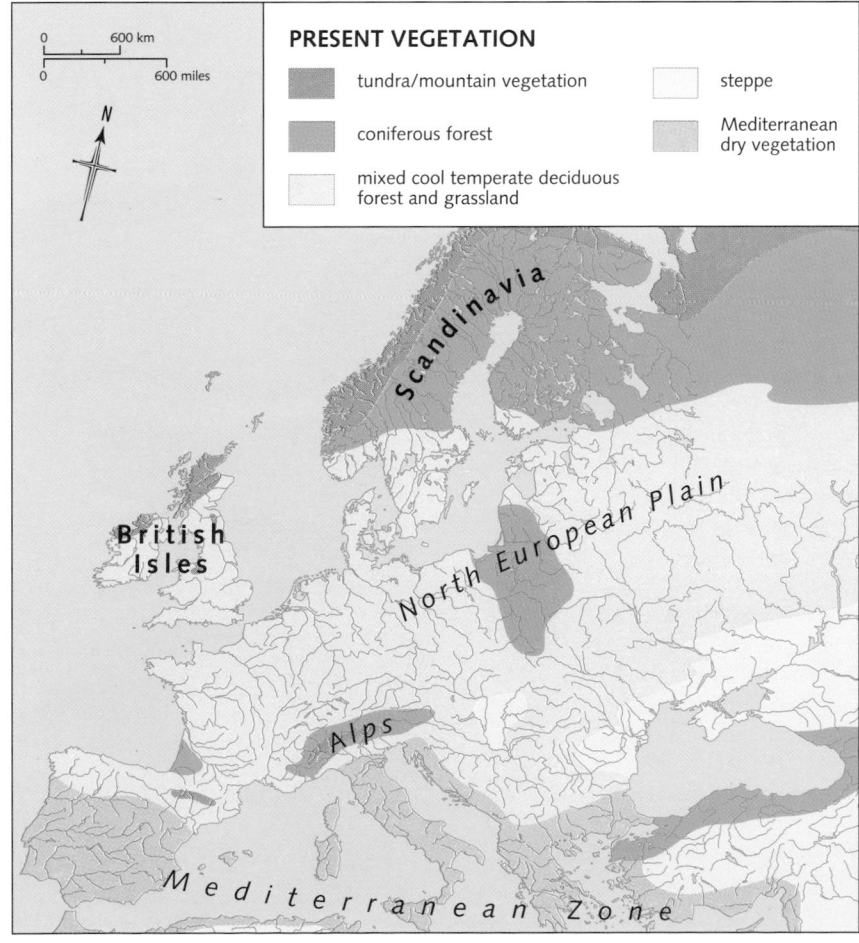

PRESENT VEGETATION

- tundra/mountain vegetation
- coniferous forest
- mixed cool temperate deciduous forest and grassland
- steppe
- Mediterranean dry vegetation

0   600 km
0   600 miles

N

Scandinavia

British Isles

North European Plain

Alps

Mediterranean Zone

Montane Faunas of the Southern Rockies
Islands in the sky

Islands are of enormous interest to evolutionary biologists. The opportunity to study plants and animals within defined geographical limits has allowed many aspects of evolution to be better understood. But islands come in many different types, and any area that is surrounded by territory hostile to particular groups of plants or animals is effectively an island for those organisms.

In many upland regions of the world, islands have been created by virtue of altitude. Certain plants will only live in the conditions that exist above a certain altitude, and certain animals will live only in regions that are occupied by those particular plants. These so-called montane floras and faunas therefore occupy the highest reaches of many mountain ranges, remaining isolated in their territories by the low-lying country around them.

In order to understand the distribution of life forms in these 'islands in the sky' we need to look at their history. The ranges of the southern Rockies are separated from one another and from the 'mainland' range by distances of over 100 kilometres. No range contains a unique set of fauna, and many species are common to most or all of them. It is unlikely that the smaller mammals that survive only on the high ranges could have crossed tracts of hostile territory to colonize other montane regions. It is important not to discount an explanation just because it is unlikely, but a more probable explanation is available.

In this theory the ranges must have been in contact with one another at some point, allowing the movement of animals between them. What seems to have happened is that, as the world cooled down during the Pleistocene ice ages, the plants that occupied the highest regions crept down the mountains, finding suitable conditions over a much wider area. The mountain animals followed, and ranged over the whole of the Rocky Mountain region. The same thing happened in other upland regions of the world. A flora and fauna common to the entire region therefore developed. Then, as the ice withdrew and the Earth began to warm up again, the montane life forms retreated to their mountain 'islands'.

As with all natural phenomena, it is wise to be cautious about giving sweeping explanations for complex patterns of behaviour. The understanding of montane animals takes the ice age theory as a starting point, not as its conclusion. But the study of mammals in the montane habitats of the southern Rockies has led to a greater understanding of the natural pressures to which animals are subjected. In general, larger mammals such as the lynx and the beaver are only found in those ranges with the largest area, as these mammals need more space in which to roam. When the area available to them shrinks to a critical size, the genetic diversity is insufficient for the animal to adapt continually. Diversity is then further reduced by inbreeding, and eventually the animal becomes extinct in that territory. This 'natural extinction' has undoubtedly taken place in the ranges of the Rockies, because successive generations of animals have been trapped in territories which have shrunk as the region has got warmer.

Not surprisingly, the greatest numbers of species are found in the largest montane ranges. The conservation of large mammals, including those much bigger than anything found in the Rockies, often requires vast areas of land in order to allow outbreeding and, in some cases, migration.

The general pattern of mammal distribution in the montane habitats reflects that found on oceanic islands, confirming the view that montane habitats are 'true' islands, with little leakage of species from one range to another. The changing size of the territories does, however, give an extra dimension to our understanding of the distribution of animals.

Montane mammals
The mammals that inhabit the montane zones range from tiny insect-eating rodents to the bobcat or lynx (*Lynx canadensis*). The largest mammals are carnivores and are restricted to the biggest ranges. The lynx and the wolverine (*Gulo luscus*), for example, are found only on the Rocky Mountain 'mainland', the Sangre de Cristo Mountains and the San Juan Mountains. Each of these ranges covers an area of more than 19,000 square kilometres. While this may not be the defining lower limit for the sustaining of a lynx population, it is a clear indication of the need for large mammals to have large ranges. The deer mouse (*Peromyscus maniculatus rufinus*) occurs on 26 of the 28 ranges and is among the smallest mammals found in the Rockies.

Southern Rocky Mountains
The southern end of the Rocky Mountain chain covers a wide area. This is caused by the complexities of its formation. While the highest mountains are concentrated in western Colorado, the area of upland stretches over a vast area reaching from Wyoming to southern Arizona. Mountain ranges are dotted over this upland, forming isolated pockets of high-altitude habitat. The topography of the region allowed a large continuous area to become converted to montane habitat during the Pleistocene ice ages. This was the source of the floras and faunas that now occupy the separated ranges. The map opposite shows a direct correlation between the area of the territory and the number of mammal species that it is able to support.

Eastern chipmunk
Tamias stiatus

Muskrat
Ondatra zibethica

American beaver
Castor canadensis

Lynx
Lynx canadensis

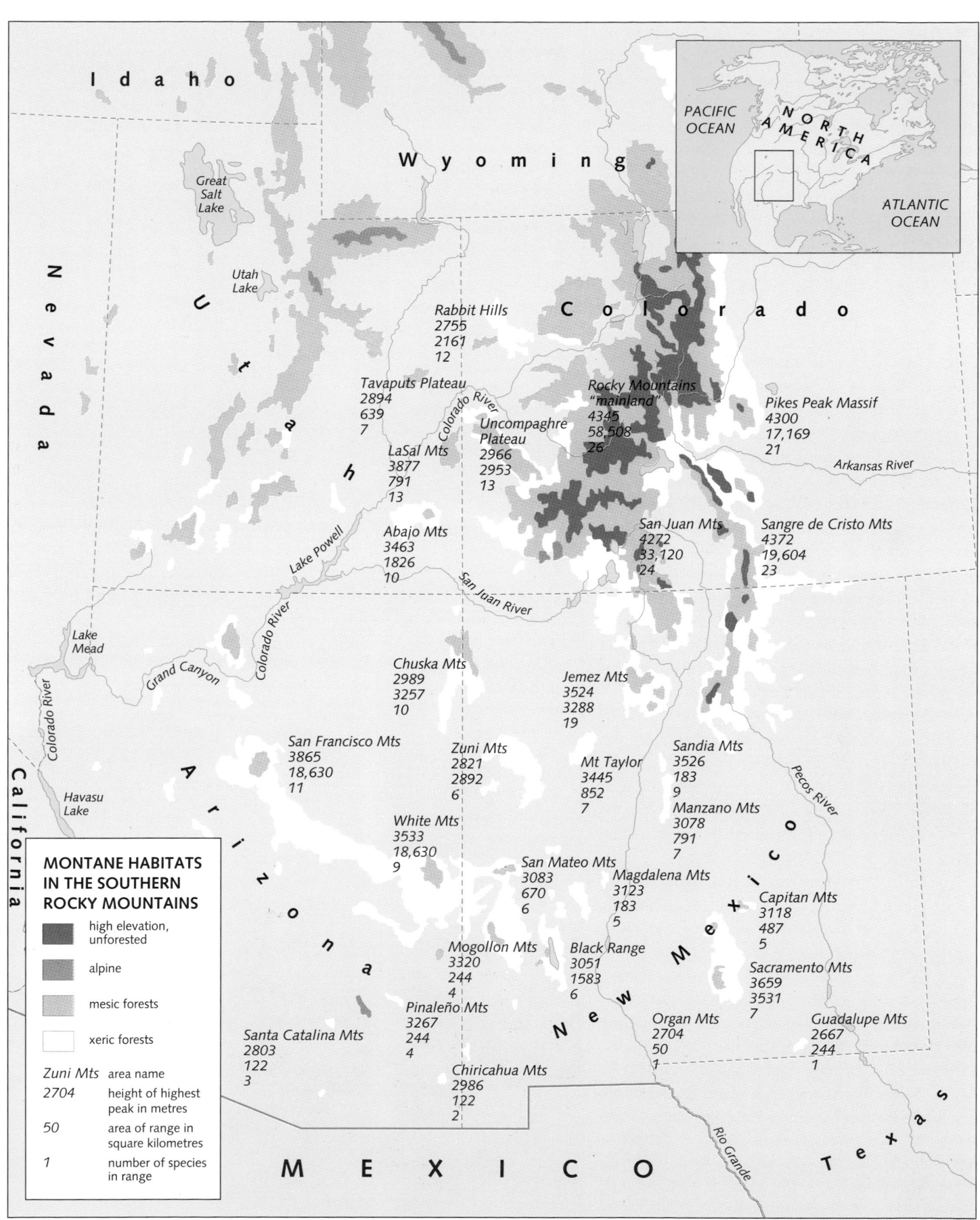

Idaho

Wyoming

Colorado

Nevada

Utah

California

Arizona

New Mexico

Texas

MEXICO

PACIFIC
OCEAN

NORTH
AMERICA

ATLANTIC
OCEAN

Great
Salt
Lake

Utah
Lake

Colorado River

Lake Powell

Lake
Mead

Colorado River

Grand Canyon

San Juan River

Colorado River

Havasu
Lake

Arkansas River

Pecos River

Rio Grande

Rabbit Hills
2755
2161
12

Tavaputs Plateau
2894
639
7

LaSal Mts
3877
791
13

**Uncompaghre
Plateau**
2966
2953
13

**Rocky Mountains
"mainland"**
4345
58,508
26

Pikes Peak Massif
4300
17,169
21

Abajo Mts
3463
1826
10

San Juan Mts
4272
33,120
24

Sangre de Cristo Mts
4372
19,604
23

Chuska Mts
2989
3257
10

Jemez Mts
3524
3288
19

San Francisco Mts
3865
18,630
11

Zuni Mts
2821
2892
6

Mt Taylor
3445
852
7

Sandia Mts
3526
183
9

Manzano Mts
3078
791
7

White Mts
3533
18,630
9

San Mateo Mts
3083
670
6

Magdalena Mts
3123
183
5

Capitan Mts
3118
487
5

Mogollon Mts
3320
244
4

Black Range
3051
1583
6

Sacramento Mts
3659
3531
7

Pinaleño Mts
3267
244
4

Organ Mts
2704
50
1

Guadalupe Mts
2667
244
1

Santa Catalina Mts
2803
122
3

Chiricahua Mts
2986
122
2

MONTANE HABITATS
IN THE SOUTHERN
ROCKY MOUNTAINS

- high elevation, unforested
- alpine
- mesic forests
- xeric forests

Zuni Mts   area name
2704   height of highest peak in metres
50   area of range in square kilometres
1   number of species in range

Paramos and Puna
Montane vegetation of the high Andes

The isolation of groups of flora and fauna in pockets of the high Andes follows a similar pattern to that seen in other highland regions. The vegetation that thrives in the highest areas is adapted to conditions which many plants, particularly trees, cannot cope with. The question of how these plants spread from one mountain 'island' to another is answered only by understanding the history of the particular region. In the Andes this is especially fascinating, since the two highland regions – the northern Andes and the Peruvian and Bolivian Andes – have separate but linked histories which are reflected in their present-day flora and fauna.

Above the tree line, at about 3,000 metres above sea level, the northern Andes are covered in dense tussocks of grass known as paramo. Further south, similar vegetation known as puna is found above the tree line in the Peruvian and Bolivian Andes. Here, the greater altitude of the region means that the puna vegetation is spread over a continuous area rather than being isolated in small islands. In both cases, certain groups of birds, butterflies and other insects are restricted to the habitats offered by the puna and paramo vegetation.

The Andes, along with the rest of the Earth's surface, became colder during the Pleistocene period. From about 3 million years ago until 12,000 years ago, the Earth was subject to a series of intense cooling periods, with ice sheets spreading outward from the polar regions and extensive glacier formation in mountainous regions. Between these glaciations, the Earth has become periodically warmer, though remaining colder than its average temperature throughout its history. For the past 12,000 years or so, we have been in an interglacial period, with the Earth's surface gradually becoming warmer.

During glaciations, the Paramos Andes became cooler, pushing the tree line down the mountains and allowing the paramo vegetation to spread over a much wider area. This in effect linked up many of the separate mountain-top 'islands'. In the Puna Andes this situation was complicated by the formation of glaciers. As these glaciers melted during the interglacial periods, the accumulation of eroded debris blocked drainage channels and created large bodies of water known as glacial lakes. Both glaciers and lakes provided barriers to the spread of montane vegetation. Glaciation therefore seems to have had an opposite effect on the two regions, causing greater chances for dispersal in the Paramos while forming barriers to dispersal in the Puna.

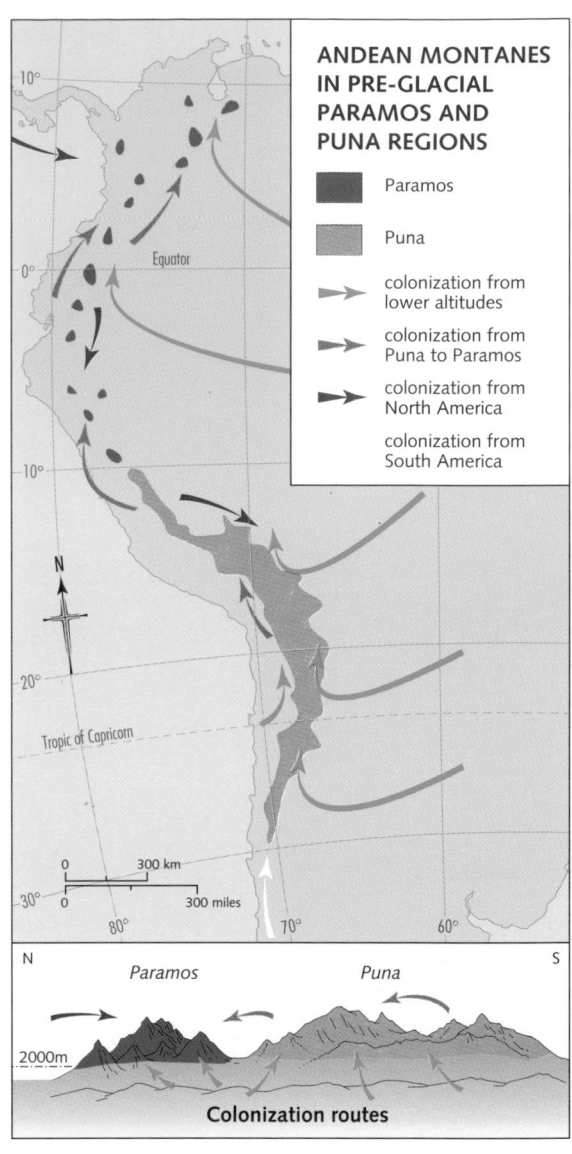

The pre-glacial Andes
The route of the pre-glacial colonization of the 'islands' of the northern Andes by paramo vegetation remains uncertain. It must have come about either by the adaptation of species from lower latitudes, or by dispersal during an early glacial episode. Colonization by birds is easier to explain. Species migrated north from southern South America up the Andean chain to the Puna and Paramos Andes. Some species came from Central America but penetrated only to the Paramos. Other groups moved, through adaptation, from lower altitudes.

ANDEAN MONTANES IN PRE-GLACIAL PARAMOS AND PUNA REGIONS

- Paramos
- Puna
- colonization from lower altitudes
- colonization from Puna to Paramos
- colonization from North America
- colonization from South America

Cross-sections through the Andes (*right*) show how changing temperature alters the range of montane vegetation and patterns of species dispersal.

Paramos   Puna

2000m

Colonization routes

Montane vegetation
A particular type of vegetation (*left*) grows above the tree line of the high Andes. Certain types of grass have developed to take advantage of the absence of competing trees and other plants. This paramo and puna vegetation is the habitat of a variety of insects, birds and mammals, some of which do not inhabit the lower reaches of the mountains. The absence of trees gives an advantage to animals that depend on the ability to see approaching predators over long distances.

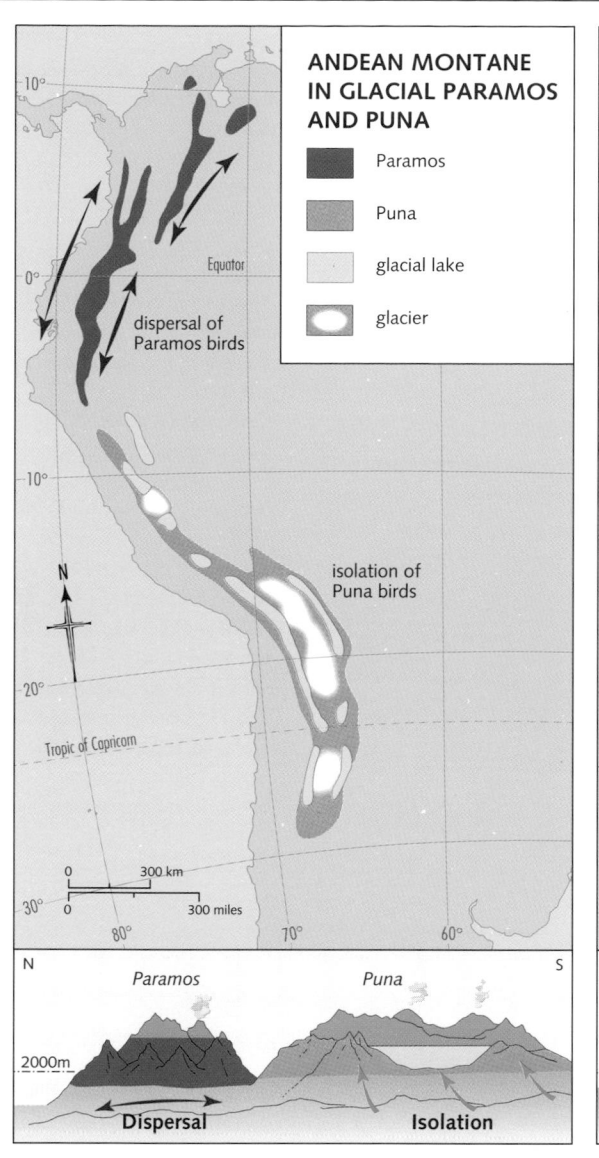

ANDEAN MONTANE IN GLACIAL PARAMOS AND PUNA

- Paramos
- Puna
- glacial lake
- glacier

dispersal of Paramos birds

isolation of Puna birds

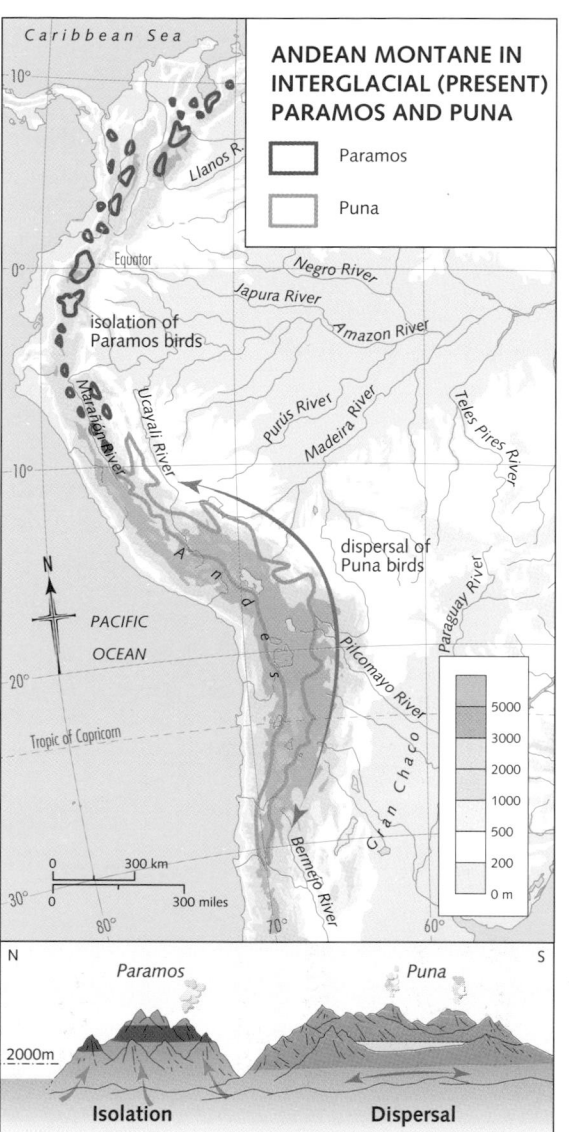

ANDEAN MONTANE IN INTERGLACIAL (PRESENT) PARAMOS AND PUNA

- Paramos
- Puna

isolation of Paramos birds

dispersal of Puna birds

Ice-age Andes *(far left)*
The cooling of the northern Andes allowed the dispersal of previously isolated Paramos species over a wide area. Plants, birds and insects came into secondary contact with species from which they had once diverged. Sympatric speciation in birds allowed different species to develop in the same area. In the Puna Andes, the continuity of the montane region was broken up by glaciers and lakes, leading to the formation of isolated communities, which evolved separate variations.

Present-day Andes *(left)*
The situation has reversed, with Paramos flora and fauna existing in isolated pockets. There will be divergence of isolated groups, until another ice age brings these communities back into contact. The continuity of the Puna has been restored, allowing the dispersal of species between previously isolated areas.

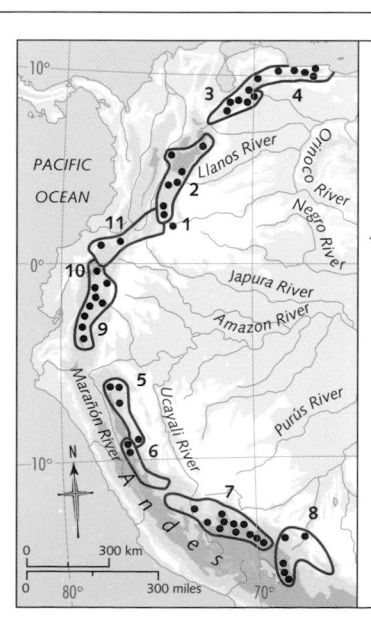

DISTRIBUTION OF THE HYALYRIS COENO BUTTERFLY

 1. *H. coeno metaensis*
 2. *H. coeno angustior*
 3. *H. coeno avinoffi*

4. *H. coeno coeno* 5. *H. coeno atrata* 6. *H. coeno schlingeri* 7. *H. coeno* ssp. nov.

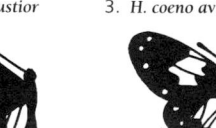

9. *H. coeno acceptabilis* 9. *H. coeno norella & H. coeno aquilonia* 10. *H. coeno norellana* 11. *H. coeno florida*

The butterflies of the Paramos Andes
The *Hyalyris coeno* group of butterflies lives at altitudes of more than 200 metres (650 feet) in the northern Andes. The separation of the populations has contributed to the diversity of the species, as each has been able to evolve in isolation.

Forest Refuges in the Amazon
Endemic centres

The tropical rainforests of South America extend over an area of about 5,300,000 square kilometres (2,050,000 square miles) and contain the greatest diversity of species found anywhere on Earth. Their position on the Equator gives little seasonal variation in temperature and this, combined with high rainfall, provides constant warmth and moisture. We have known for a long time the drastic effect of the Pleistocene ice ages on the high latitude regions of the world. More recently their effects have been discovered in tropical and subtropical regions.

During the Pleistocene period, from about 3 million to 10,000 years ago, this region became periodically cooler, along with the rest of the Earth. The lowering of sea levels and the changes in prevailing winds also brought dry conditions to the Amazon basin. This cool, arid climate proved unsuitable for rainforest plants. The forests shrank dramatically – at the height of the last glaciation 18,000 years ago, they covered less than one-tenth of their present area. The exact extent of the rainforests in the interglacial periods and before the Pleistocene is difficult to estimate, but it is thought that they were at least as extensive as they are now at some time before the last ice age.

The places where the rainforest held on are known as the forest refuges. The locations of these refuges have been pieced together from different types of evidence, not all of which gives exactly the same answer. Preserved layers of previous vegetation are the most obvious source of information, but these are difficult to identify in the rainforest. We would expect the rainforest to endure longest in those areas where

The location of forest refuges

The position of those islands of rainforest that lasted throughout the ice ages (*below*) are known only from the evidence provided by the fauna and flora of the present day (*bottom right*), supported by data from soil sampling (*top right*). Invasions by the sea during the warm interglacial periods are known to have destroyed rainforest along the Amazon River itself, so the refuges must have occupied higher ground. They were situated where there was high humidity and high moisture in the soil. Following the warming of the region, the forests spread out from the refuges – as can be seen from the high number of plant and animal species that are endemic to those places.

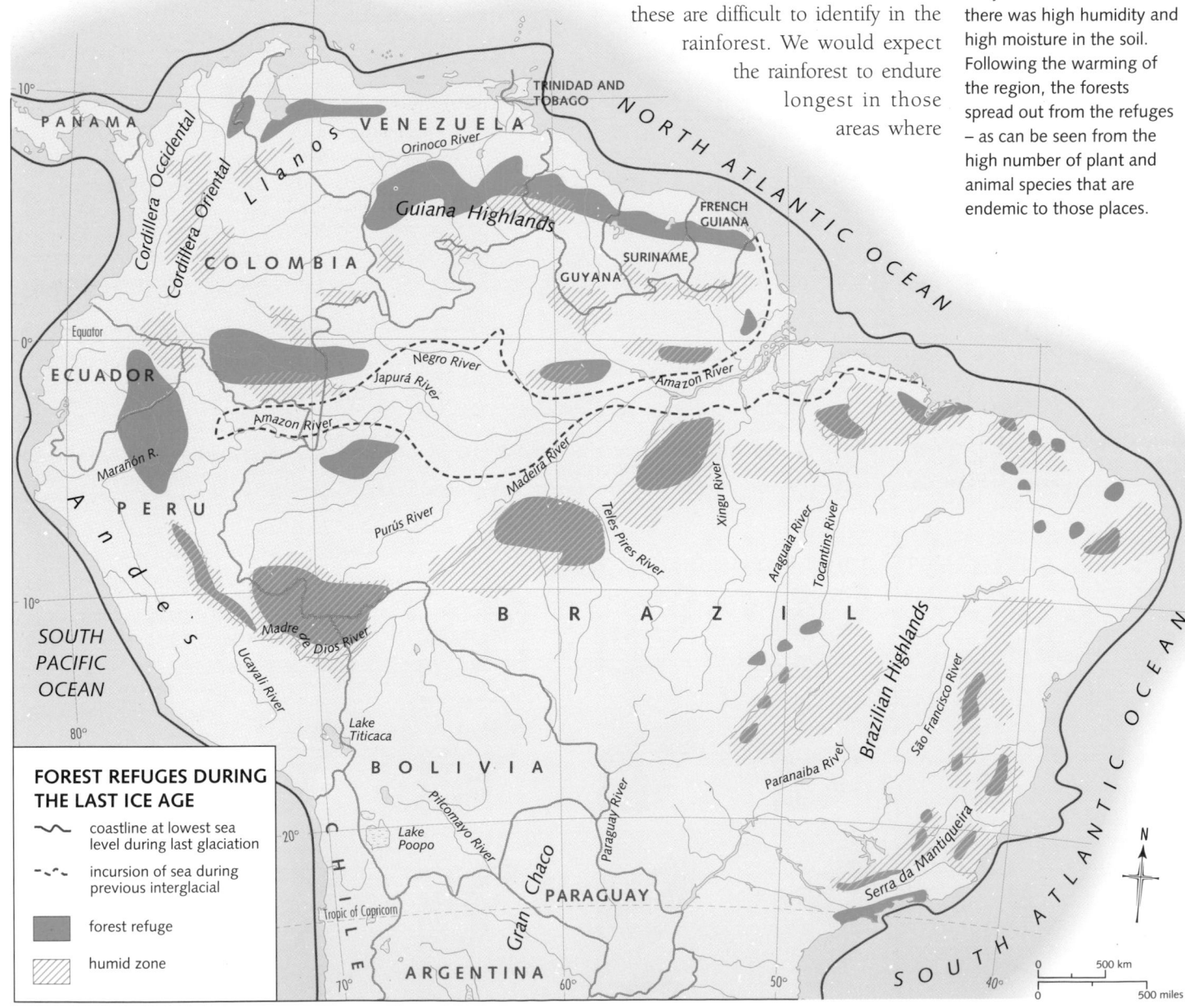

FOREST REFUGES DURING THE LAST ICE AGE

- ∿ coastline at lowest sea level during last glaciation
- ⌐⌐⌐ incursion of sea during previous interglacial
- ▬ forest refuge
- ▨ humid zone

SOIL CLASSES OF SOUTH AMERICA

- very high moisture retention
- high to moderate moisture retention

The apparent uniformity of tropical rainforests is shown, on close inspection, to be illusory. Species diversity is high in some areas, and within the range of each individual species there are areas of abundance and areas in which individuals are rare. The number of species of fern (*above*) in tropical South America, for example, is unlikely ever to be known.

the soil shows the highest level of moisture retention (*see map above*). This level varies a great deal, even within the Amazon basin. Evidence from soil sampling does coincide well with information obtained from other sources.

More crucially, biologists have calculated the number of endemic species in different regions of the Amazon, and have found that they are concentrated in specific areas. These endemic centres vary slightly for different types of flora and fauna, but there is remarkable overlap between those for birds, those for flowering plants and those for insects such as butterflies.

Endemic centres are areas in which there are a greater than normal number of species that are unique to that particular area. In the case of the Amazon region, it is assumed that a large number of species became concentrated in the forest refuges during the cold dry periods, and died out in other areas. As the region has warmed up and become wet, these species have gradually expanded outwards again, although they are still proportionally concentrated in the areas that remained rainforest throughout the ice ages.

ENDEMIC CENTRES

- centre of endemic plants, butterflies and birds

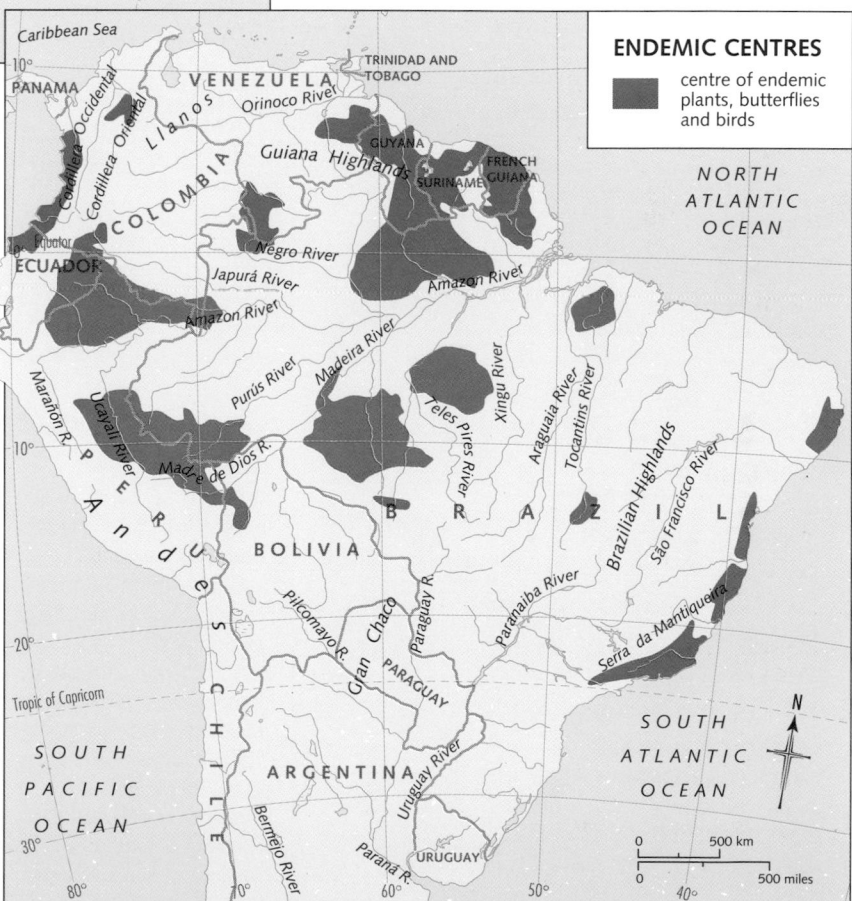

Pleistocene Extinctions

Climatic change and the death of cold-adapted mammals

Extinction is a normal part of evolution – new species evolve and others disappear all the time. Extinction rates have often been high, most notably during the 'big five' mass extinctions (*see pages 68–73*). The Late Pleistocene extinction event of about 10,000 years ago was hardly a mass extinction, since it apparently affected only large mammals, but the species that disappeared were impressive, and fossil remains and other evidence are still relatively fresh.

The Pleistocene epoch began 2 million years ago with a major cooling episode. The northern hemisphere became cold and ice sheets extended far south several times. The last ice age, the Wisconsinan (in North America) or Devensian (in Europe), ended only 10,000 years ago. At its height 18,000 years ago, the north polar ice cap extended over Canada, the northern United States, and northern Europe and Asia. The icy north became home to a variety of cold-adapted plants and animals. Tundra and alpine plants extended their natural ranges enormously (*see pages 106–109*) and heavily insulated mammals evolved to exploit the new habitats.

In North America, the ice sheets advanced and retreated, and major floral zones moved north and south at the same time. At the height of the Wisconsinan Glacial, the ice sheet extended 200

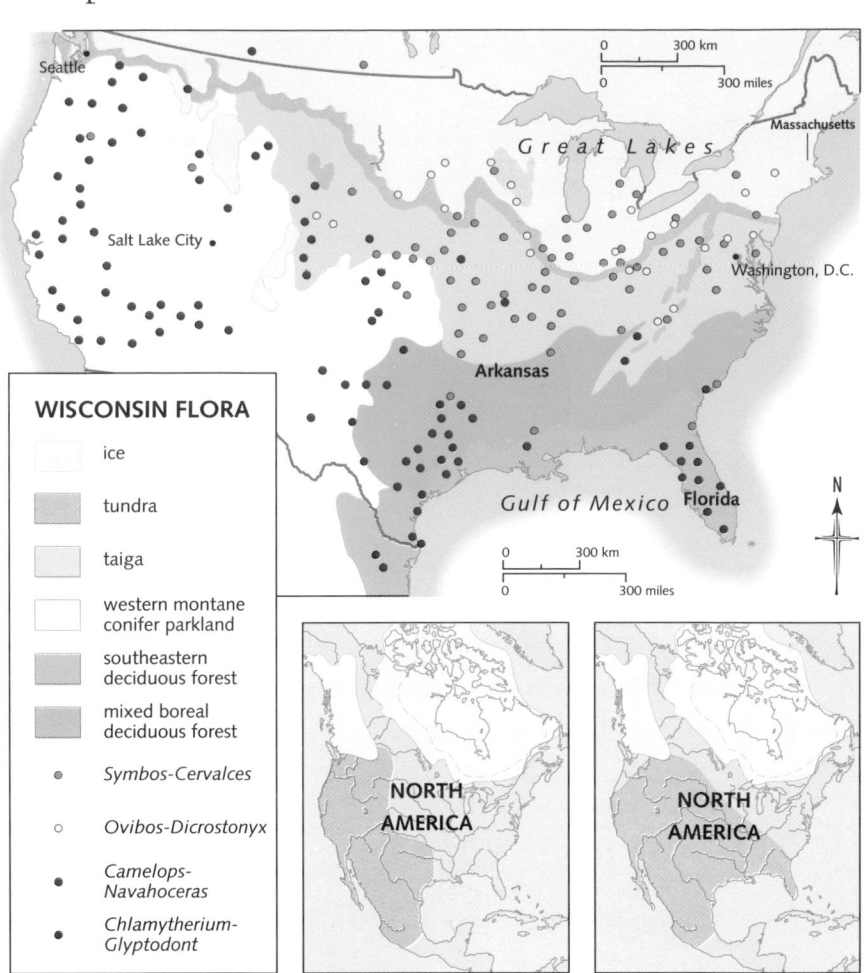

WISCONSIN FLORA

- ice
- tundra
- taiga
- western montane conifer parkland
- southeastern deciduous forest
- mixed boreal deciduous forest

- ● *Symbos-Cervalces*
- ○ *Ovibos-Dicrostonyx*
- ● *Camelops-Navahoceras*
- ● *Chlamytherium-Glyptodont*

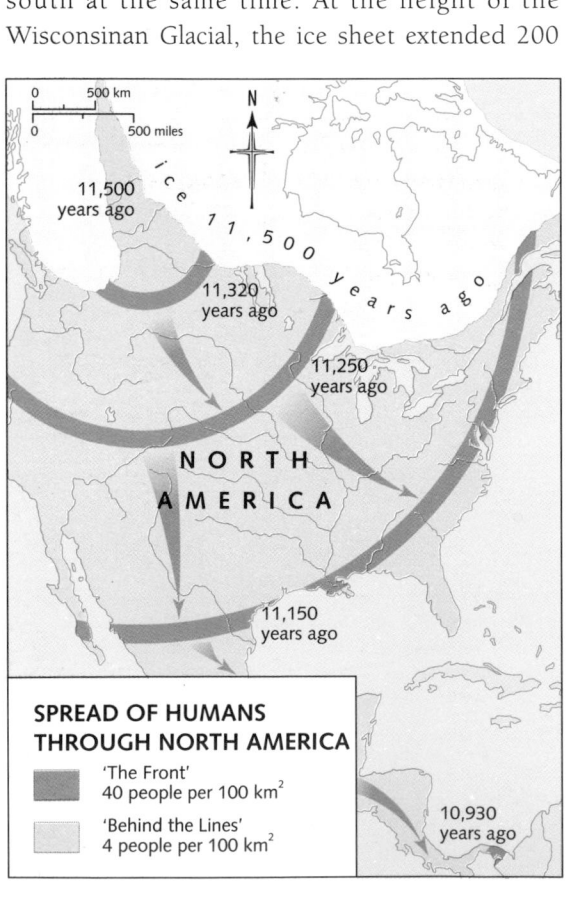

SPREAD OF HUMANS THROUGH NORTH AMERICA

- 'The Front' 40 people per 100 km²
- 'Behind the Lines' 4 people per 100 km²

11,500 years ago
11,320 years ago
11,250 years ago
11,150 years ago
10,930 years ago

Smilodon

Mammal extinctions

Several large mammals had wide distributions in North America during the Wisconsinan ice advance (*above*), such as the Shasta ground sloth, *Nothrotheriops*, the sabre-toothed cat, *Smilodon*, and the mastodon, *Mammut*. The mastodon lived right up to the tundra belt. Were these mammals killed by warming climates or by the advance of humans across North America (*left*)?

Changing geography

During the Wisconsinan glacial phase, up to 10,000 years ago, ice sheets extended well south of the Great Lakes (*top*). A belt of tundra ran from Massachusetts to Seattle. Low-vegetation taiga extended down to Washington, D.C., in the east and Salt Lake City in the west. The rest of North America consisted of north-temperate conifer and deciduous forests.

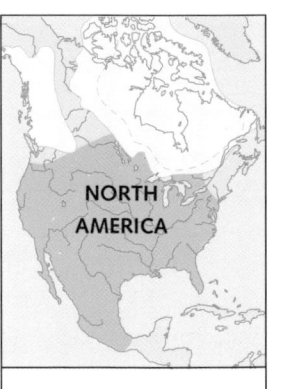

UPPER PLEISTOCENE DISTRIBUTIONS

- ice sheets 12,000 BP
- ice sheets 10,500 BP
- Shasta ground sloth (*Nothrotheriops shastensis*)
- sabre-toothed cat (*Smilodon fatalis*)
- mastodon (*Mammut americanum*)

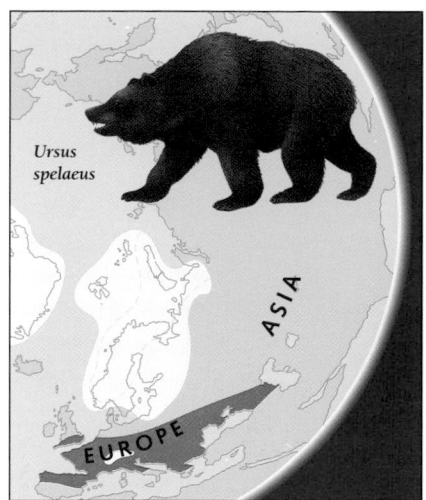

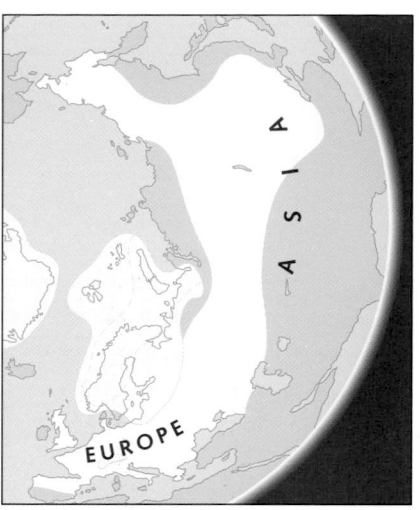

European ice age
In Late Pleistocene times, ice sheets covered northern Europe and the region around the Alps. Cold-adapted mammals had wide distributions over Europe and western Asia (*left* and *below*). The cave bear, *Ursus spelaeus*, was found abundantly over central and southern Europe, as was the giant deer, *Megaloceros giganteus*, which also lived further north and east. Rhinoceroses included species that were adapted to both cold and hot conditions. The woolly rhinoceros, *Coelodonta antiquitatis*, was found in a wide strip over Europe and Asia during the last ice age. The rhinoceros *Dicerorhinus hemitoechus* was a warm-adapted species, like modern rhinos, and was found widely in Europe and Asia during the last interglacial (40,000 years ago), but its distribution shrunk rapidly to Mediterranean regions when the cold phase began.

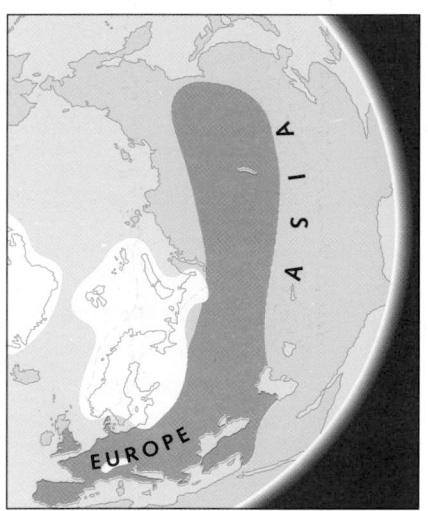

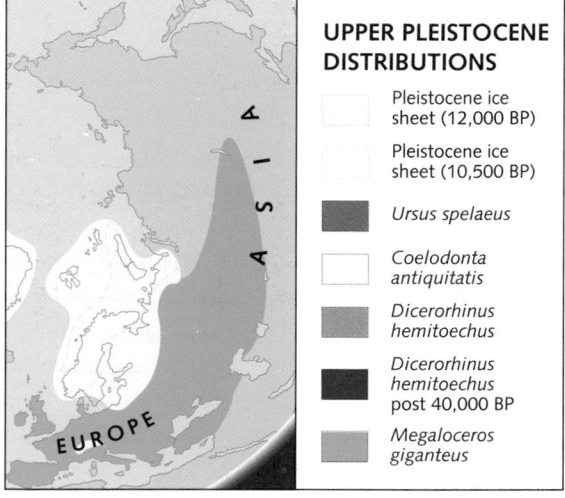

UPPER PLEISTOCENE DISTRIBUTIONS

- Pleistocene ice sheet (12,000 BP)
- Pleistocene ice sheet (10,500 BP)
- *Ursus spelaeus*
- *Coelodonta antiquitatis*
- *Dicerorhinus hemitoechus*
- *Dicerorhinus hemitoechus* post 40,000 BP
- *Megaloceros giganteus*

MAMMOTH DISTRIBUTION

- Pleistocene ice sheet (12,000 BP)
- *Mammuthus primigenius*
- *Mammuthus columbi*
- 12.2 latest survival dates in thousands of years BP

kilometres south of the Great Lakes, and most North American flora was arctic or cold-temperate in adaptations. Mammals included muskox, yak, lemming, moose, bear and giant beaver as far south as Florida and Arkansas. Warmer-climate plants and animals, with South American glyptodonts and giant rodents (*see pages 100–101*), were found only in Florida and a thin strip along the Caribbean coastline.

Many large mammals had wide distributions over North America in the Late Pleistocene but disappeared 11,000–10,000 years ago, a date that corresponds not only with the beginning of the retreat of the Wisconsinan ice sheet, but also with the arrival of humans in North America (*see pages 116–117*). Extinction patterns were similar in Eurasia, where many large mammals were widespread during the Devensian phase. Were they wiped out by changing climates or by hunting overkill? The overkill hypothesis might hold for North America, but climatic change was more important in Europe where humans had been present during the last ice age.

Mammoths had a wide distribution in the north during the Late Pleistocene (*below*). The Columbian mammoth, *Mammuthus columbi*, occupied all of North America south of the ice sheets. The woolly mammoth, *Mammuthus primigenius*, inhabited Europe, Asia and parts of Alaska and northern Canada.

Human Migrations

Entry into North America and Australia via land bridges

The origin and spread of humans coincided with the latter part of the Pleistocene ice ages, and the geographical distribution of early humans was influenced by these ice ages in a number of ways. As with all work on our early ancestors, research into migration patterns is severely hampered by the scarcity of preserved remains. Our current state of knowledge is therefore precarious, but we have made some interesting findings about how and why our forebears spread over the globe.

The normal Darwinian explanation for the expansion of a species range is pressure of numbers. As pointed out by Malthus in his celebrated *Essay on the Principle of Population* (1798), which was a great influence on Darwin's thinking (*see page 34*), population will naturally increase exponentially, while food resources can at best only increase in a linear fashion. Species therefore outgrow their resources and are forced to expand their territory, or to move on.

Early humans probably remained in Africa for the first 50,000 years of their existence. Groups then migrated north through Arabia and east across Asia. Though slow in modern terms, this migration was rapid on an evolutionary time-scale. But why did humans spread across the globe? Pressure of numbers does not seem to provide an answer – the population of Africa was tiny at the time humans left. It is known that the first migrations coincided with an interglacial period and warming of all parts of the Earth. Northeast Africa and the Middle East became fertile regions suited to human habitation. This has led to the theory that humans were trapped in Africa until the end of a particular ice age. They then burst out of the northeast corner and pushed eastwards.

This 'cork in a bottle' theory does not explain *why* this nomadic animal needed to spread so far. It seems clear that early *Homo sapiens* lived in extended family groups like other primates. As their main physical difference from their ape relatives is brain size, recent research has looked at the main use of the human brain – as a tool of social interaction – as the key to understanding human migration. Groups established territories in which they could find food and safety. When food diminished, advance parties were sent out to seek the best places to move to, and the information they brought back was shared and acted on by the group as a whole. The human brain, with a greater capacity for storing and processing information, gave the species a tool for rapid movement and the exploitation of new situations, and this may explain the huge area over which it spread.

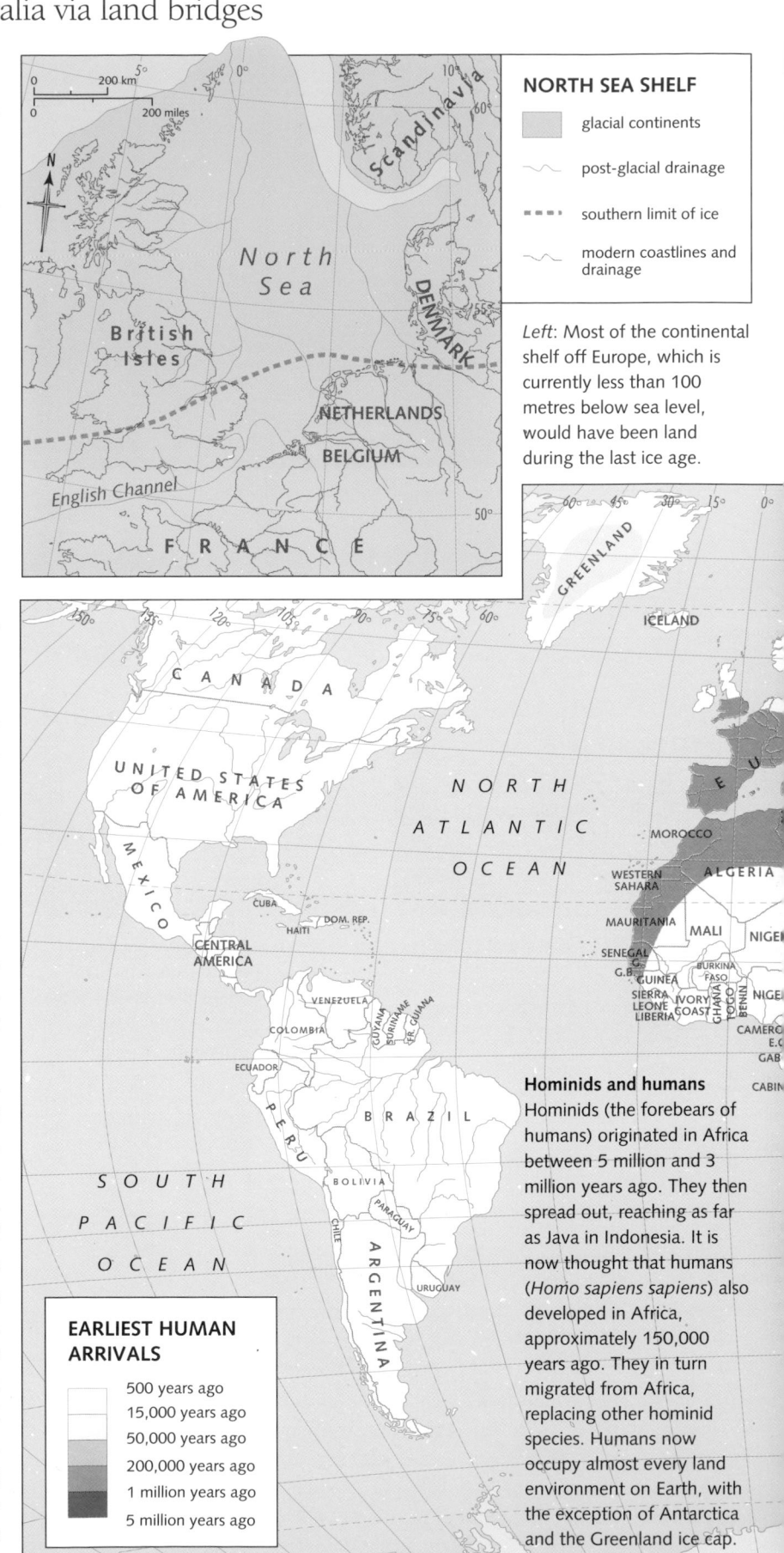

NORTH SEA SHELF

- glacial continents
- post-glacial drainage
- ---- southern limit of ice
- modern coastlines and drainage

Left: Most of the continental shelf off Europe, which is currently less than 100 metres below sea level, would have been land during the last ice age.

EARLIEST HUMAN ARRIVALS

- 500 years ago
- 15,000 years ago
- 50,000 years ago
- 200,000 years ago
- 1 million years ago
- 5 million years ago

Hominids and humans
Hominids (the forebears of humans) originated in Africa between 5 million and 3 million years ago. They then spread out, reaching as far as Java in Indonesia. It is now thought that humans (*Homo sapiens sapiens*) also developed in Africa, approximately 150,000 years ago. They in turn migrated from Africa, replacing other hominid species. Humans now occupy almost every land environment on Earth, with the exception of Antarctica and the Greenland ice cap.

At times of maximum glaciation, enormous quantities of seawater were frozen in polar ice sheets, reducing the overall world sea level by up to 150 metres. This created bridges between separate land masses, allowing human beings and other animals to migrate. Human entry into North America and Australia came via land bridges. Migration to North America took place over the wide piece of land known as Beringia (*right*). The first phase was between 30,000 and 18,000 years ago, when the whole of North America, apart from a strip down the west coast, was under ice.

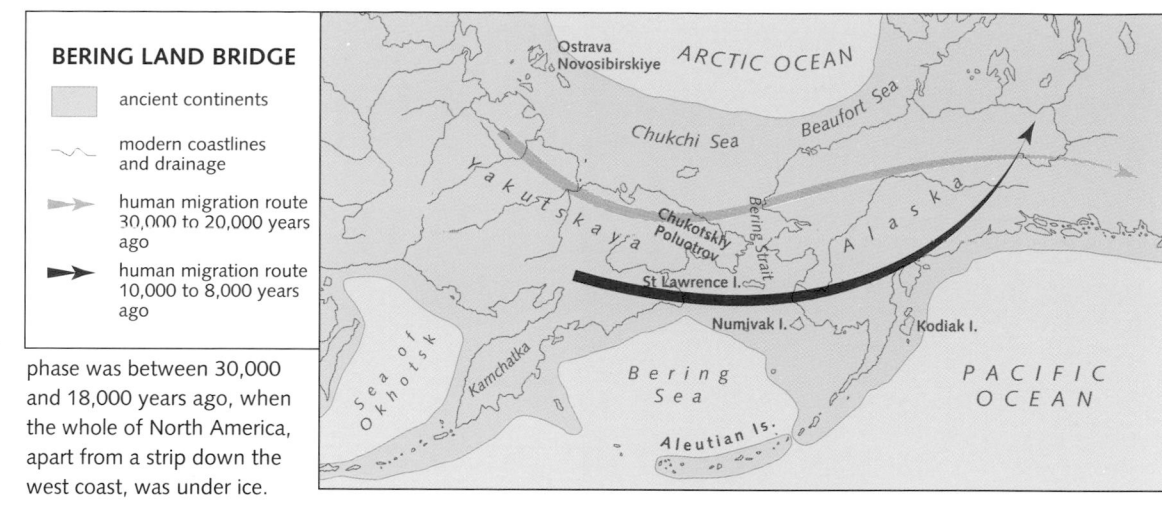

BERING LAND BRIDGE

ancient continents

modern coastlines and drainage

human migration route 30,000 to 20,000 years ago

human migration route 10,000 to 8,000 years ago

SUNDA LAND BRIDGE

ancient continents

modern coastlines

probable human land migration routes

SAHUL LAND BRIDGE

ancient continents

modern coastlines

earlier Pleistocene migrations

late Pleistocene migrations

post-glacial migrations

Far right: Australia was probably reached via the Sunda and Sahul land bridges. There was a wave of migration with every glaciation, each following a different route.
Right: Cave paintings from Finnmark Alta show the presence of humans in northern Europe around 32,000 years ago.

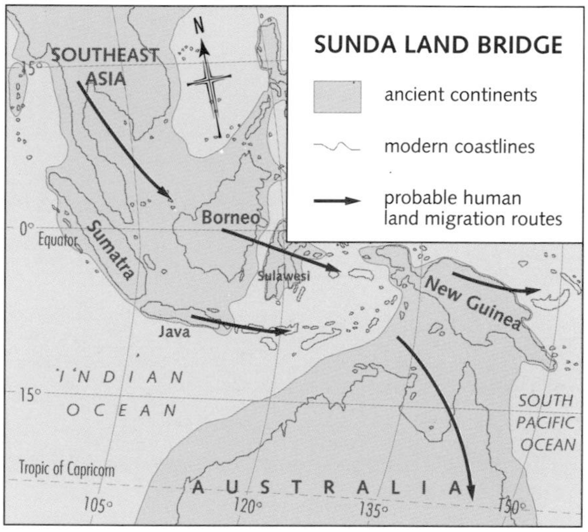

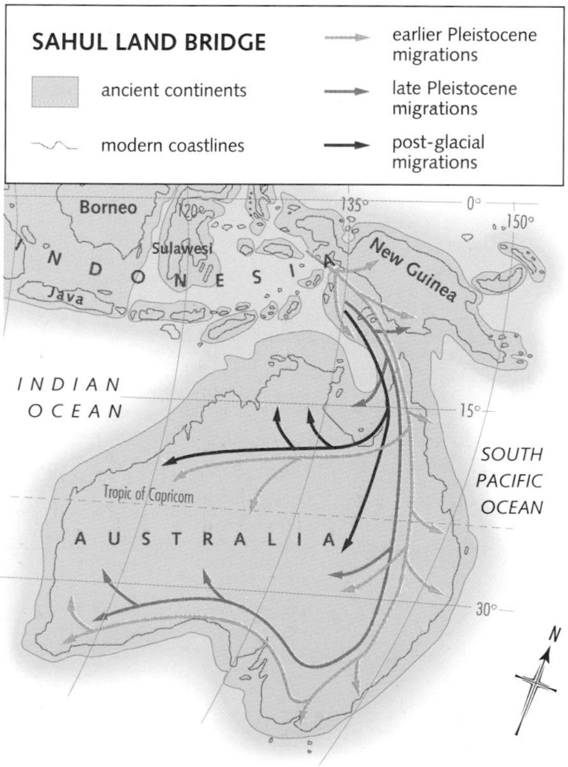

Part Three
The Pattern of Life

Diversity and Environment

The Distribution of Life on Earth

Islands and Endemics

Human Intervention

Section VII: Diversity and Environment

The total number of species in any place varies enormously over the Earth. On average, there is much greater species diversity in the tropics than there is in colder regions. Variability of habitat encourages diversity, as does the abundance of food.

Victorian naturalists who travelled from Europe to equatorial regions were astonished by the diversity of life in the tropics. Charles Darwin saw it in his visits to South America and the Pacific. His contemporary Alfred Russell Wallace was similarly impressed by this phenomenon during his work in Southeast Asia, as was the famous German explorer-naturalist Alexander von Humboldt. These naturalists presented their results in a series of successful books, and the vast diversity of tropical birds, insects and plants became known to a wide reading public. Darwin, Wallace and Humboldt all asked why it was that the tropics support such high diversities. It is a question that has still not been completely answered.

The high diversity of the tropics is part of a broader gradient in species diversity extending from polar to equatorial regions. This has been demonstrated for numerous groups of organisms. Lizards, for example, show a dramatic change in diversity from fewer than five genera in northern parts of the United States, Canada and Eurasia from France to northern China, to more than 40 genera in western parts of the Amazon basin. This increase may be tracked from the temperate belt to the Equator – there are five genera in Washington State and Virginia, ten in northern California and Florida, 20 in southern Mexico, 30 in Colombia and northern Brazil and 40 in the Amazon basin. A similar pattern spans Eurasia and Africa.

Patterns of increasing diversity are also shown by many marine groups. Cowrie shells, which inhabit shallow seawater, are rare in temperate latitudes, occurring only sporadically as far north as the south coast of England and northern Japan, and as far south as northern Chile and New Zealand. The gradient of genera rises from two on the Baja California coast, off northern Florida, northern Spain, northern Italy and Japan, to four off Mexico, north Africa, southern France and Korea, eight off most of the west African coast, northern India and southern China, 16 in the Red Sea, central India and Vietnam, and 32 off the east African coast, southern India and Thailand. The cowrie hot spot – oceanic zones with more than 32 genera in any location – covers a broad equatorial belt of the Indian Ocean and the Pacific.

These trends are repeated for whole habitats: tropical rainforests and coral reefs are much more species-rich than their temperate equivalents. There are exceptions to the rule, of course. Penguins are clearly more abundant at the South Pole than the Equator, where only a single species is found, in the Galapagos Islands. Cacti occur only in arid zones north and south of the equatorial belt.

Tropical species richness

There have been many attempts to explain the rich species diversity of the tropics. One suggestion is that tropical habitats are more stable than others. If this is correct, then there has been a long-term predictability about climatic variables and food supplies that has allowed species to divide repeatedly and occupy ever-narrower niches. If temperature and rainfall fluctuate and food supplies vary, it is hard for many specialist species to establish themselves. In temperate zones, there are dramatic seasonal fluctuations, and species must either have broad adaptations to survive through the year, or they must be able to migrate long distances. Other non-tropical zones, such as deserts, tundra and polar regions, are also subject to unpredictable conditions and dramatic catastrophes. Storms, droughts and major freezing episodes present great challenges to the survival of many species, so it may be that diversity can never expand beyond a great extent. Perhaps, then, tropical environments on land have large species diversities because the habitats are predictable. This, however, does not readily explain the high diversity in tropical oceans.

A second reason for high diversity in the tropics, particularly on land, may be high productivity. Productivity refers to the amount of food available to support life, and is based on the primary production of carbohydrates by plant photosynthesis. More sunlight is available for longer times of the day, and for more of the year, in the tropics, so plants can capture more of the Sun's energy and produce more tissue. This supports more herbivores, which in turn support more carnivores, than in temperate and polar zones. This may also be the case in the oceans, where the base of the food chain is photosynthesizing plankton. However, productivity in the sea depends more on the availability of certain nutrients, and although the high points of such nutrients occur in some tropical areas, high productivity in the oceans occurs more in cold temperate and polar waters.

Complexity of habitats may be a further reason for the latitudinal gradient of species diversity. Complex communities are particularly common in tropical forests, where thousands of plant species may provide niches for similarly high numbers of animal species. The same is true of tropical coral reefs, which themselves consist of a wide diversity of skeleton-building organisms, and which provide homes for a huge number of species of shellfish, fish and arthropods. The complexity of tropical habitats may relate partly to the current environmental stability of these zones, but perhaps also to longer-term stability in geological terms. The tropics were not as affected as the polar and temperate belts by the Pleistocene glaciations. On a longer time-scale, the tropics have probably maintained their environmental constancy, and even during phases of generalized global warming and cooling, species relations may have been maintained. It may be the long history of the tropics that lends them their high species diversity.

Diversity and Geography

Global and local variation

The term diversity is used throughout this and many other books on evolution to mean the number of species. A diverse environment is therefore one where there are a large number of species in a given area. On this count we find that there is a trend, over the surface of the Earth, for diversity to increase towards the Equator and to decrease towards the poles (*see pages 28–29*).

The rainforests of the tropics contain the greatest number of species per square kilometre and are therefore said to show the richest diversity of species on Earth. It is important to note, however, that a greater number of species does not necessarily mean a greater diversity of life forms as this is generally understood. Hundreds of species of fungi may inhabit a small area in the rainforest, for instance, but there may be little physical difference between them. They will therefore add to the technical diversity without increasing the forms of life.

In fact, the tendency of life forms to follow a relatively small number of basic patterns has become of increasing interest in the study of evolution. This phenomenon, known as stereotypy (as in the adjective stereotypical), is a product of the pattern of life over the past 600 million years. It seems likely that early multicellular organisms took on a wide variety of body plans. Many of these life forms became extinct before they developed a large number of species. The surviving forms were left to 'diversify' and produce many species, though all following the body form of their ancestor. It had been thought that the many alternative body forms died out because they somehow failed the test of life – that they were experimental life forms that did not work. This view is increasingly called into question. The alternative view – that some forms survived because of a lucky combination of circumstances – seems more likely.

NORTHERN LIMITS OF VARIOUS GROUPS OF COLD-BLOODED VERTEBRATES

freshwater fish
frogs
salamanders
snakes
lizards
turtles
crocodilians
EUROPE
EASTERN ASIA
WESTERN NORTH AMERICA
EASTERN NORTH AMERICA
Arctic Circle
Tropic of Cancer
Equator

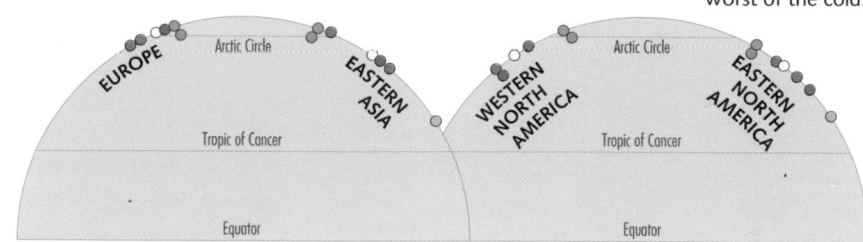

The distribution of cold-blooded animals such as reptiles is restrained by cold climates. Some can inhabit temperate regions by hibernating to avoid the worst of the cold.

Early this century, Russian scientists carried out extensive work into the productivity of different regions. The map below shows the estimated production of dry vegetation matter measured in metric tonnes per hectare – thought to be the most basic measure of the productivity of an environment because all life depends on vegetation production.

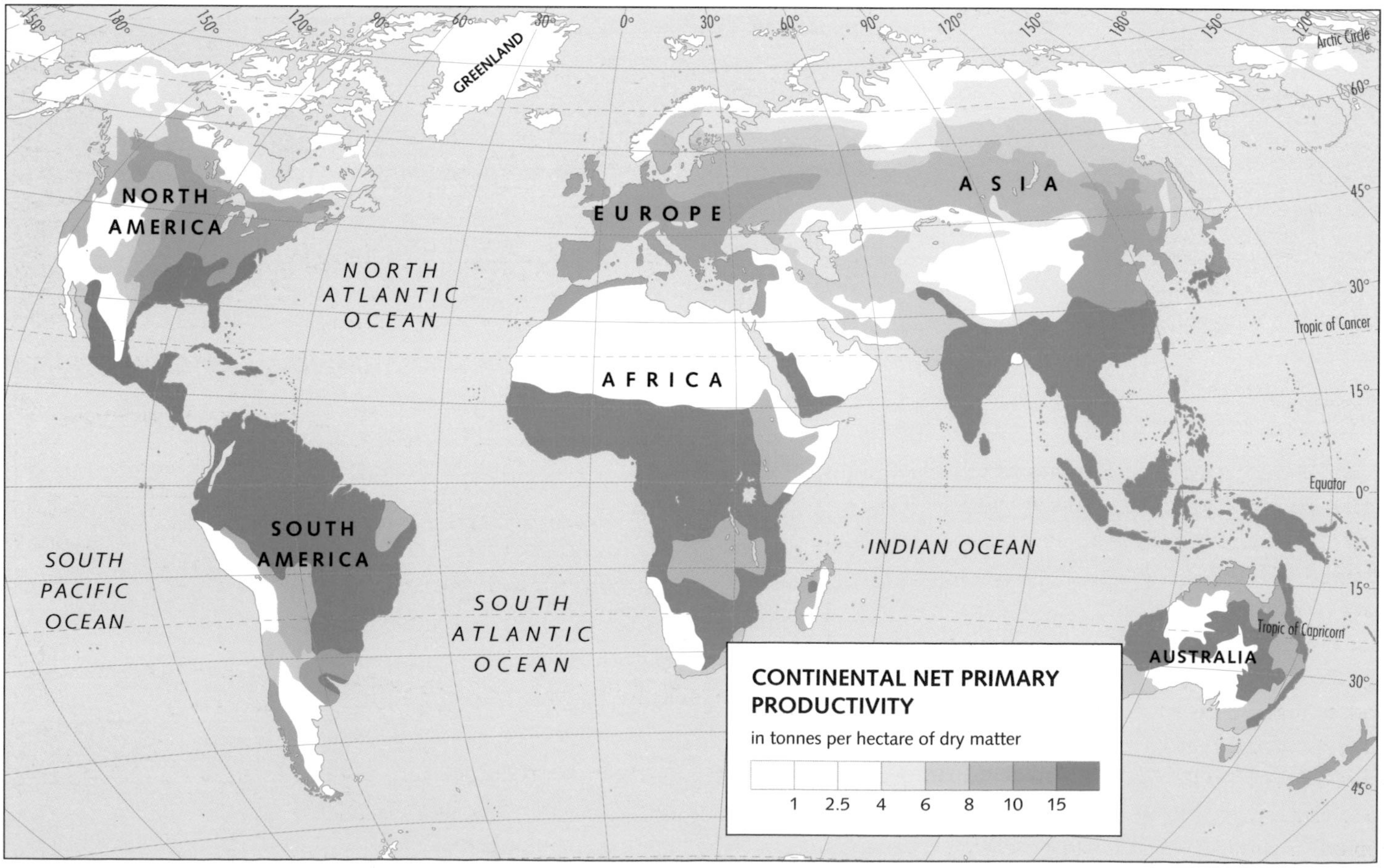

GREENLAND
NORTH AMERICA
NORTH ATLANTIC OCEAN
EUROPE
ASIA
AFRICA
SOUTH AMERICA
SOUTH PACIFIC OCEAN
SOUTH ATLANTIC OCEAN
INDIAN OCEAN
AUSTRALIA
Arctic Circle
Tropic of Cancer
Equator
Tropic of Capricorn
CONTINENTAL NET PRIMARY PRODUCTIVITY
in tonnes per hectare of dry matter
1
2.5
4
6
8
10
15

BIRD SPECIES RICHNESS IN NORTH AMERICA

—50— number of species

Beaufort Sea

Hudson Bay

NORTH PACIFIC OCEAN

NORTH ATLANTIC OCEAN

Gulf of Mexico

N

0 1000 km
0 1000 miles

The diversity of both bird species (*left*) and tree species (*below left*) in North America shows a general increase from north to south. But there are significant differences in the patterns which are not easily explained. There is a much greater richness of tree species in the southeastern USA than in California for example, while the situation is reversed in the case of bird species. Plants take longer to migrate than animals, and it seems that trees are still recovering from the impact of the last ice age, which affected the high country west of the Rockies more than the southeast. Trees are almost absent from the desert regions of the southwest.

The maps below show how the diversity of marine animals, in this case cowrie shells, follows a similar pattern to land animals such as lizards (*above*). The diversity of marine animals is influenced by water temperature, sunlight and the availability of nutrients (*see page 136*), which is often extremely variable locally. The richest areas for numbers of marine animals, though not necessarily species diversity, have cold water moving up from the depths to the warm surface waters, which encourages the growth of plankton. Extremely rich feeding areas are usually at the junction of warm and cold currents.

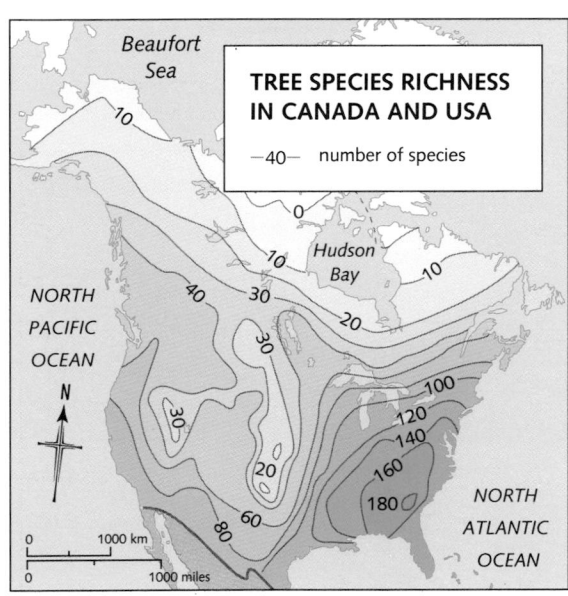

TREE SPECIES RICHNESS IN CANADA AND USA

—40— number of species

Beaufort Sea

Hudson Bay

NORTH PACIFIC OCEAN

NORTH ATLANTIC OCEAN

N

0 1000 km
0 1000 miles

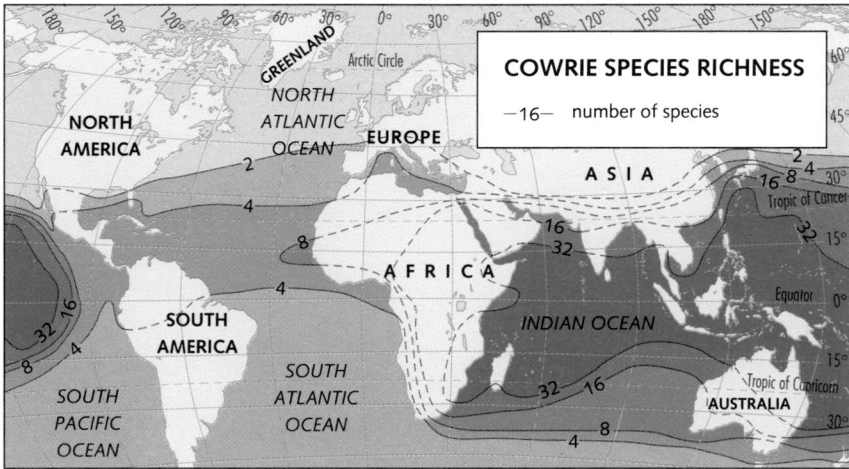

COWRIE SPECIES RICHNESS

—16— number of species

GREENLAND Arctic Circle
NORTH ATLANTIC OCEAN *EUROPE*
NORTH AMERICA *ASIA* Tropic of Cancer
AFRICA *INDIAN OCEAN* Equator
SOUTH AMERICA *SOUTH ATLANTIC OCEAN* Tropic of Capricorn
SOUTH PACIFIC OCEAN *AUSTRALIA*

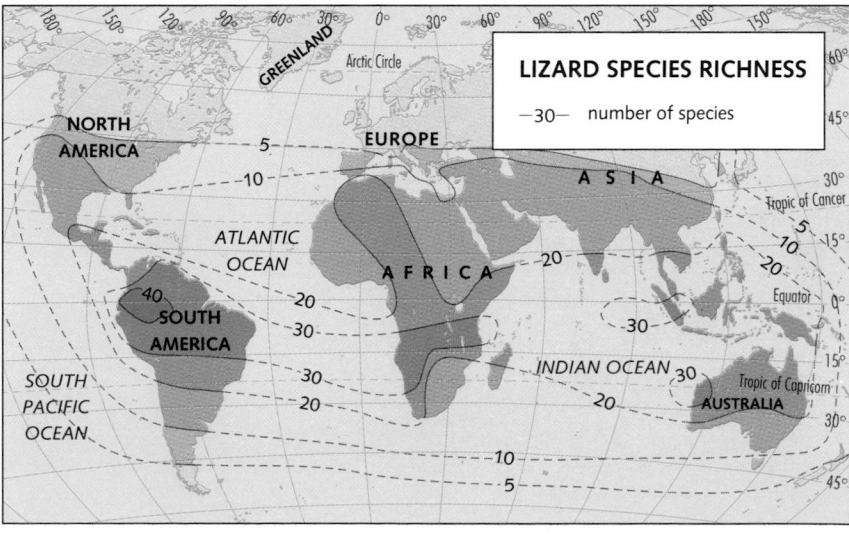

LIZARD SPECIES RICHNESS

—30— number of species

GREENLAND Arctic Circle
NORTH AMERICA *EUROPE* *ASIA* Tropic of Cancer
ATLANTIC OCEAN *AFRICA* Equator
SOUTH AMERICA *INDIAN OCEAN* Tropic of Capricorn
SOUTH PACIFIC OCEAN *AUSTRALIA*

Different conditions encourage diversity in different groups of organisms, so while the general pattern is for more species nearer the Equator, this is reversed within some groups. Local conditions also have an effect. Mountainous areas, for example, provide a wide variety of habitats and therefore encourage diversity among many different groups of plants and animals. Even environments with few species, such as tundra regions, compensate by showing a greater variety within species. This allows organisms to fulfill a variety of roles while retaining the ability to reproduce within a large enough population.

Tropical Rainforests
Hotbeds of diversity

The tropical rainforests are the richest environments on the planet, containing a greater diversity of species per cubic metre than any other ecosystem. The Earth's gradient of diversity (*see pages 122–123*) reaches its peak in these vast forests, where 20–40 per cent of all plant and animal species live. Most of these species will never be identified and classified – they are simply too numerous.

But why are the tropical rainforests so diverse? High rainfall and a warm year-round climate seem to be obvious encouragements to plants but do not necessarily encourage diversity. Variations in habitat are likely to be a dominant factor. Although the massed treetops look uniform from above, the forests are a highly variable environment. This is confirmed by the apparently small ranges of many species – it is possible to map the total apparent range of a species of butterfly or orchid to within a few hundred metres. As well as conditions on the ground, rainforests have developed a vertical stratification of environments that is not seen in other ecosystems. This adds an extra dimension to the variability of habitat (*far right*).

Some species are common in the rainforests while many others are extremely rare. One theory is that each species has a source base where conditions are ideal for it. From this source area it sends out individuals through normal dispersal methods. These may become established in new areas, but most of the time they find that conditions are not quite right or that another species got there first. In these cases, a few may hang on for a while before dying out. Any given area will therefore be a source for some species, which will be common, and an outlying area for others, which will appear to be extremely rare.

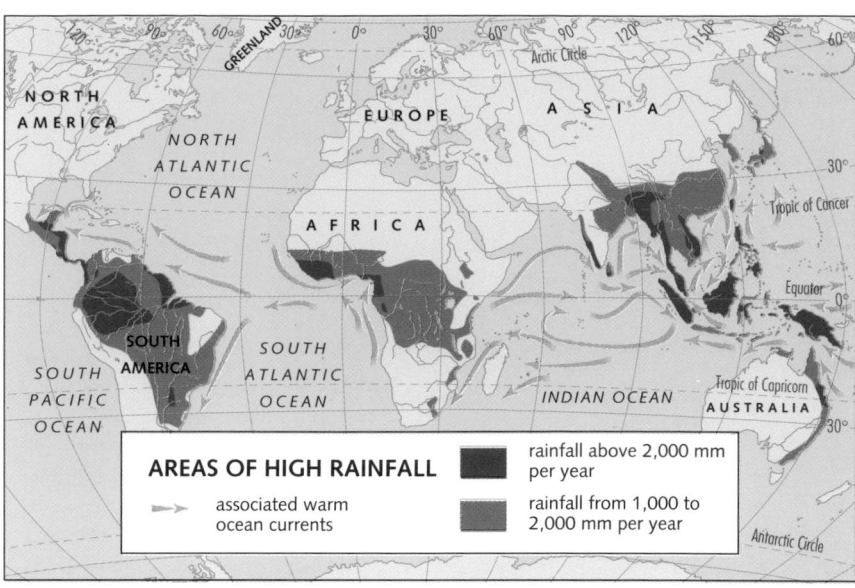

AREAS OF HIGH RAINFALL

→ associated warm ocean currents

■ rainfall above 2,000 mm per year

■ rainfall from 1,000 to 2,000 mm per year

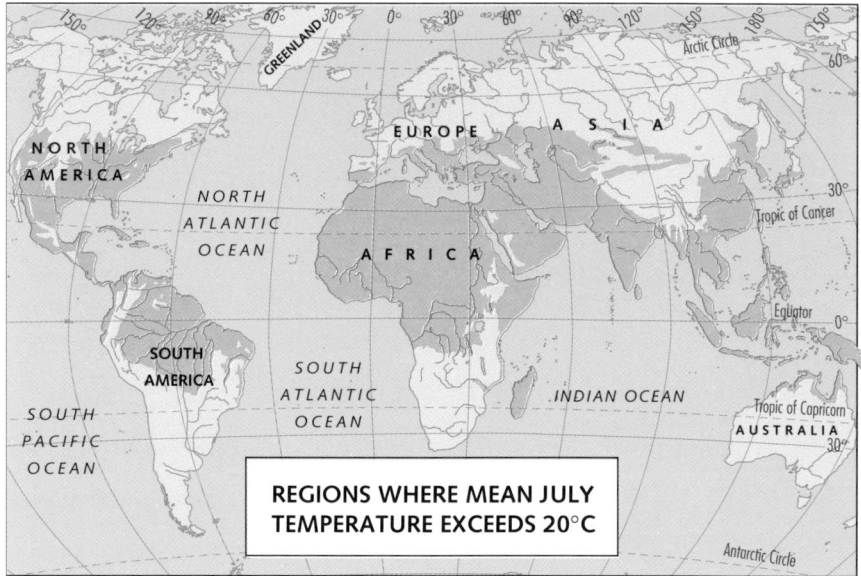

REGIONS WHERE MEAN JULY TEMPERATURE EXCEEDS 20°C

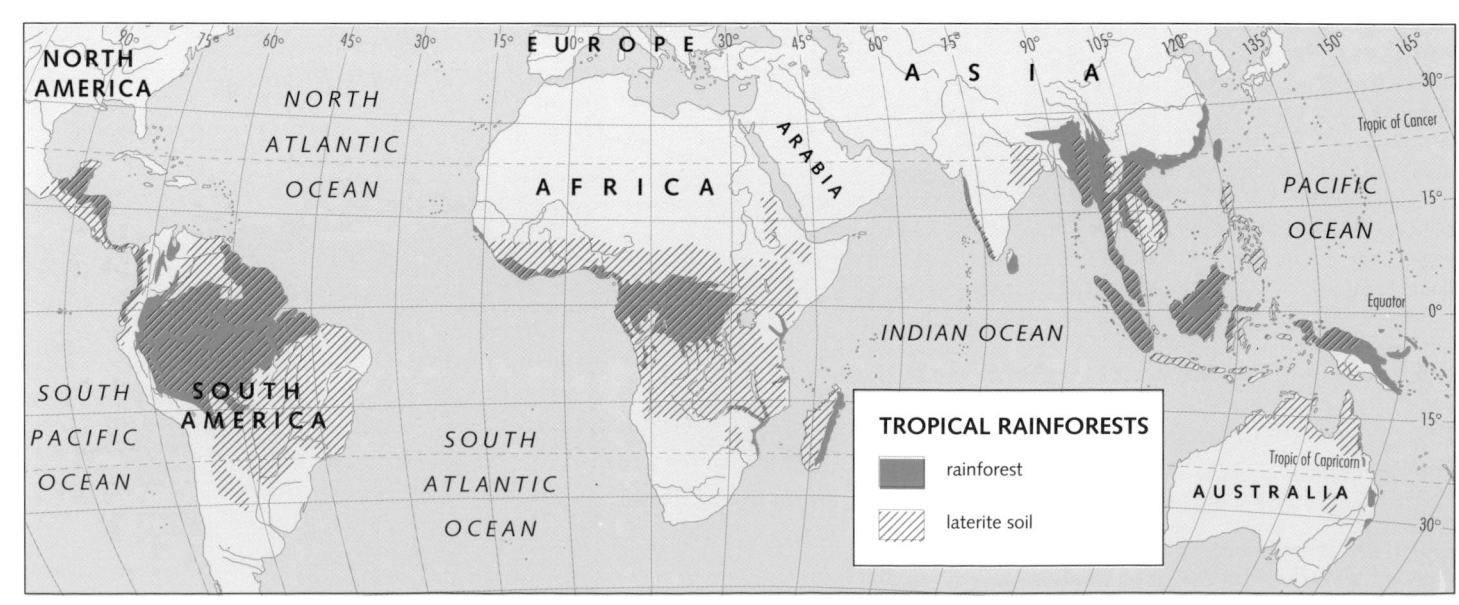

TROPICAL RAINFORESTS

■ rainforest

▨ laterite soil

Above: Some of the variety of plants found in tropical rainforests can be seen on Diego Garcia, an island in the Indian Ocean.

The maps opposite show the coincidence of tropical rainforest with areas of high temperature and rainfall. The bottom map shows the distribution of laterite, a soil type associated with rainforests. The relationship between vegetation and soil type is two-way. Dying plants help to form the soil, which in turn provides nutrition for the living flora. But despite the apparently inexhaustible vigour of the vegetation, rainforests have fragile foundations. Nutrient elements in the soil are washed out by the heavy rains, and what survives remains in the top 5 centimetres. Dead vegetation on the forest floor is consumed by fungi and the nutrients are returned to the living vegetation. The soil contains almost no vegetative matter. This system is so efficient that in some areas the composition of streams originating in tropical rainforests is close to that of distilled water.

The vertical rainforest (*right*)
The stratification of vegetation is more flexible than was once thought, with gradation and movement between levels.
Upper emergent level: Mature trees are in full sunlight. Animal life consists of birds and insects, which live at all levels.
Canopy level: Almost continuous cover is provided by treetops, that absorb 70–80 per cent of sunlight. Monkeys and flying squirrels use lianas for added mobility. Animals from this level rarely visit the floor.
Under-canopy: Epiphytes (plants that live on tree trunks without extracting nutrients from them) are common. Trees bear fruit and flowers directly on their trunks and branches. The vegetation consists of immature trees which will grow to canopy level or reach maturity at a lower height. Animals such as tree frogs visit the floor regularly.
Forest floor: Sunlight is poor so vegetation grows slowly. The temperature stays fairly constant. Litter is consumed by fungi and insects, so little gets into the soil. Animals include large mammals and flightless birds.

Butterflies of Tropical America
Speciation and replacement

The study of the distribution of plant and animal groups is generally accompanied by another enquiry – the question of how and why the present distribution has come into existence. In many parts of the world this second part of the study has been answered before the first has been properly understood. In other words, researchers have often been prone to fit assumed distributions of organisms into current theories, rather than do the extensive (and sometimes extraordinarily difficult and painstaking) fieldwork that is needed to underpin the theories.

Nowhere is this more true than in the tropical regions of South America. Access to many areas is difficult, if not impossible, and the extraordinary diversity of species makes the classification of organisms into evolutionary relationships a mind-boggling task. The use of modern genetic techniques has shown many previous classifications of insects, particularly Lepidoptera (butterflies), to be wrong, and this makes the usual process of building on earlier fieldwork a precarious pursuit.

One striking pattern that has emerged in the mapping of the ranges of particular subspecies of tropical butterfly is the presence of so-called endemic centres for species and subspecies (*see pages 112–113*). These endemic centres contain a great concentration of species that occur only within that area. Endemic centres for butterflies coincide with the small areas of forest that persisted through the ice ages (*see pages 110–111*), and this reinforces the idea that the formation of new species has occurred in parallel with movement out of these continuously forested areas. The fact that the ranges of related species and subspecies do not generally overlap lends weight to the theory that new species come into existence by movement away from territory occupied by their forebears.

The astonishing variety of wing patterns that can be seen in tropical butterflies raises some very interesting questions about evolution. Some butterflies can successfully mimic others for evolutionary advantage, convincing predators that they are poisonous (*see pages 42–43*). But other slight variations in pattern are assumed to give advantage in particular locations through differences in habitat. There is also an argument that such characteristics sometimes vary by accident – that is, without any evolutionary advantage being conferred – and that they persist because the species or subspecies that bears them has other characteristics that are advantageous.

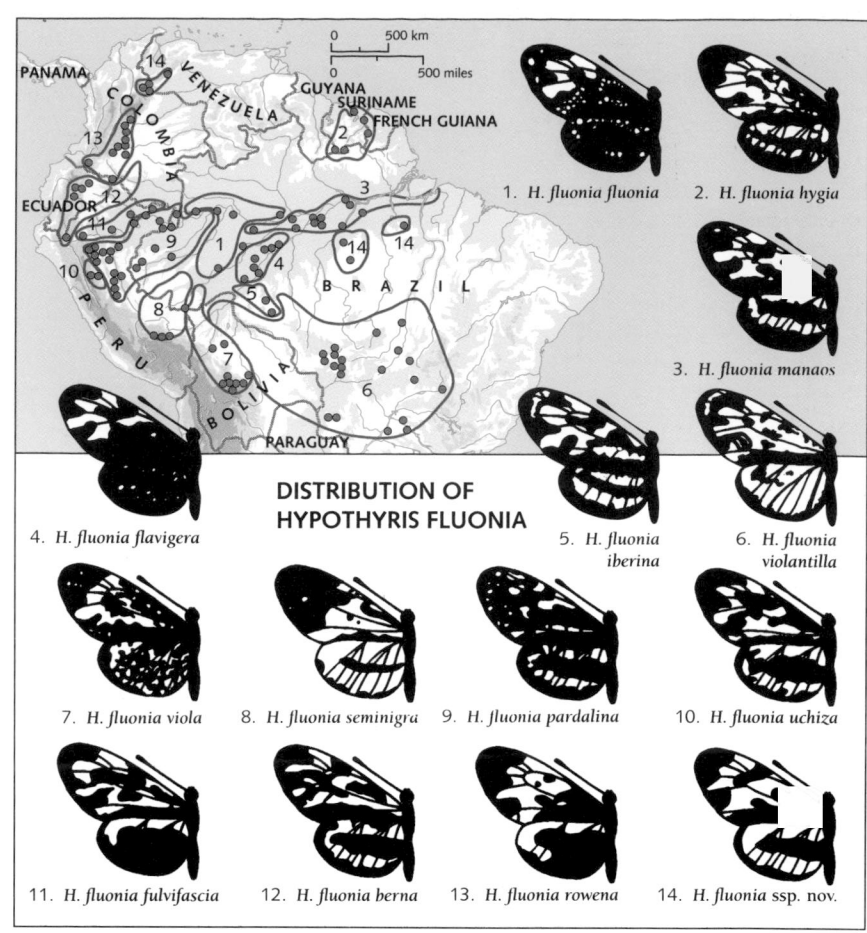

DISTRIBUTION OF HYPOTHYRIS FLUONIA

1. *H. fluonia fluonia*
2. *H. fluonia hygia*
3. *H. fluonia manaos*
4. *H. fluonia flavigera*
5. *H. fluonia iberina*
6. *H. fluonia violantilla*
7. *H. fluonia viola*
8. *H. fluonia seminigra*
9. *H. fluonia pardalina*
10. *H. fluonia uchiza*
11. *H. fluonia fulvifascia*
12. *H. fluonia berna*
13. *H. fluonia rowena*
14. *H. fluonia ssp. nov.*

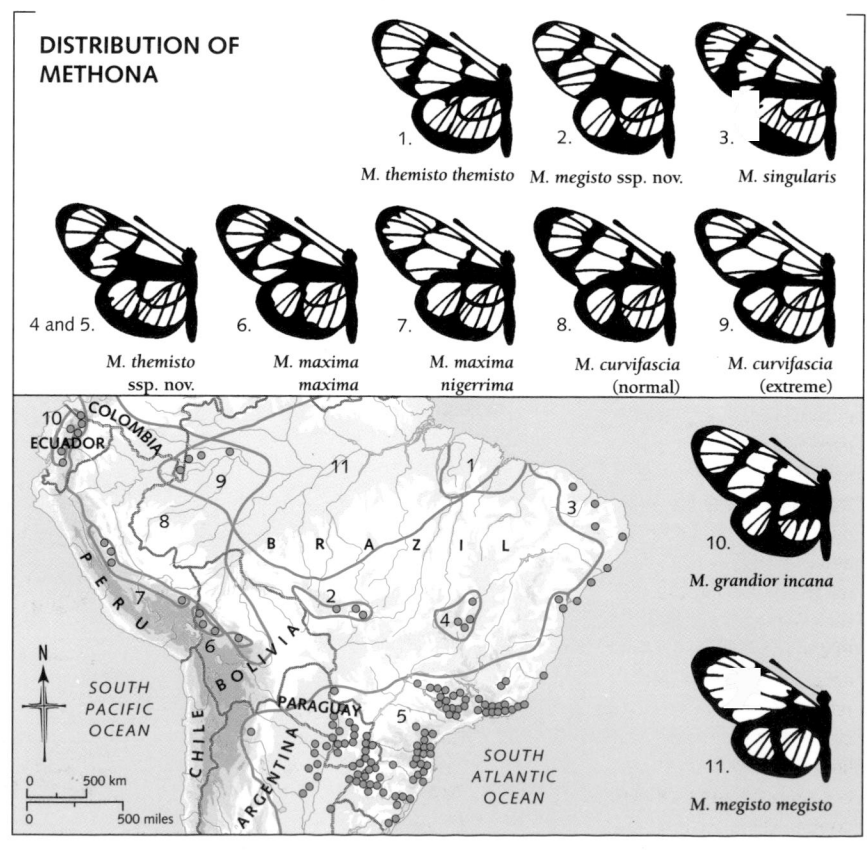

DISTRIBUTION OF METHONA

1. *M. themisto themisto*
2. *M. megisto ssp. nov.*
3. *M. singularis*
4 and 5. *M. themisto ssp. nov.*
6. *M. maxima maxima*
7. *M. maxima nigerrima*
8. *M. curvifascia (normal)*
9. *M. curvifascia (extreme)*
10. *M. grandior incana*
11. *M. megisto megisto*

Tropical butterflies

The maps on this page and on the opposite page show the ranges of various species and subspecies of tropical butterfly. Researchers have used maps in their attempts to understand the spatial relationships of different types of butterfly. To do this they have employed the concept of a phenotype. By this we mean an organism that has a distinct morphology: that is, a distinct set of physical characteristics. A phenotype may not be the same as a species, since one species cannot breed with another. Phenotypes may be subspecies, in which case they are able to breed with other phenotypes.

Heliconiidae

The occurrence of different phenotypes is indicated on the maps by the dots, and the deduced range of a particular phenotype is shown by the outlined areas. Indications of the physical features that are likely to form barriers between phenotypes are used to construct these ranges. Some species of tropical butterfly are particularly sensitive to altitude, so that even low ranges of hills will prevent their migration.

In common with other plant and animal groups (*see pages 140–141*), it becomes clear that phenotypes are limited in their ranges by certain restraints that are continually in operation.

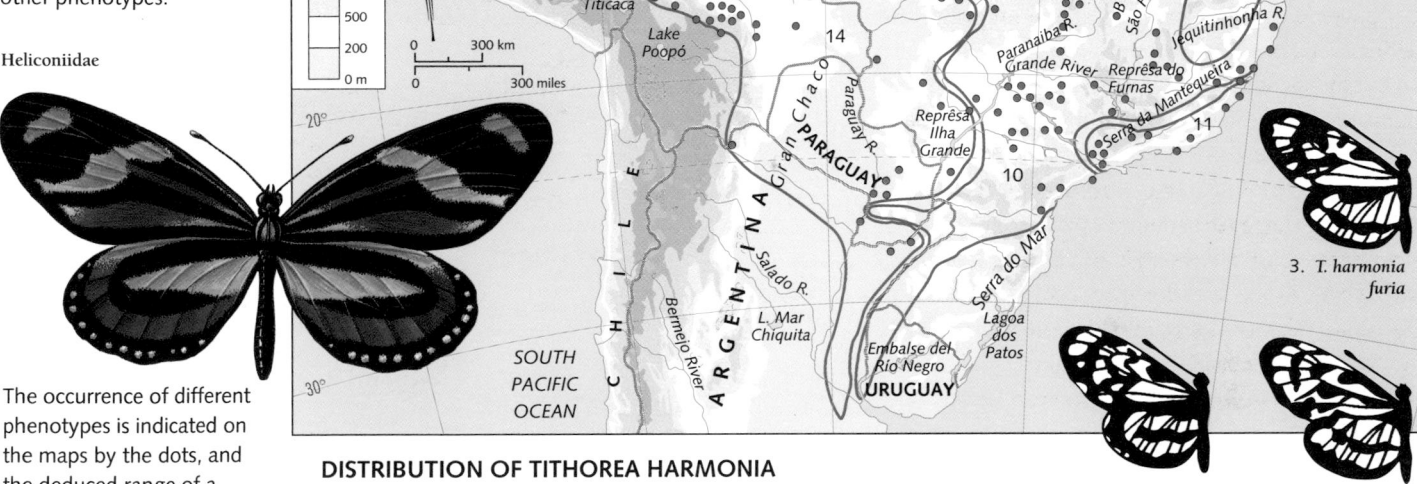

DISTRIBUTION OF TITHOREA HARMONIA

1. *T. harmonia furina*
2. *T. harmonia sulphurata*
3. *T. harmonia furia*
4. *T. megara*
5. *T. dorada*

6. *T. ssp. nov.*

7. *T. harmonia*

8. *T. moppa*

9. *T. cuparina*

10. *T. pseudethra*

11. *T. caissara*

12. *T. harmonia egaensis*

13. *T. harmonia lateflava*

14. *T. harmonia pseudonyma*

15. *T. harmonia brunnea*

16. *T. harmonia melanina*

17. *T. harmonia neitha*

18. *T. harmonia martina*

19. *T. harmonia gilberti*

20. *T. harmonia hermias*

21. *T. harmonia napona*

22. *T. harmonia manabiana*

23. *T. harmonia ssp. nov.*

Antarctic Seas
Cold waters and a rich marine environment

The ocean that surrounds the continent of Antarctica appears to be inhospitable to human life. The written accounts of early explorers describe it as a desolate region. It might therefore come as a surprise to find that the Southern Ocean provides a varied and productive environment for marine life.

The key to this productivity is the movement of ocean water both horizontally (around on the surface) and vertically (through the layers of the ocean depths). This vertical and horizontal movement of waters is caused in the Southern Ocean mainly by differences in temperature. The seawater immediately around the continent of Antarctica is cooled by the polar temperatures. This cold water continually sinks to the ocean floor and is pushed away from the polar region by more cold water descending behind it. Deep ocean currents follow a general pattern that is dictated by the flow of cold water away from the poles towards the Equator.

On the surface of the ocean, cold water is also pushed away from the pole, as well as being continually circulated westwards by the West Wind Drift. The cold surface water meets warm water that is moving southwards at the Antarctic Convergence. This marks the northernmost boundary of the Southern Ocean. Because the two currents are pushing together, the cold Antarctic waters are displaced downwards. To replace this water, and the water that is sinking around the margins of the continent, water is continually welling up from the depths of the ocean in various places in the Southern Ocean. It is this upwelling water that brings nutrients to the surface and helps to create a rich environment for marine life.

The food chain of the oceans begins with the microscopic phytoplankton and zooplankton (plant and animal plankton). These form the diet of many marine creatures, including invertebrates such as squid, fish and seabirds. In the Southern Ocean krill (*Euphasia superba*), another small life form, provides a crucial link in the ecosystem. Krill are tiny shrimp-like animals and form the main food source for baleen whales, penguins and seals. Whales migrate to the Antarctic each summer to feed off the vast shoals of krill (*see pages 148–149*). It is estimated that, before whaling decimated their numbers, baleen whales consumed about 200 million tonnes of krill every year. The food chain of the Southern Ocean – from plankton to krill to baleen whale – is a remarkably short one.

ENDEMIC SPECIES FROM THE SOUTHERN OCEAN

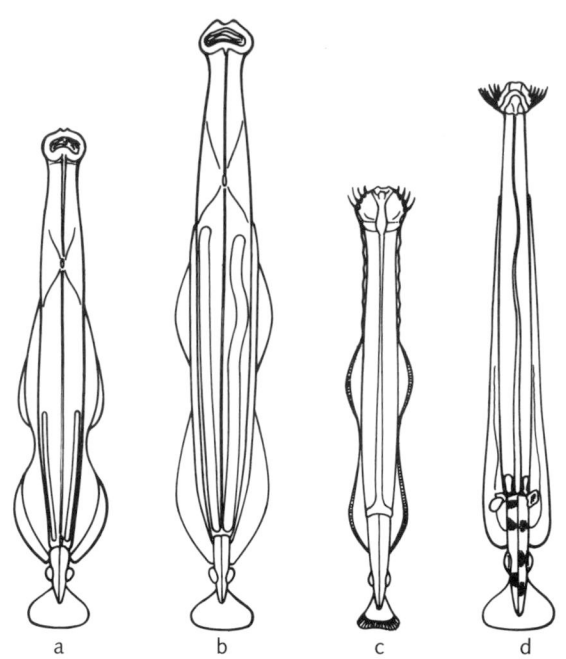

a) *S. gazellae*, small northern race   b) *S. gazellae*, large southern race   c) *S. marri*   d) *E. bathyantarctica*

Arrow worms (*S. gazellae*) are small free-floating animals found in all oceans. The three species on the left are endemic to the Southern Ocean. The map opposite shows sampling of these species by a research vessel. The worms are sensitive to temperature and salinity, and gradually increase in density away from Antarctica before tapering off above a latitude of about 40°S. There is a notable decrease at the Antarctic Convergence where salinity and temperature variation may not suit them. Distribution of crustacea species in the Southern Ocean (*below*) is influenced by the direction of the prevailing westerly current, which is known as the West Wind Drift or Circum-Antarctic Current.

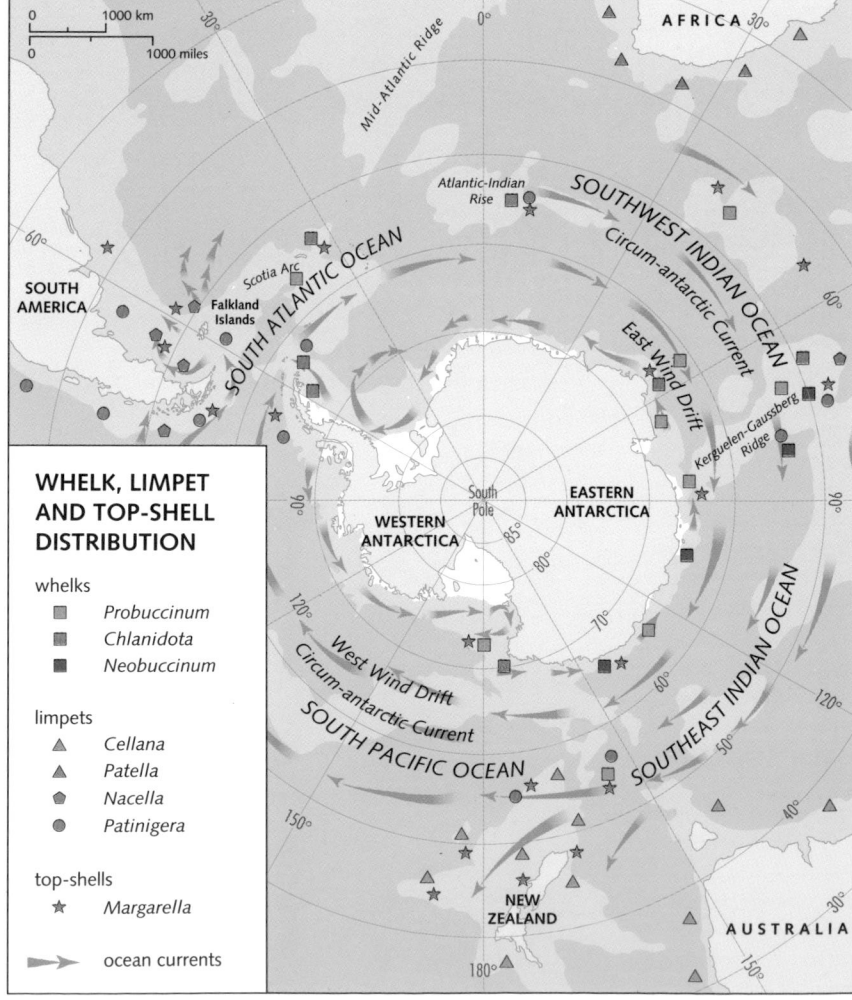

WHELK, LIMPET AND TOP-SHELL DISTRIBUTION

whelks
- ▪ *Probuccinum*
- ▪ *Chlanidota*
- ■ *Neobuccinum*

limpets
- ▲ *Cellana*
- ▲ *Patella*
- ⬠ *Nacella*
- ● *Patinigera*

top-shells
- ★ *Margarella*

➤ ocean currents

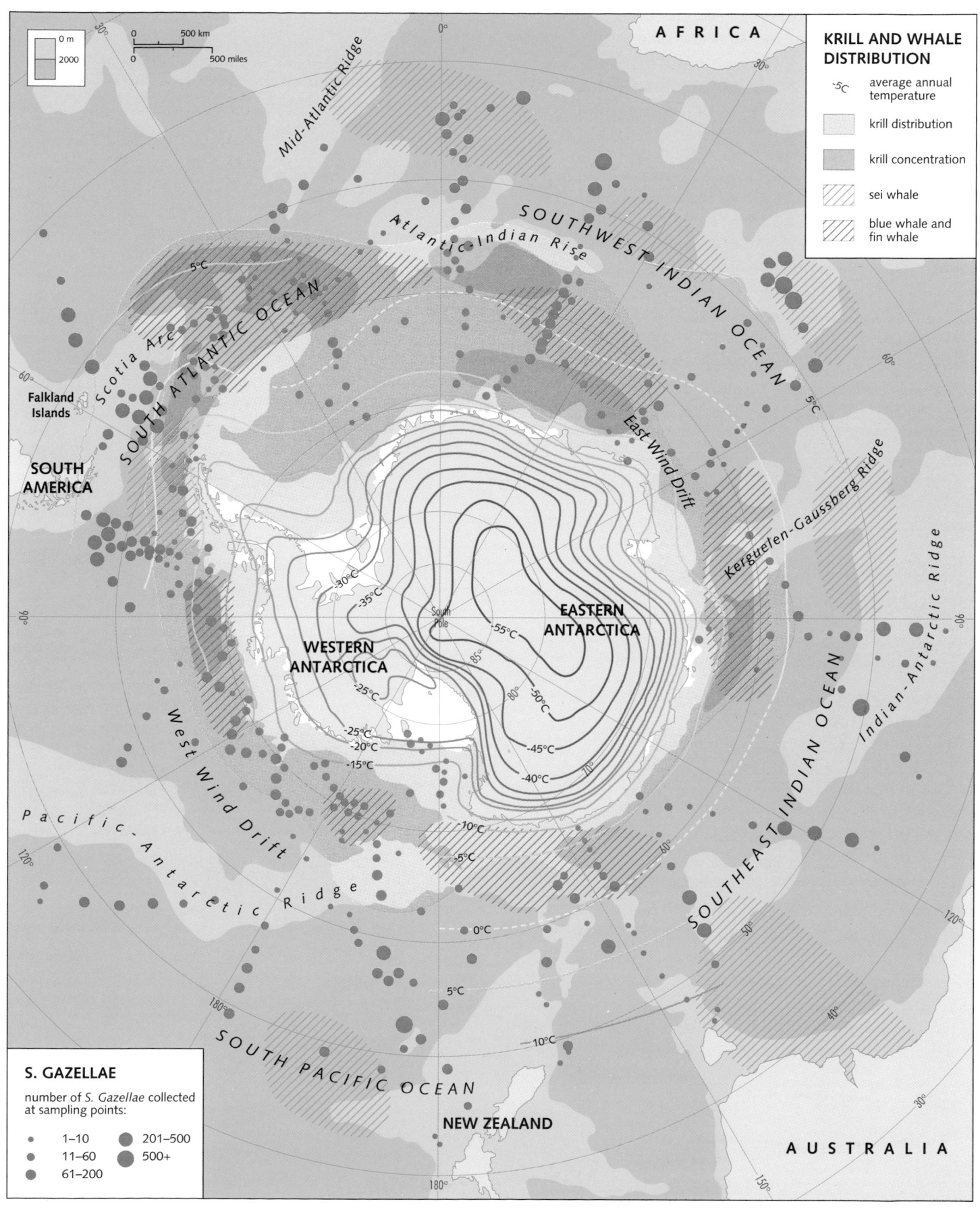

KRILL AND WHALE DISTRIBUTION

-5°C average annual temperature

krill distribution

krill concentration

sei whale

blue whale and fin whale

0 m
2000

0   500 km
0   500 miles

AFRICA

SOUTHWEST INDIAN OCEAN

Mid-Atlantic Ridge

Atlantic-Indian Rise

Scotia Arc

SOUTH ATLANTIC OCEAN

Falkland Islands

SOUTH AMERICA

East Wind Drift

Kerguelen-Gaussberg Ridge

Indian-Antarctic Ridge

5°C

60°

30°

0°

30°

60°

90°

5°C

60°

WESTERN ANTARCTICA

EASTERN ANTARCTICA

South Pole

-30°C

-35°C

-55°C

-25°C

-25°C

-20°C

-15°C

-50°C

-45°C

-40°C

85°

80°

70°

West Wind Drift

Pacific - Antarctic Ridge

SOUTHEAST INDIAN OCEAN

-10°C

-5°C

0°C

5°C

10°C

120°

90°

120°

60°

50°

40°

30°

S. GAZELLAE

number of S. Gazellae collected at sampling points:

● 1–10   ● 201–500
● 11–60   ● 500+
● 61–200

SOUTH PACIFIC OCEAN

NEW ZEALAND

AUSTRALIA

180°

150°

New World Monkeys
Separation and distribution

South American monkeys are entirely distinct from those of Africa and Asia and are known as New World monkeys. The key physical differences between Old and New World monkeys give a strong indication of when the two types separated, but the scarcity of fossils prevents an accurate fixing of the date or method of separation.

Primates are divided into groups on the basis of differences in nostril shape. New World monkeys (Platyrrhini or 'broad-nosed') have completely separated nostrils. Old World monkeys (Catarrhini or 'downward-nosed') have nostrils separated by a thin membrane. All primate groups descended from an extinct ancestor. Splitting into groups is thought to have taken place in the early Tertiary.

New world monkeys are also divided into two groups: the marmosets and tamarins (Callithricidae), and the capuchins (Cebidae). Both differ from Old World monkeys and must therefore have developed from a common ancestor, a proto-platyrrhini that was already isolated, presumably by migration to South America. Biogeographers see this as a case of continental drift separating land-based groups and leading to the development of separate evolutionary branches. The splitting of South America from Africa as the South Atlantic opened must have been the mechanism by which separation took place,

PRIMATE PHYLOGENY

Omomyidae Tarsiidae Cebidae Parapithecidae Cercopithecidae Hominoidae

Adapidae Lemuridae

Strepsirhini   Haplorhini   Platyrrhini   Catarrhini

Primate ancestor   ⌐ ⌐ New World monkeys

but unfortunately the dates do not quite match up. By the Early Tertiary, South America was already 200 kilometres from Africa. This can be verified by measuring the ages of the rocks on the Atlantic seafloor. Rocks older than Tertiary occur along the margins of South America and Africa, which are each 100 kilometres wide at their narrowest points. Our best guess is that some monkeys did cross, maybe by island-hopping and travelling accidentally on driftwood. The chances of a breeding pair making it are slight, but only one pair would have to get there in several million years, which reduces the odds.

Primates are thought to have originated in Africa in the Early Tertiary, around 60 million years ago (*below*). From there they spread into Asia and Europe. Migration routes to the Americas are less certain. Primates were present in North America in the Tertiary, but South America was separated from both North America and Africa through most of the Tertiary. Island-hopping from Africa is still regarded as the most likely route.

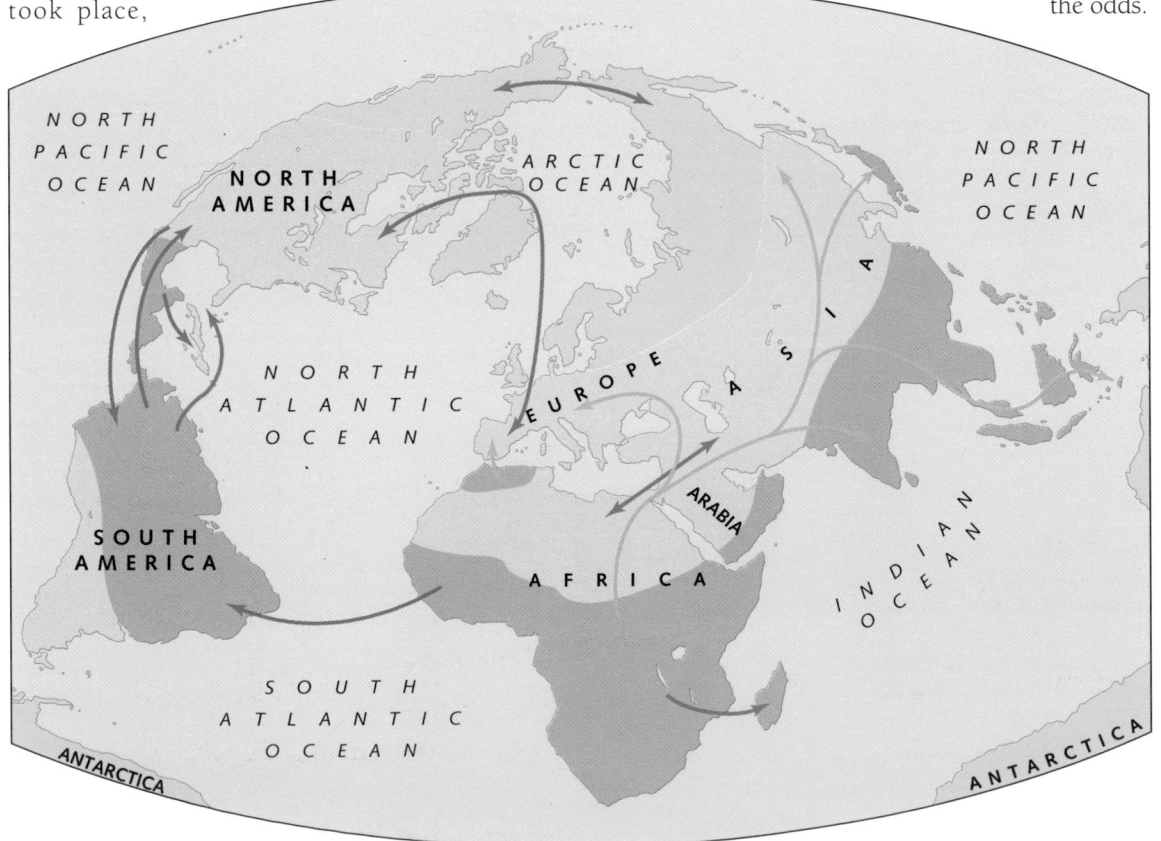

DISTRIBUTIONAL HISTORY AND MIGRATION ROUTES OF PRIMATES (EXCLUDING HUMANS)

→ exchange of prosimian fauna in the Early Tertiary

→ possible migration of proto-platyrrhini, a New World progenitor, in the Early Tertiary

→ migration of New World primates in the Late Tertiary and Quaternary

→ migration of Old World primates in the Late Tertiary and Quaternary

former distribution

modern distribution

or will not cross wide rivers. In the case of the latter, different subspecies have developed within each group. The range of each subspecies in the moloch group is defined by the southern tributaries of the Amazon. The monkeys can cross smaller rivers in the south, but as groups have pushed north they have become isolated from one other by the widening rivers. This has led to the evolution of separate characteristics and therefore separate subspecies.

Tamarin monkeys are thought to have spread northward along the margin of the Andes. Some then branched eastwards, others crossed the Andes to reach Panama. The eastern branch must have crossed the Amazon River several times, probably at bends where islands and cut-offs sometimes develop.

MIGRATION AND DISTRIBUTION OF TAMARINS

— boundary of genus *Saguinus*

---- interspecies boundaries

➤ probable migrations/river crossings

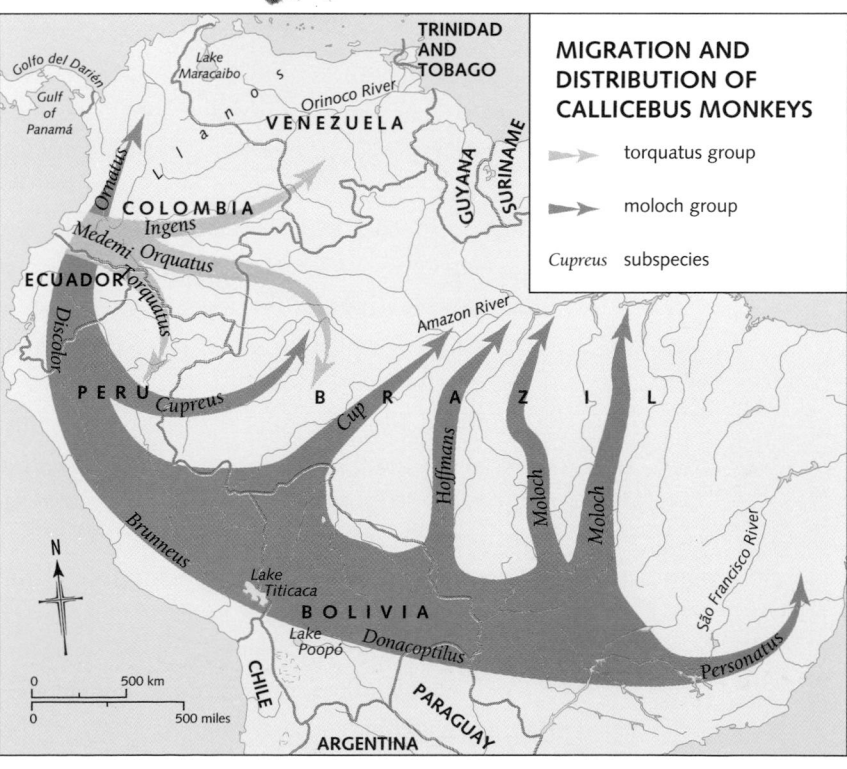

Studies of monkey distribution in the rainforests of South America have enabled us to derive the likely history of their dispersal throughout the region. The most striking feature of these maps is the role that rivers play in distribution patterns. Both tamarins (*above*) and titis monkeys (*below right*) cannot

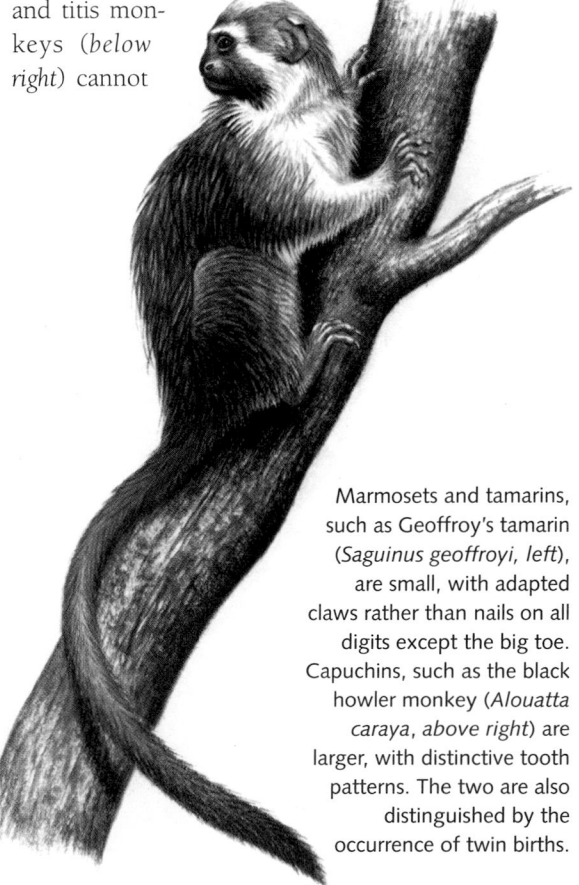

Marmosets and tamarins, such as Geoffroy's tamarin (*Saguinus geoffroyi, left*), are small, with adapted claws rather than nails on all digits except the big toe. Capuchins, such as the black howler monkey (*Alouatta caraya, above right*) are larger, with distinctive tooth patterns. The two are also distinguished by the occurrence of twin births.

MIGRATION AND DISTRIBUTION OF CALLICEBUS MONKEYS

➤ torquatus group

➤ moloch group

Cupreus subspecies

Section VIII: The Distribution of Life on Earth

Life on land occurs in six biogeographic regions. These are separated by natural barriers that prevent species from passing from one to the other. Modern biogeographic realms evolved over the course of the past 100 million years, as the continents moved to their present positions.

One of Darwin's keen interests was biogeography. His observations of small-scale geographic variation among the finches and land tortoises of the Galapagos Islands were instrumental in shaping his ideas on evolution. He also devoted a great deal of effort to the study of dispersal in an attempt to find plausible ways in which organisms of all kinds could cross wide expanses of sea. But he was not particularly interested in regional biogeography – the attempt to define floral and faunal zones, and to interpret their origins – perhaps because he could not see any major contribution from such studies to an understanding of the processes of evolution.

Regional biogeography had, however, fascinated generations of naturalists before Darwin, including such notable names as the Comte de Buffon and Linnaeus (*see pages 20–21*). These naturalists had noted the great variations in plants and animals from region to region and observed some major regional patterns. This patchy work was formalized by Philip Sclater, an energetic British ornithologist, in 1858. Sclater divided the world into six zoogeographic realms based on bird distributions: Nearctic (North America), Palearctic (Europe, Asia and North Africa), Neotropical (South America), Ethiopian (central and southern Africa), Oriental (India and Southeast Asia) and Australasian. These realms have subsequently been confirmed by studies of other groups, and are easy to distinguish. Pouched marsupials, egg-laying monotremes, unique birds, and plants such as the eucalyptus clearly set Australia, New Guinea and neighbouring areas apart from all others. South America has a similar assemblage of endemic plants and animals: sloths and armadillos, opossums, New World monkeys and unique tropical forest plants and butterflies. Africa south of the Sahara is set apart by its elephants, giraffes, apes, plants and insects. These distinctive features could be catalogued at great length.

The descriptive approach to regional faunas and floras formed a focus for the work of biogeographers for the next 60 years. Their bible was Alfred Russell Wallace's *The Geographical Distribution of Animals* (1876), in which the author presented copious amounts of data on regional distributions of animals, concentrating in particular on the southeast Asian region where he had spent years engaged in field work. He tried to refine the definition of the boundary between the distinctive Oriental realm and the Australasian realm, and the line of demarcation, the Wallace Line, is still set where he proposed it.

Biogeographers continued to debate the origins of the six realms. Mammal experts noted great similarities between the mammals of

North America and Eurasia, while ornithologists found greater similarities between the birds of North America and South America. Botanists found quite different patterns. They regarded the plants of Southeast Asia and the Pacific islands as one flora, and this spanned the Wallace Line that is so important for animals. Discussions on how to further subdivide the six major biogeographic realms continued among zoologists and botanists but did not produce any useful results. What was more important from the 1920s onward was to understand the origins of present-day faunas and floras in a geological and historical context.

Biogeographic evolution

Three approaches to the problem of the origins of modern biogeographic patterns flourished in the early 20th century. First were the biologists who postulated the previous existence of land bridges between all the regions that appeared to show floral and faunal links. Darwin, Wallace and many later biogeographers were uncomfortable with any proposal that went beyond the bounds of geological knowledge, however, and could not accept land bridges across the oceans. Land links from offshore islands to the mainland, and between Siberia and Alaska, were supported by the present configurations of seafloors, and by evidence for lowered sea levels in the Pleistocene, but anything beyond that was fantasy. The postulating of land bridges was too often pure guesswork, the imposition of a *deus ex machina* to rescue an inexplicable problem in biogeography.

Darwin and Wallace believed that most terrestrial species could disperse readily over wide oceans. Birds, bats and insects could fly or be blown off course. Many seeds could also travel huge distances over oceans, as could spiders and other small animals if caught up in unusual winds. Other plants and animals could be rafted hundreds of miles on matted pieces of timber and undergrowth. The Pacific islands were clearly of recent volcanic origin, and it must have been chance dispersal that populated them with species. Similar rafting, lofting and island-hopping could explain distribution from continent to continent.

The third dispersalist school came into being with the proposal of continental drift by Alfred Wegener in 1912. These biogeographers saw that many of the profound patterns of floral and faunal distribution related to the long-term movements of continents. There was no need for a land bridge from India to Madagascar, because the two land masses were in contact 150 million years ago. The Americas were linked to the Old World before the Atlantic Ocean opened up, and Australia may have acquired its complement of marsupials from South America by a direct Gondwana link. Present opinions in biogeography explain the major biogeographic realms in terms of the breakup of the supercontinent Pangea over the course of the past 200 million years, combined with sporadic long-distance dispersal to offshore islands by the means favoured by Darwin and Wallace.

Biogeographic Regions
Biology, geography and history

Plants and animals occur in particular places for a wide variety of reasons – most of this book is taken up with explaining what those reasons are. The distribution of flora and fauna is studied at every point of the scale, from single habitats to the whole world, and different factors can be seen operating at varying intensities at each level.

Large areas of the world contain floras and faunas that occur only within that region. Within these biogeographic regions, plant and animal groups co-exist in often highly complex relationships which are the result of evolutionary forces at work in specific environments. In the rainforests of South America, for instance, plants evolved to take advantage of the special climatic and soil conditions. Insects, reptiles, birds and mammals then emerged that thrive in this environment. Other plants then evolved that use these animals for reproduction. A complex web of interdependency is thus built up over millions of years. These webs always lie within a biogeographic region, which is why such areas form the basis for the study of the distribution of life on Earth.

The distribution of particular organisms also depends on the history of continental movements (*see Section 5*). Some animals from Asia are now moving across the Malay archipelago to Australia, for instance, and vice versa. For millions of years, however, Australasia was totally isolated, allowing a separate flora and fauna to develop, so the two areas form two distinct biogeographic regions.

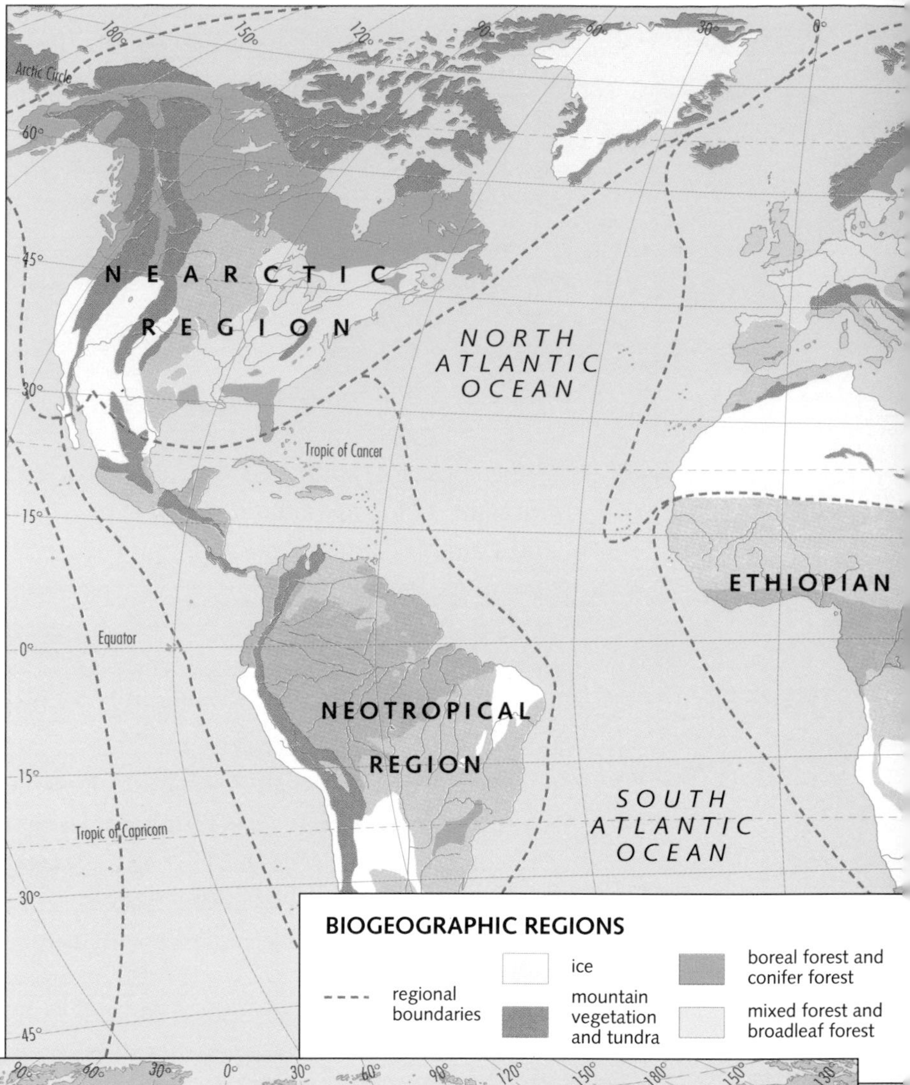

Climatic regions

The main factor in the distribution of organisms is climate. Much of the world has been warming up since the last ice age, so many plants are not in equilibrium with their environment but are constantly adapting to keep up with change.

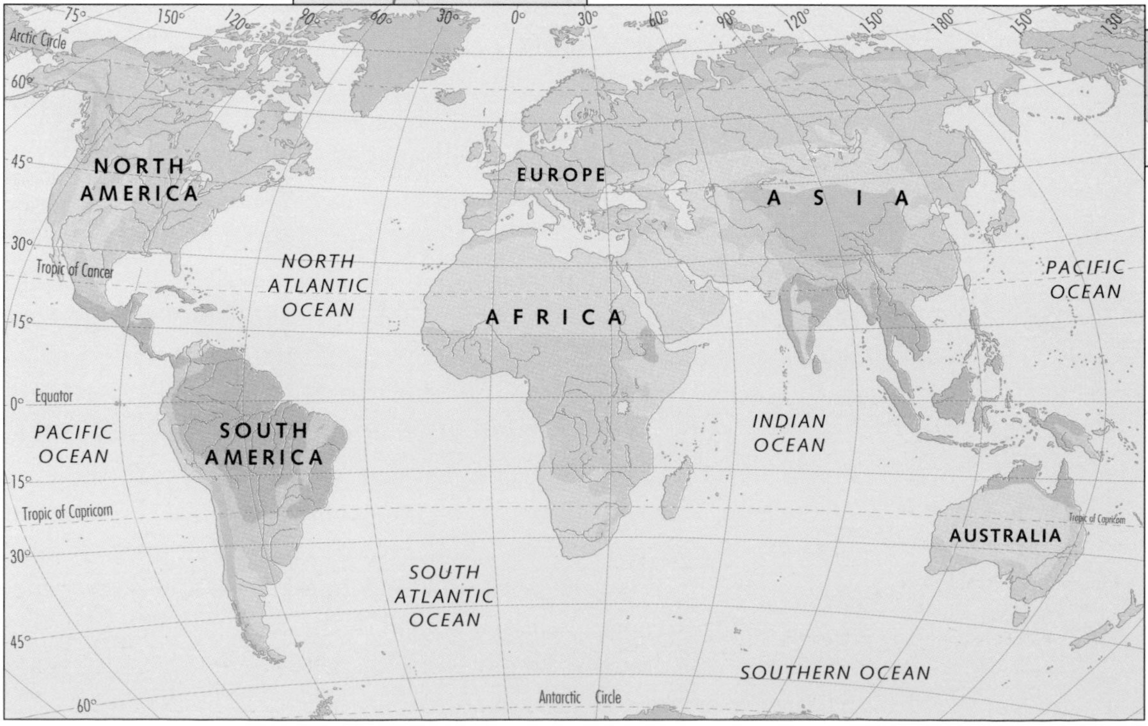

BIOGEOGRAPHIC REGIONS

- - - - regional boundaries

ice

mountain vegetation and tundra

boreal forest and conifer forest

mixed forest and broadleaf forest

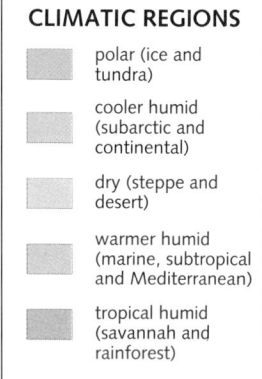

CLIMATIC REGIONS

- polar (ice and tundra)
- cooler humid (subarctic and continental)
- dry (steppe and desert)
- warmer humid (marine, subtropical and Mediterranean)
- tropical humid (savannah and rainforest)

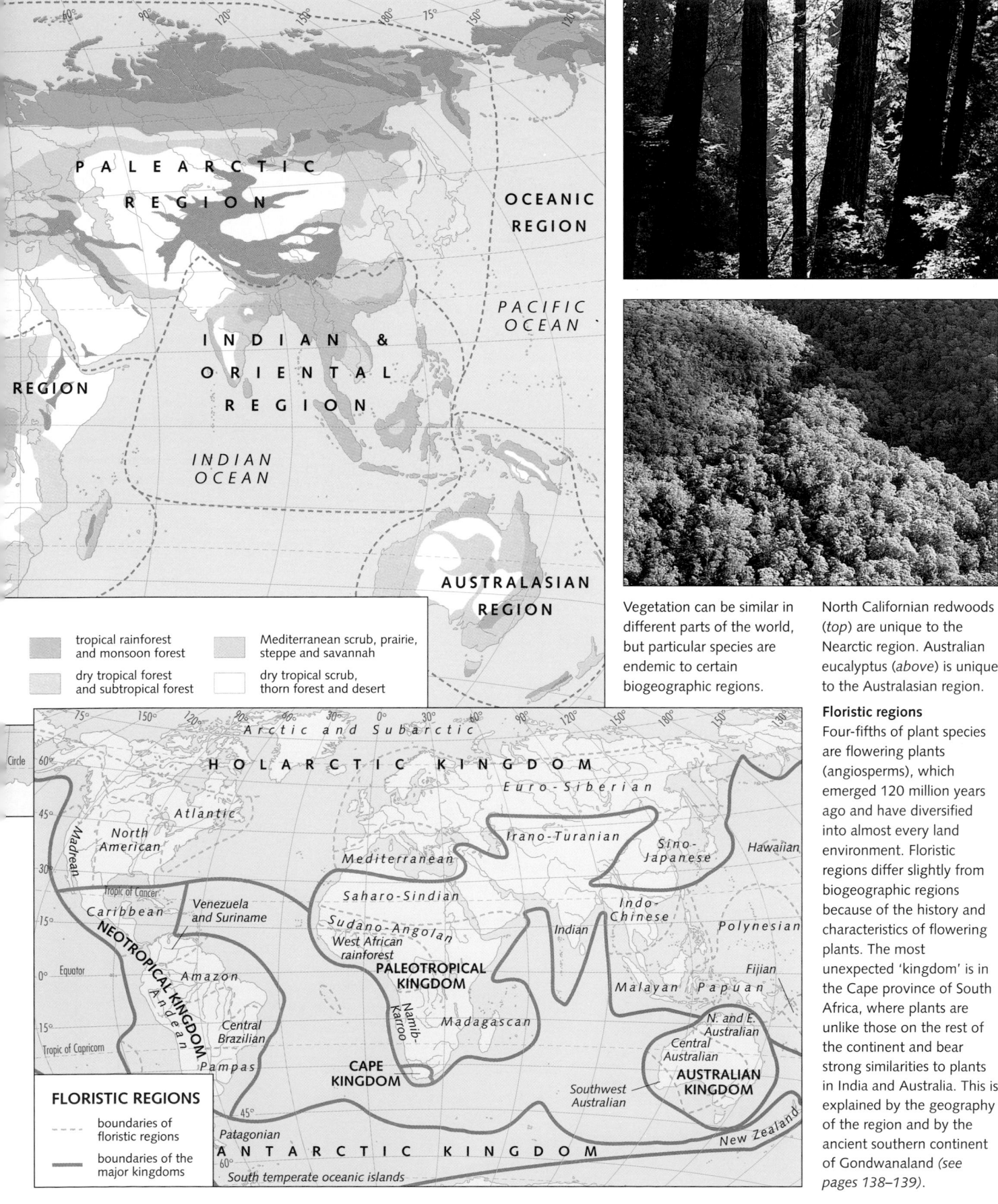

tropical rainforest
and monsoon forest

dry tropical forest
and subtropical forest

Mediterranean scrub, prairie,
steppe and savannah

dry tropical scrub,
thorn forest and desert

Vegetation can be similar in different parts of the world, but particular species are endemic to certain biogeographic regions.

North Californian redwoods (*top*) are unique to the Nearctic region. Australian eucalyptus (*above*) is unique to the Australasian region.

FLORISTIC REGIONS

boundaries of
floristic regions

boundaries of the
major kingdoms

Floristic regions

Four-fifths of plant species are flowering plants (angiosperms), which emerged 120 million years ago and have diversified into almost every land environment. Floristic regions differ slightly from biogeographic regions because of the history and characteristics of flowering plants. The most unexpected 'kingdom' is in the Cape province of South Africa, where plants are unlike those on the rest of the continent and bear strong similarities to plants in India and Australia. This is explained by the geography of the region and by the ancient southern continent of Gondwanaland (*see pages 138–139*).

Biogeography of the Oceans
The zones of ocean life

The oceans of the world can, like the continents (*see pages 134–135*), be usefully divided into distinct biogeographic regions. The boundaries between these are dictated by the surface temperature of the sea, and the broad belts are similar in the northern and southern hemispheres.

In the Arctic and Antarctic regions, the surface temperature of the ocean is between 0°C and 5°C. In the temperate and boreal regions it ranges from 5°C to 18°C, and in the tropics it is between 18°C and 20°C. These figures are for the mean annual temperature. The ocean currents influence the positions of these zones, and there are less rigid boundaries between them than there are between the biogeographic regions of the land.

In colder zones the diversity of species tends to be low, while in the seas of the tropics the number of species is very high, although the total amount of living matter in the ocean – the biomass – may be relatively low. In the temperate regions, the number of species and the amount of biomass are influenced by changes in water temperature and sunlight according to the seasons.

Life in the oceans depends on photosynthesis. The initial stage in the food chain, the phytoplankton, can therefore only live at or above the depth to which sunlight is capable of penetrating. This range is known as the euphotic zone, and normally extends to a maximum of 100 metres below the surface of the ocean. All other life forms in the sea are ultimately dependent on the microscopic photosynthesizing plants which live in this zone. There is therefore a general relation between the amount of phytoplankton and the total amount of life forms or biomass in any one area.

Although higher temperatures tend to lead to greater productivity, plankton also depend on nutrients circulating through the different levels of the ocean to the surface. In some tropical areas the surface temperatures may be high but the circulation is poor, resulting in a lower production of plankton, and therefore of other life, than in cooler but more active systems. The Sargasso Sea – a calm area in the central Atlantic – is a good example of this. The same phenomenon is seen at a more local level in parts of the central Mediterranean.

Plankton and biomass
(*maps right*)
The abundances of phytoplankton (microscopic photosynthesizing plants), zooplankton (microscopic free-floating animals) and benthic biomass (total mass of seafloor life) are closely related due to the ultimate reliance of all oceanic life on phytoplankton. Seafloor life is most abundant in shallow seas around the continents, particularly where seabed animals are in contact with the fish above. In very deep oceans this contact is broken and the abundance of seafloor life diminishes rapidly. In the most isolated parts of the Pacific, where there is little movement between depth layers, seabed life is less common than in the deeper trenches bordering the continents.

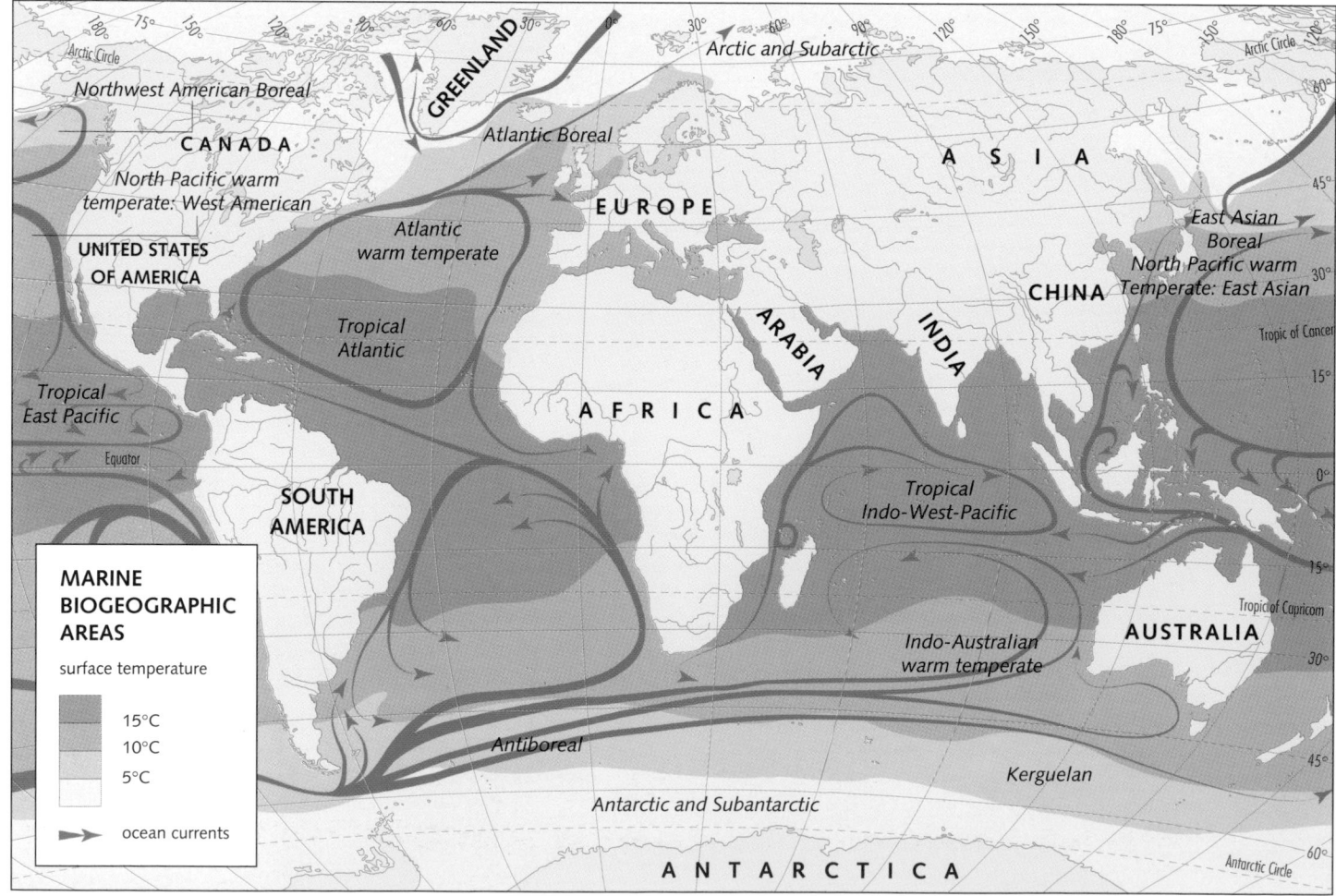

MARINE BIOGEOGRAPHIC AREAS

surface temperature

15°C
10°C
5°C

ocean currents

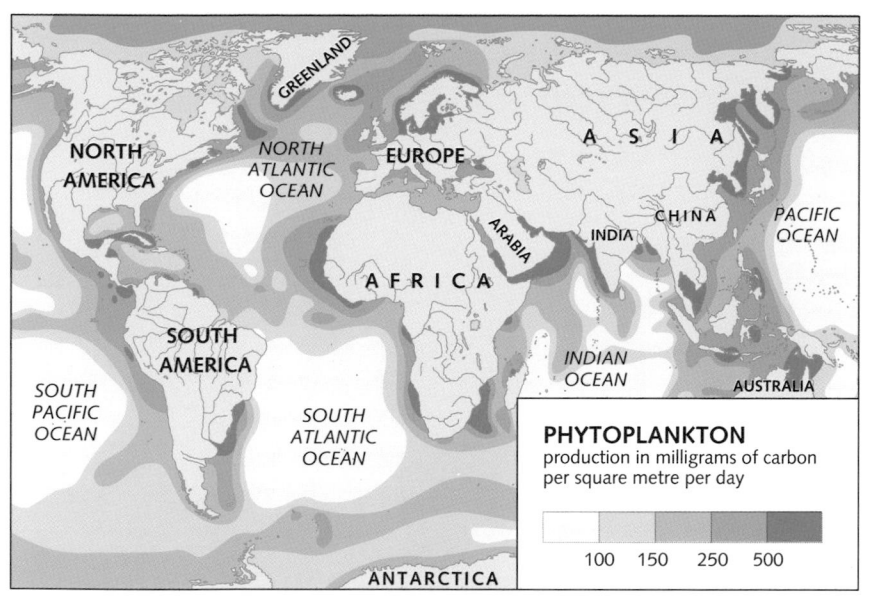

PHYTOPLANKTON
production in milligrams of carbon
per square metre per day

100 150 250 500

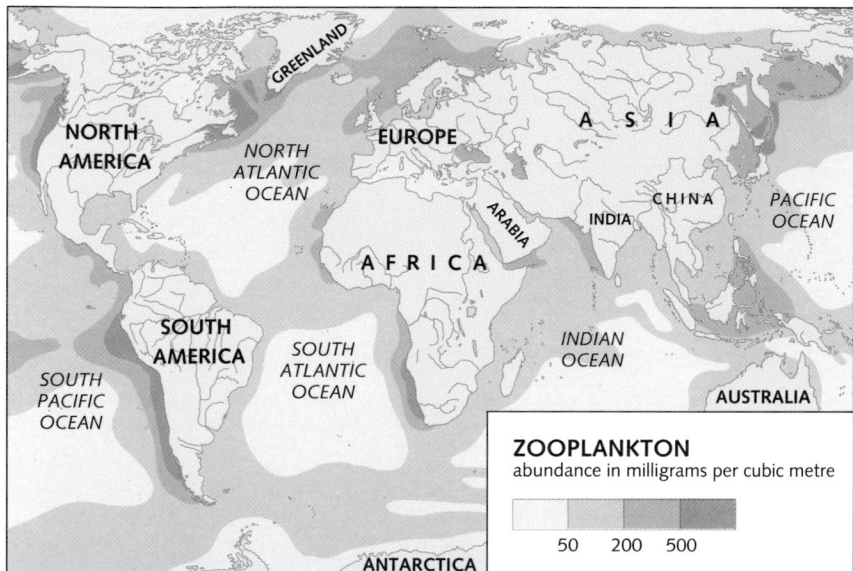

ZOOPLANKTON
abundance in milligrams per cubic metre

50 200 500

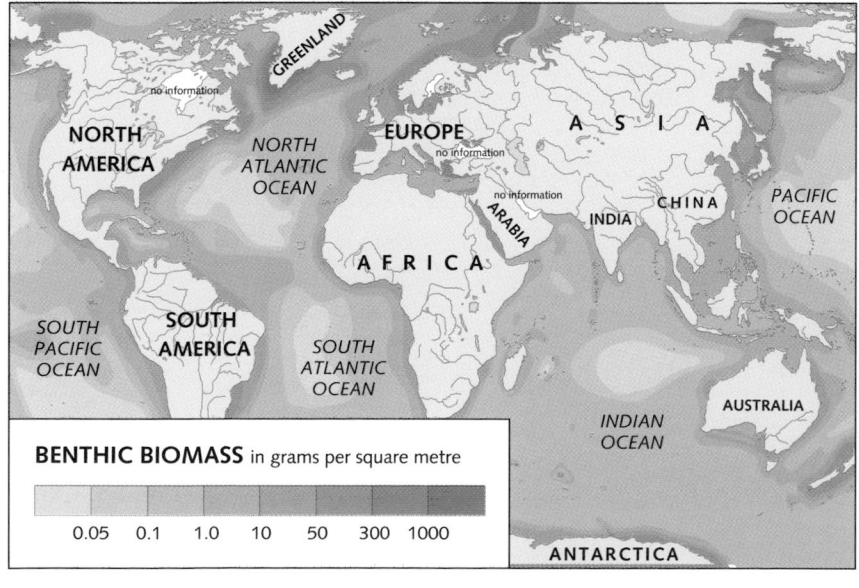

BENTHIC BIOMASS in grams per square metre

0.05 0.1 1.0 10 50 300 1000

Oceanic zones

The vertical stratification of the oceans depends on sunlight, temperature and pressure. The euphotic zone extends to a maximum of 100 metres below the ocean surface – the depth to which sunlight is capable of penetrating. The pelagic zone extends from 100 metres to 3,000 metres below the surface. Many life forms in this zone make regular vertical migrations to the euphotic zone in order to search for food. The benthic or abyssopelagic zone varies in depth according to the depth of the ocean. It contains those life forms that are adapted to seafloor dwelling, which may exist either in the deepest oceans or in relatively shallow continental seas.

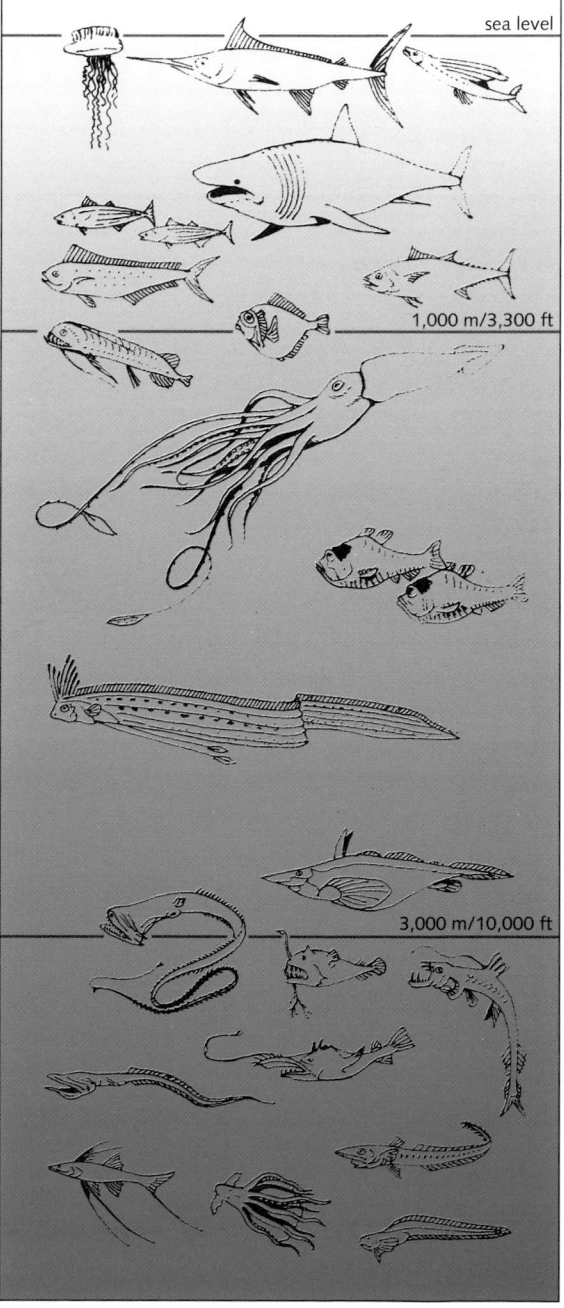

sea level

1,000 m/3,300 ft

3,000 m/10,000 ft

Flora of the Cape
Biogeographic barriers on the continent of Africa

Biogeographic regions broadly follow the boundaries that we might expect. Continental land masses form their basis, and boundaries are formed by physical barriers such as seas, mountain ranges and deserts. Within the plant kingdom, the flowering plants can themselves be divided into what are known as floristic regions (*see pages 134–135*). These again follow expected boundaries, with a notable exception – the flora of the Cape of Good Hope area on Africa's southern tip is unlike that on the rest of the continent. The tiny sliver of land occupies a floristic region of its own.

Africa straddles the Equator from latitude 35°N to 35°S, so the climate ranges from subtropical Mediterranean in the extreme north and south to equatorial heat in the centre. The most crucial influence on the vegetation is rainfall. Coastal areas around the Gulf of Guinea have the world's highest annual rainfall at over 500 millimetres, and there is dense tropical rainforest there. A thousand kilometres north the annual rainfall is less than 25 millimetres, causing the world's largest desert area. Between these extremes there is dry forest, monsoon forest, scrubland, mountain vegetation and grasslands.

At the southern end of the continent many of these vegetation types occur within a relatively small area, partly as a result of variations in altitude, but also of the climatic system around the Cape. But these do not explain the affinities that Cape flora has to that of other southern continents. In the wake of the theory of continental drift, it is now thought that many of the Cape plants are descendants of plants that occurred across the ancient southern continent of Gondwanaland. These plants have been eradicated further north in Africa because of the unsuitable climate, but their descendants have persisted on this small sliver of land with its own micro-climate.

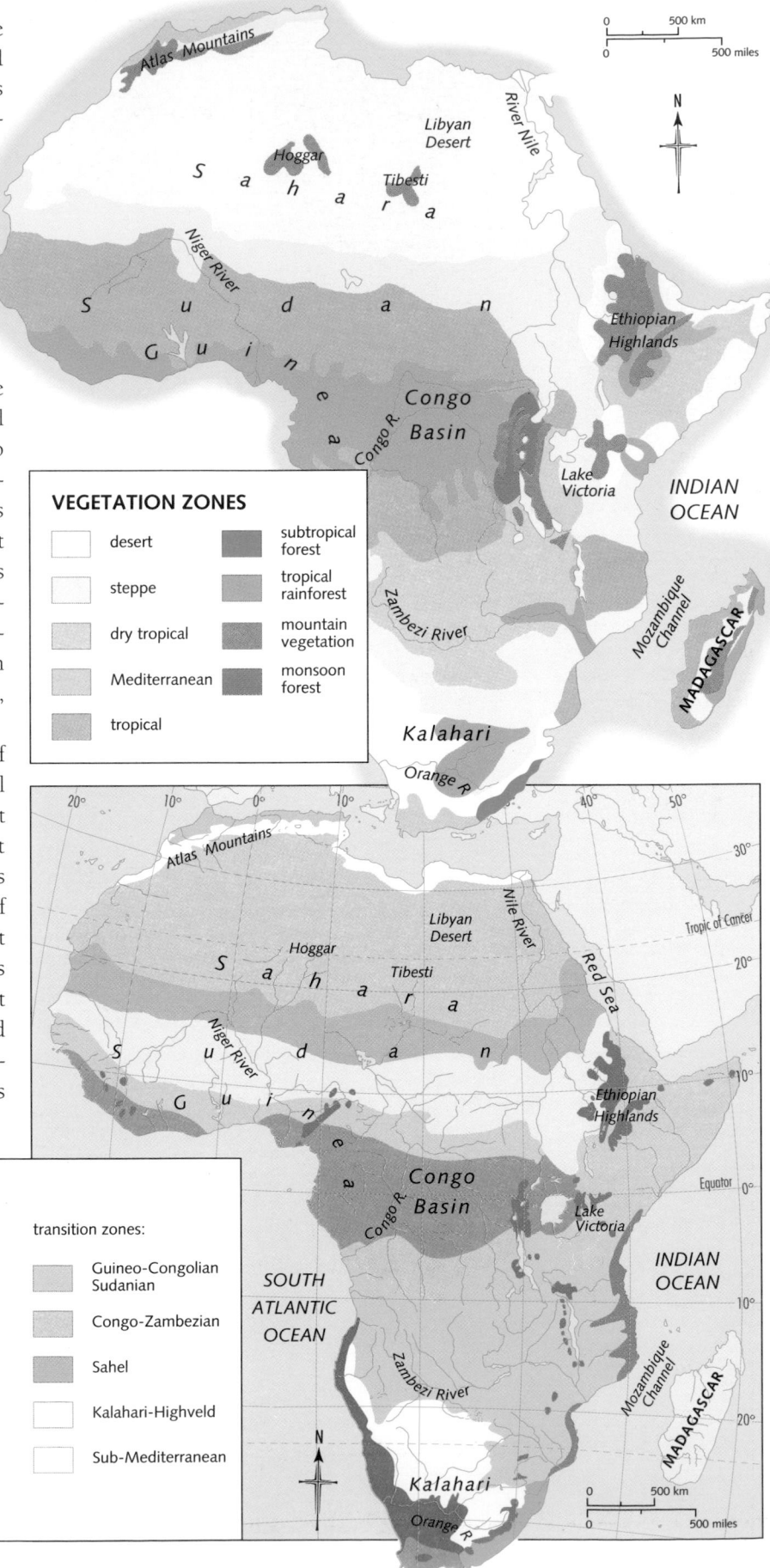

VEGETATION ZONES

- desert
- steppe
- dry tropical
- Mediterranean
- tropical
- subtropical forest
- tropical rainforest
- mountain vegetation
- monsoon forest

REGIONS OF ENDEMISM

centres of endemism:

- Guineo-Congolian
- Sudanian
- Zambezian
- Somalia-Masai
- Saharan
- Karoo-Namib
- Cape
- Mediterranean
- Afromontane

regional mosaics:

- Lake Victoria Basin
- Zanzibar-Inhambane
- Tongaland-Pondoland

transition zones:

- Guineo-Congolian Sudanian
- Congo-Zambezian
- Sahel
- Kalahari-Highveld
- Sub-Mediterranean

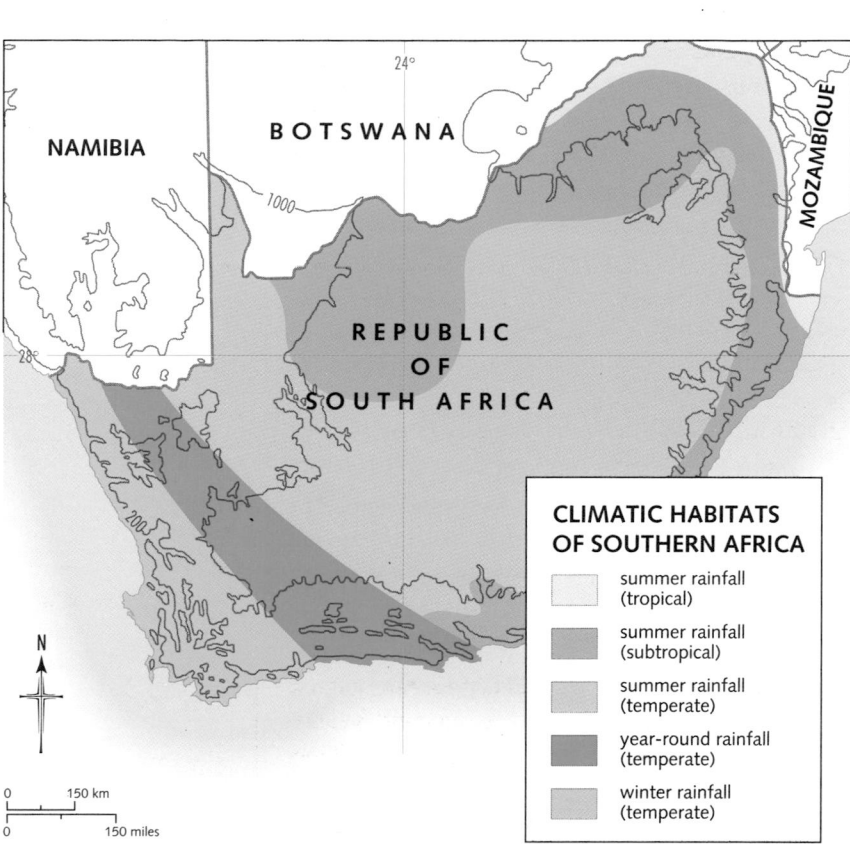

Africa's southwest tip enjoys all-year or winter rainfall and a temperate climate – conditions unknown in the rest of southern Africa. The coastal strip is cut off by uplands behind. Climate and relief combine to continue its isolation. Cape plants, such as the aloe and wild flowers above, are thought to have descended from plants that were once spread across the southern continents. The isolation has never been complete and is increasingly broken by human intervention.

CLIMATIC HABITATS OF SOUTHERN AFRICA

- summer rainfall (tropical)
- summer rainfall (subtropical)
- summer rainfall (temperate)
- year-round rainfall (temperate)
- winter rainfall (temperate)

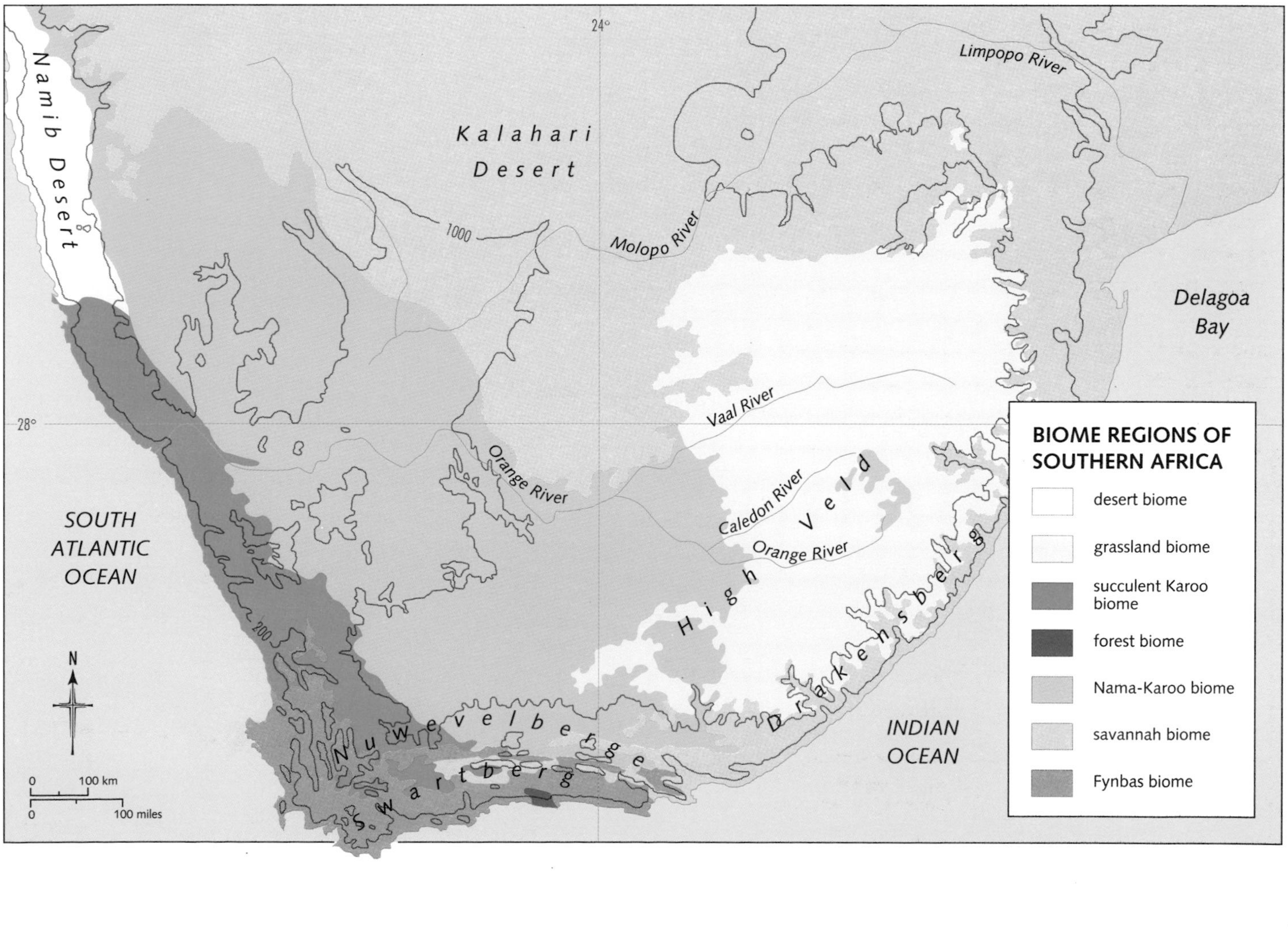

BIOME REGIONS OF SOUTHERN AFRICA

- desert biome
- grassland biome
- succulent Karoo biome
- forest biome
- Nama-Karoo biome
- savannah biome
- Fynbas biome

Counterpart and Replacement Species
Conditions for the limitations of a species range

The distribution of plant and animal species bears an obviously strong relation to environmental conditions. These conditions are themselves partly created by the floras and faunas that live in a particular place. Each organism can thus exist only by fitting in with those around it. Distribution is also affected by history and by chance. A species may emerge in a particular place and spread in order to take advantage of a particular ecological niche. Another species that is adapted to the same conditions but that arrives at a later date will find it more difficult, if not impossible, to become established. In fact a species will tend to spread until it finds that it can go no further, either because the habitat has changed, or because another species got there first and has taken all the space.

When the distribution of different bird species is mapped, it becomes clear that both of these conditions for the limitation of a species range are constantly in operation. Most commonly a bird species will occupy a region with a fairly uniform habitat. This really applies to breeding ranges, as birds are mobile seasonal migrants (although even then they are essentially finding uniform conditions by chasing the seasons around the globe). Once the habitat changes, the bird will venture no further. What is apparent, however, is that a relative of the first bird will often be found occupying a habitat that is adjacent but environmentally slightly different. This second species is sometimes referred to as a replacement species for the first.

A closely related group of species can evolve to fit into a range of habitats. The boundaries between the species are well defined by difference in habitat, and there is little overlap between them. The species are therefore not in competition. The evolution of a separate species would probably follow the classic model, where mutation allows a population to explore a novel but nearby habitat with success. Differences in habitat need not be a general difference but simply a difference in the range of vegetation or climate that that group of species occupies.

The mirror image of replacement distribution is the ecological counterpart. In these instances, different species from the same family occupy the same ecological niche, but in different areas. Here there is no overlap between the species, as the boundary is defined by the primacy of one species over the other. The emergence of a separate species is more complex in this situation, with separation being followed by the abuttal of each others' territories.

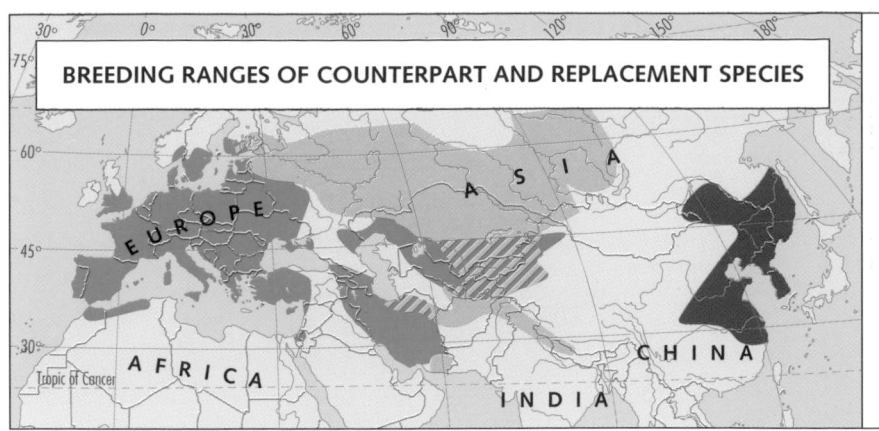

BREEDING RANGES OF COUNTERPART AND REPLACEMENT SPECIES

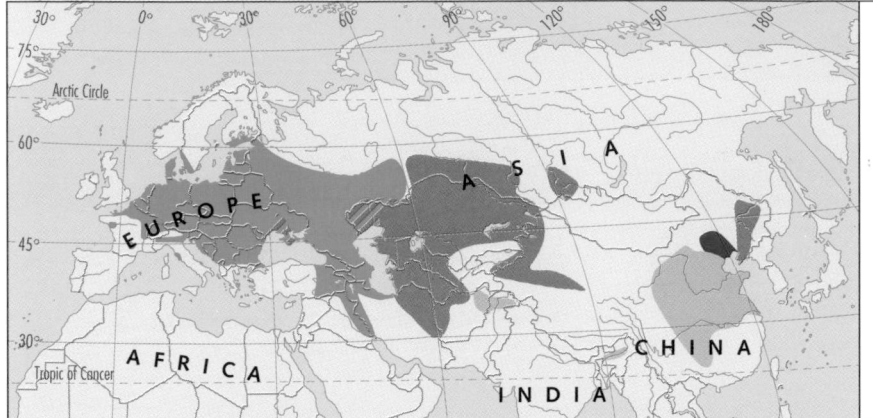

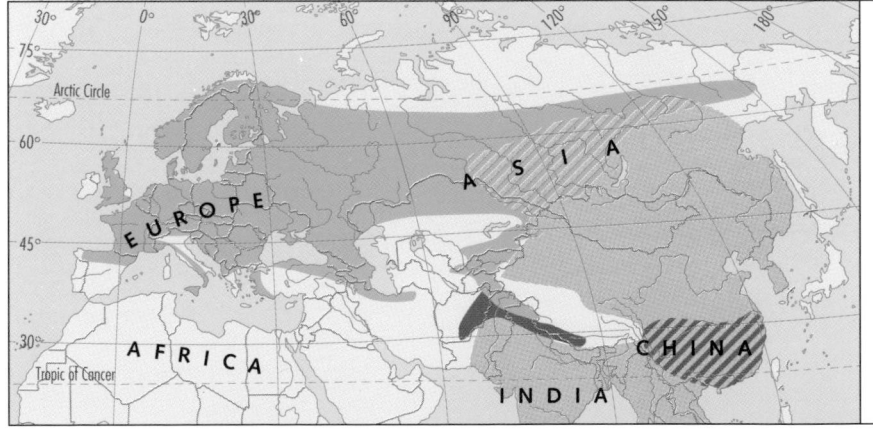

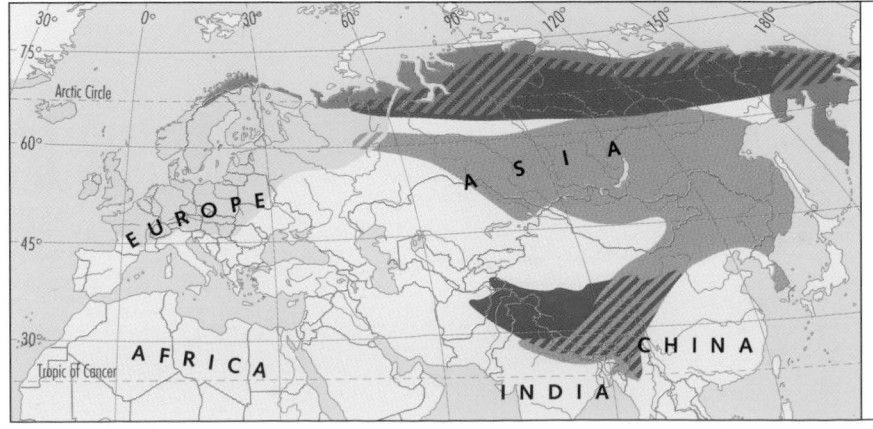

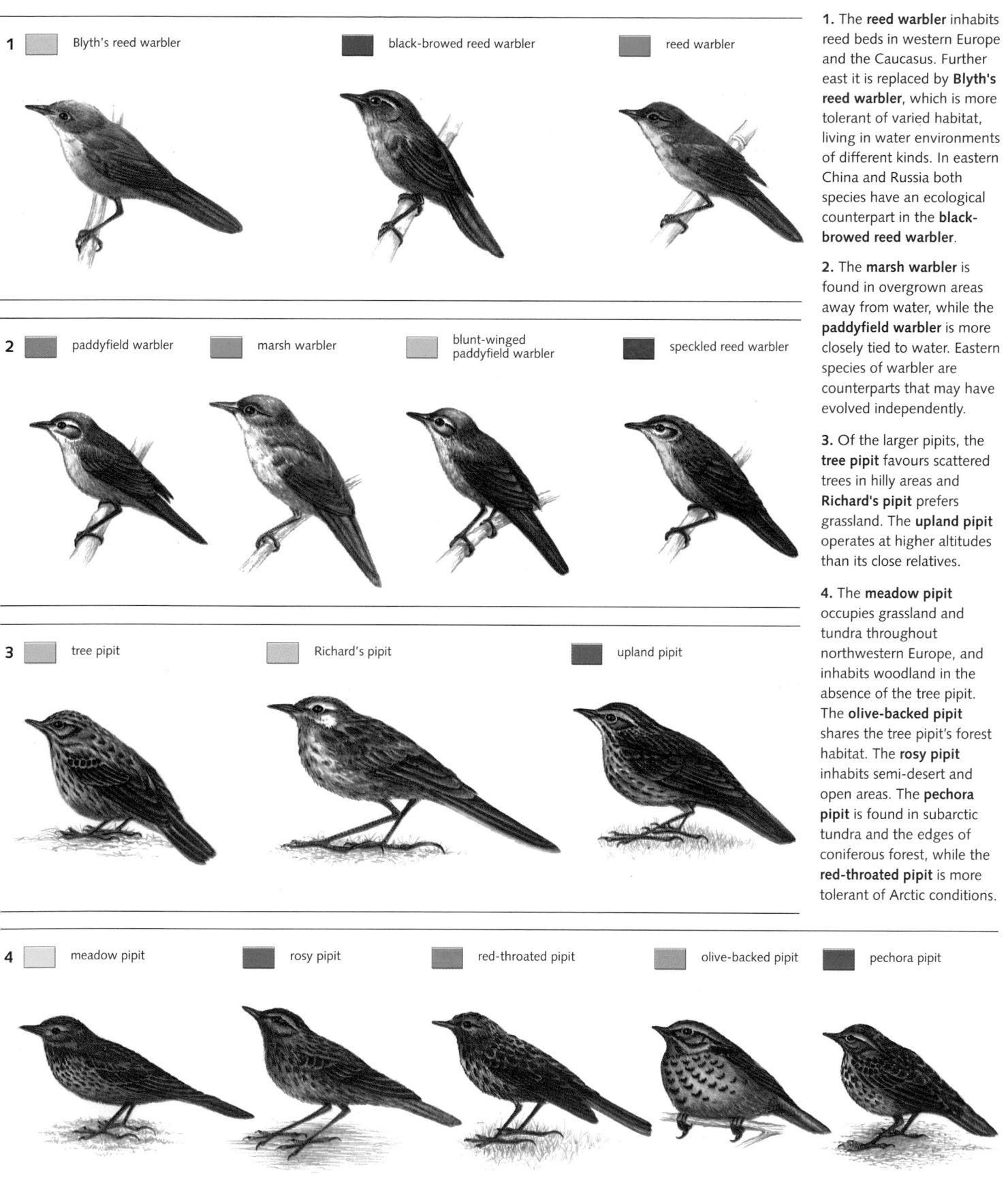

1 Blyth's reed warbler   black-browed reed warbler   reed warbler

2 paddyfield warbler   marsh warbler   blunt-winged paddyfield warbler   speckled reed warbler

3 tree pipit   Richard's pipit   upland pipit

4 meadow pipit   rosy pipit   red-throated pipit   olive-backed pipit   pechora pipit

1. The **reed warbler** inhabits reed beds in western Europe and the Caucasus. Further east it is replaced by **Blyth's reed warbler**, which is more tolerant of varied habitat, living in water environments of different kinds. In eastern China and Russia both species have an ecological counterpart in the **black-browed reed warbler**.

2. The **marsh warbler** is found in overgrown areas away from water, while the **paddyfield warbler** is more closely tied to water. Eastern species of warbler are counterparts that may have evolved independently.

3. Of the larger pipits, the **tree pipit** favours scattered trees in hilly areas and **Richard's pipit** prefers grassland. The **upland pipit** operates at higher altitudes than its close relatives.

4. The **meadow pipit** occupies grassland and tundra throughout northwestern Europe, and inhabits woodland in the absence of the tree pipit. The **olive-backed pipit** shares the tree pipit's forest habitat. The **rosy pipit** inhabits semi-desert and open areas. The **pechora pipit** is found in subarctic tundra and the edges of coniferous forest, while the **red-throated pipit** is more tolerant of Arctic conditions.

Animal Territories
Individual and group boundaries

In other parts of this atlas we have seen how groups and species of plants and animals are distributed over the Earth. This distribution is largely the result of the ways in which new species have evolved. But within the ranges of many species of animals another pattern of distribution is at work – the establishment of territory by individuals or groups of animals.

Territories work to the advantage of animals in a variety of ways. Typically, males compete for territories which they then use as a base for display to attract female mates. The strongest males will be able to command the largest territories. Females choose their mates on the basis that the male with the largest territory should be able to produce the strongest offspring. The setting up and defence of territories involves mock fights, posturing, threatening behaviour and real battles. These rarely end in death, although complete exhaustion is common.

The size and occupation of territories changes over time, as was shown by pioneering studies on the robin that were carried out by

Right: The robin (*Erithacus rubecula*) is found throughout Europe, Africa, western Asia and the Azores. Its red breast is used to warn intruders against incursions into its territory.

British biologist David Lack in the 1930s. The seven maps on the right show changes in the territories of robins in a part of Devon in southwest England over a period of three years. In each year, the territories are occupied by male and female pairs in the spring, and by solitary males and, less commonly, solitary females in the winter. But each year new males and females arrive and take over, altering the boundaries of territories.

It seems that mature males have a limited time in which to establish a territory and a short time in which they will be strong enough to hold it against rival males. In some birds the male display is taken to apparently destructive lengths. The peacock's tail, for instance, is a definite physical handicap. However, it appears that the peahen chooses the male with the grandest tail on the basis that he must be strong (and therefore produce strong offspring) to be able to survive such a burden.

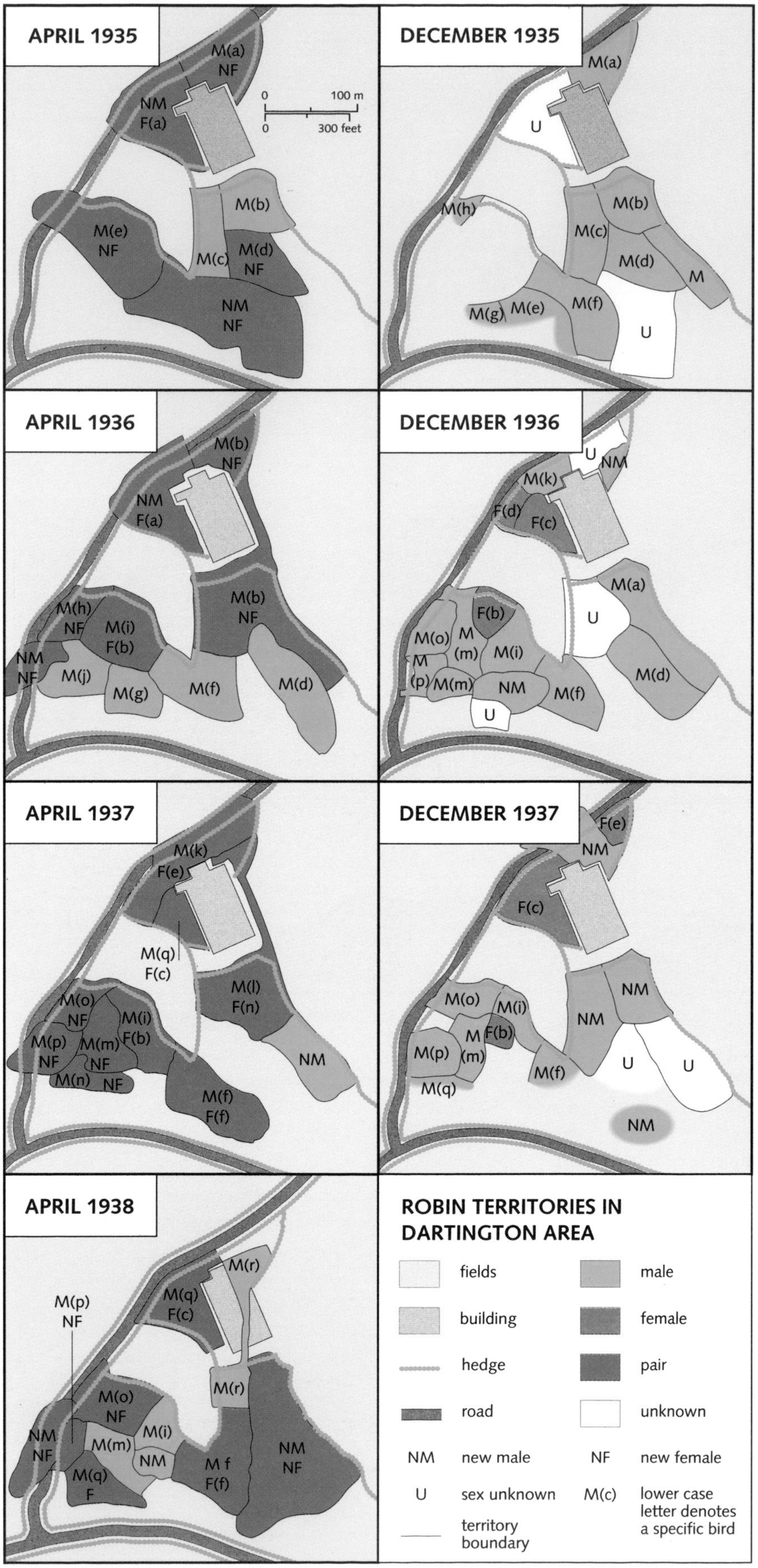

APRIL 1935

M(a) NF
NM F(a)
M(b)
M(e) NF
M(c) M(d) NF
NM NF

0 100 m
0 300 feet

DECEMBER 1935

M(a)
U
M(h)
M(b)
M(c)
M(d)
M
M(g) M(e) M(f)
U

APRIL 1936

M(b) NF
NM F(a)
M(h) NF M(i) F(b)
M(b) NF
NM NF M(j)
M(g) M(f) M(d)

DECEMBER 1936

U NM
M(k)
F(d) F(c)
M(a)
F(b)
M(o) M (m)
M M(i) U
(p) M(m) NM M(d)
U M(f)

APRIL 1937

M(k) F(e)
M(q) F(c)
M(o) M(l) F(n)
NF M(i)
M(p) M(m) F(b)
NF NF
M(n) NF NM
M(f) F(f)

DECEMBER 1937

F(e) NM
F(c)
M(o) M(i) NM
NM
M F(b)
M(p) (m) U U
M(q) M(f)
NM

APRIL 1938

M(r)
M(p) NF
M(q) F(c)
M(r)
M(o) NF
M(i)
NM NF M(m) NM
M(q) F NM F(f) NM NF

ROBIN TERRITORIES IN DARTINGTON AREA

fields		male	
building		female	
hedge		pair	
road		unknown	

NM new male
U sex unknown
territory boundary

NF new female
M(c) lower case letter denotes a specific bird

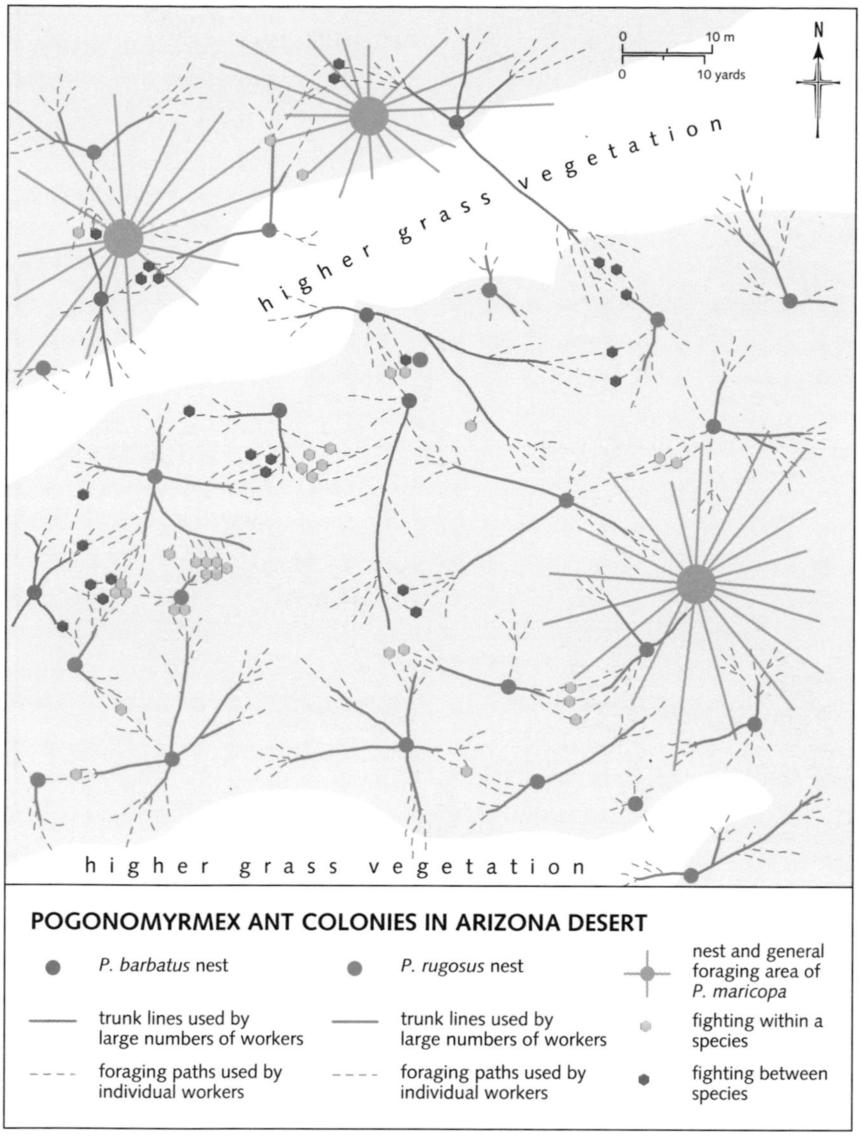

The formation of territories by individuals may seem to invite conflict, whereas the formation of territories by groups may have the opposite effect – they can help to establish behaviour which avoids unnecessary conflict. Studies of animals as diverse as ants and wolves have revealed some of the ways in which this is achieved. Separate colonies of ants, including those from different species, divide areas up into territories and establish routes along which worker ants will travel and areas in which they will forage. These territories seem to be marked in some way, and are defended against intruders from other species or colonies. Wolf packs mark the edge of their territories with scent, and rarely cross into another pack's territory. To avoid encounters the packs have another device – they howl at one another. The concerted howling of a pack of wolves carries over a distance of around 10 kilometres (6 miles) – roughly the radius of a typical wolf pack territory.

Timber wolf (*Canis lupus*)

POGONOMYRMEX ANT COLONIES IN ARIZONA DESERT

higher grass vegetation

higher grass vegetation

- ● *P. barbatus* nest
- ● *P. rugosus* nest
- ┼ nest and general foraging area of *P. maricopa*
- —— trunk lines used by large numbers of workers
- —— trunk lines used by large numbers of workers
- ○ fighting within a species
- - - - foraging paths used by individual workers
- - - - foraging paths used by individual workers
- ● fighting between species

If food remains in sufficient supply, a wolf pack's territory may remain constant for over ten years, but shortage of prey will drive the pack outside its territory. *Right*: Packs will risk crossing one another's territories to reach winter deer yards.

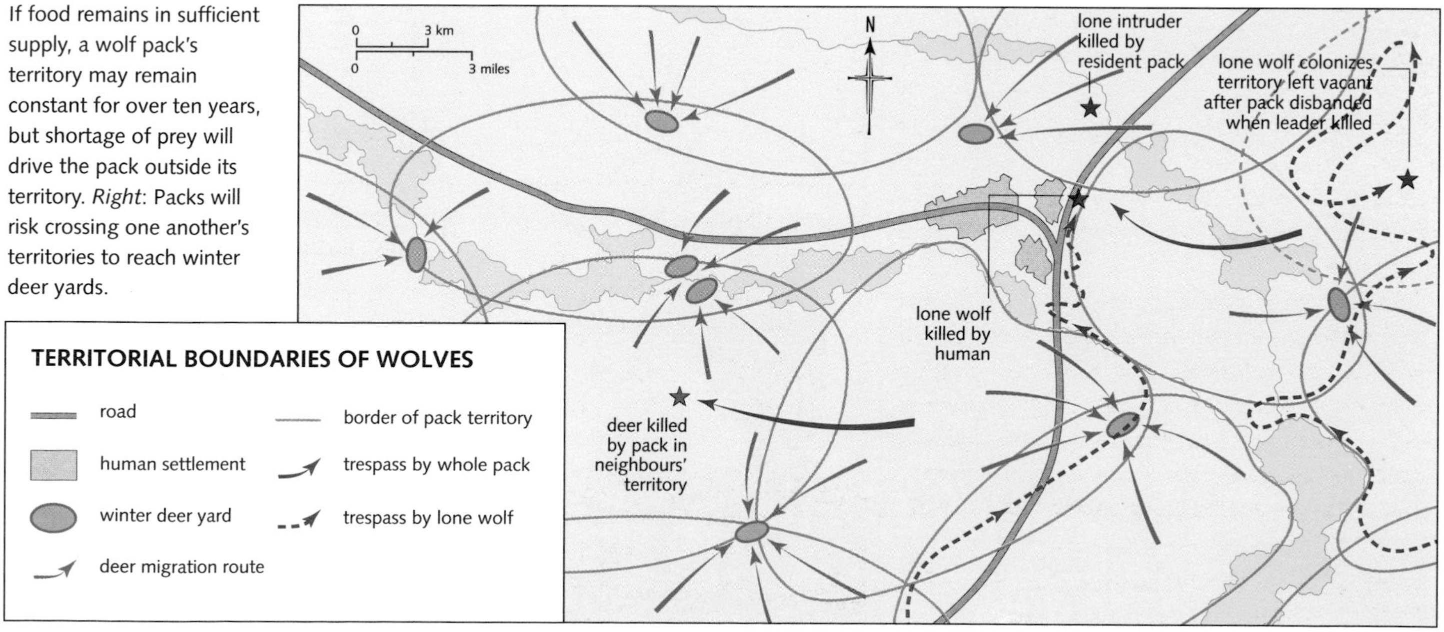

lone intruder killed by resident pack

lone wolf colonizes territory left vacant after pack disbanded when leader killed

lone wolf killed by human

deer killed by pack in neighbours' territory

TERRITORIAL BOUNDARIES OF WOLVES

- ══ road
- —— border of pack territory
- ▨ human settlement
- ⤙ trespass by whole pack
- ⬭ winter deer yard
- ⤙ trespass by lone wolf
- ⤙ deer migration route

Bird Migration
Chasing the seasons

Migratory birds are a familiar sight to people in every part of the world. Temperate countries have summer visitors that disappear in the autumn, such as swallows, while the prodigious feats of birds such as the arctic tern, which flies 40,000 kilometres (25,000 miles) from the Arctic to the Antarctic and back again each year, are well-documented. The main reason for these migrations is quite simple – food. Swallows feed off gnats and mosquitoes which are abundant in temperate countries only in the summer months, so the birds must make the trek from one side of the Equator to the other in order to find a constant supply of these insects. The mechanisms of navigation and the evolution of migratory habits are more complex, and in many cases remain unknown.

Overall, the Earth's climate has become more seasonal over the past 5 million years or so. In particular, the cooling of the Earth during the Pleistocene ice ages has left a legacy of cold polar regions and warm equatorial regions, with temperate regions in-between where there is much greater variation through the year than in the past. Some birds that originally lived in one region all year round found that they had to move a short distance with the changing seasons in order to maintain their food supply. As the annual temperature cycle became more extreme they had to go further and further, and in some cases the summer and winter feeding grounds have become entirely separate, forcing the birds on long journeys, often with no feeding en route.

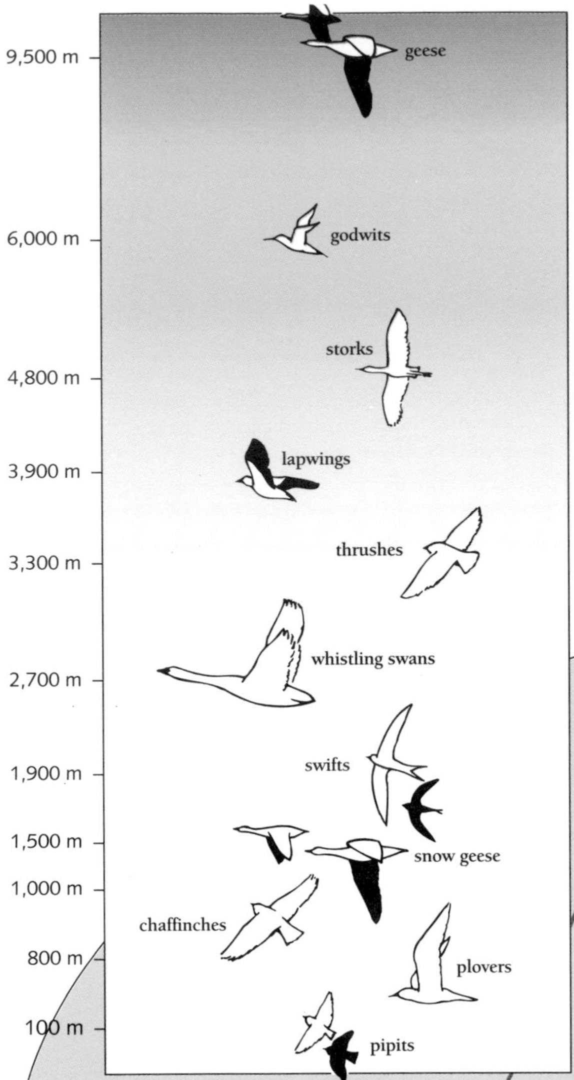

High fliers
Most migrating birds fly at heights of between 100 and 1,500 metres (300 and 3,300 feet) above sea level, with much migration taking place well beyond the range of human vision. The maximum recorded height of certain birds is shown left. Most small birds fly at about 35 kilometres (20 miles) per hour, and starlings and blackbirds can manage 55 kilometres (35 miles) per hour. Measurements of these speeds enable researchers to work out the migratory range of the different species.

9,500 m — geese

6,000 m — godwits

4,800 m — storks

3,900 m — lapwings

3,300 m — thrushes

2,700 m — whistling swans

1,900 m — swifts

1,500 m — chaffinches / snow geese

1,000 m

800 m — plovers

100 m — pipits

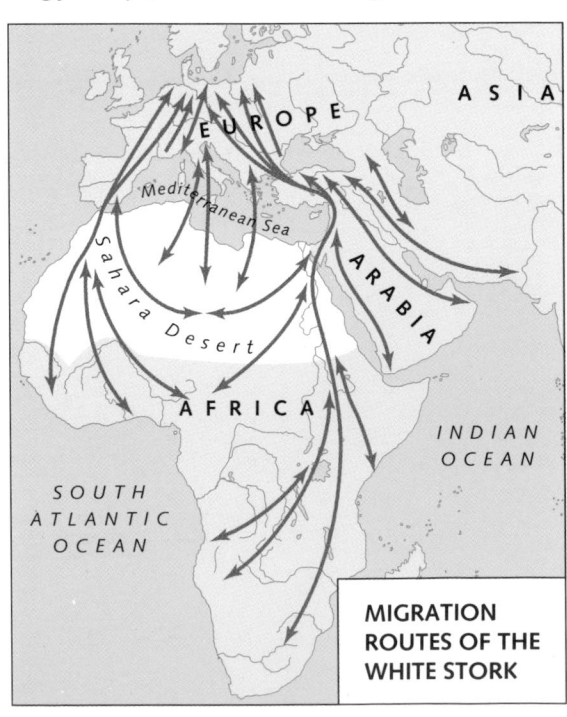

MIGRATION ROUTES OF THE WHITE STORK

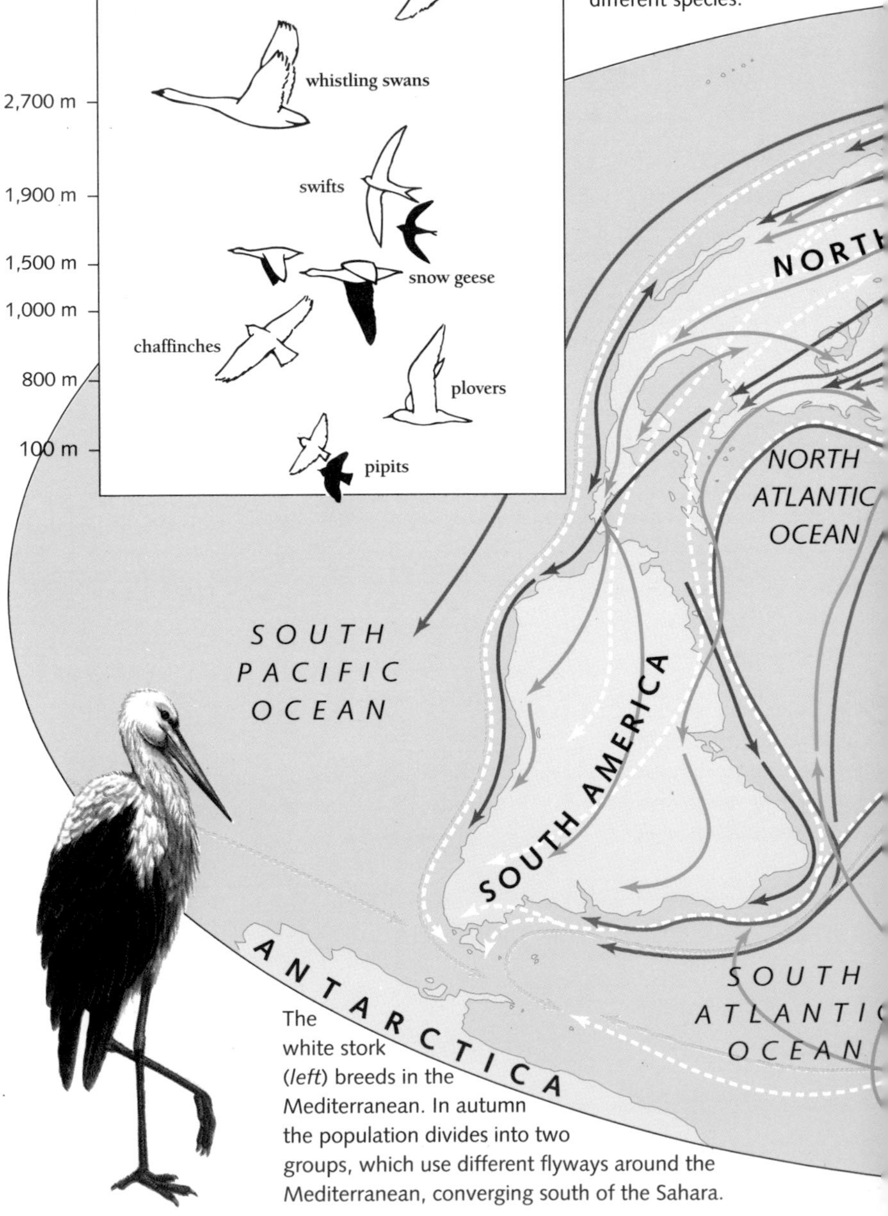

EUROPE
ASIA
Mediterranean Sea
Sahara Desert
ARABIA
AFRICA
INDIAN OCEAN
SOUTH ATLANTIC OCEAN

NORTH
NORTH ATLANTIC OCEAN
SOUTH PACIFIC OCEAN
SOUTH AMERICA
ANTARCTICA
SOUTH ATLANTIC OCEAN

The white stork (*left*) breeds in the Mediterranean. In autumn the population divides into two groups, which use different flyways around the Mediterranean, converging south of the Sahara.

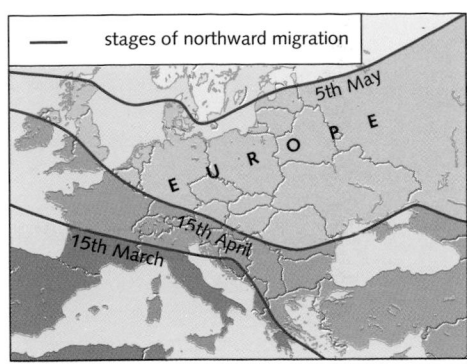

Spring swallows

The appearance of the barn swallow (*Hirundo rustica*) in Europe announces the coming of spring. Its arrival is regular and is followed by that of its fellow aerial insect-feeders, the house martin and the swift.

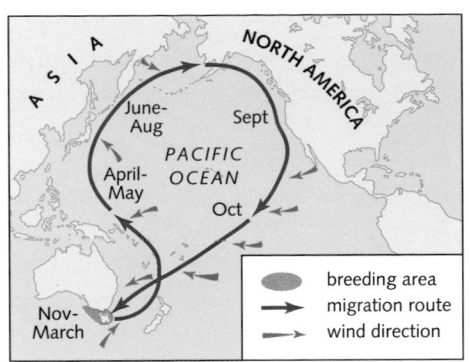

Short-tailed shearwater

Shearwaters seem to have a figure-of-eight migration pattern. The short-tailed shearwater, *Puffinus tenuirostris*, uses the prevailing winds of the Pacific to help it to cover the 30,000 kilometres (20,000 miles) of its annual journey.

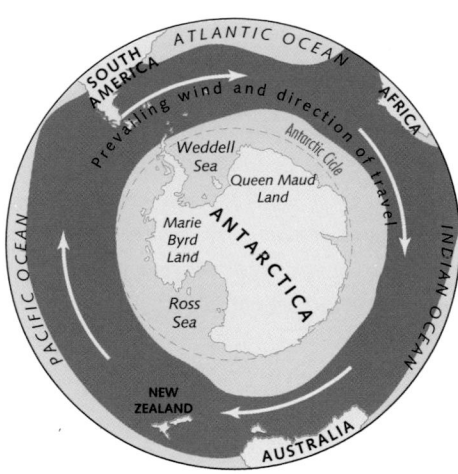

Above: the range of the wandering albatross (*Diomedea exulans*), which circumnavigates the globe on the southern trade winds and nests on remote islands, to which it returns after many years of wandering.

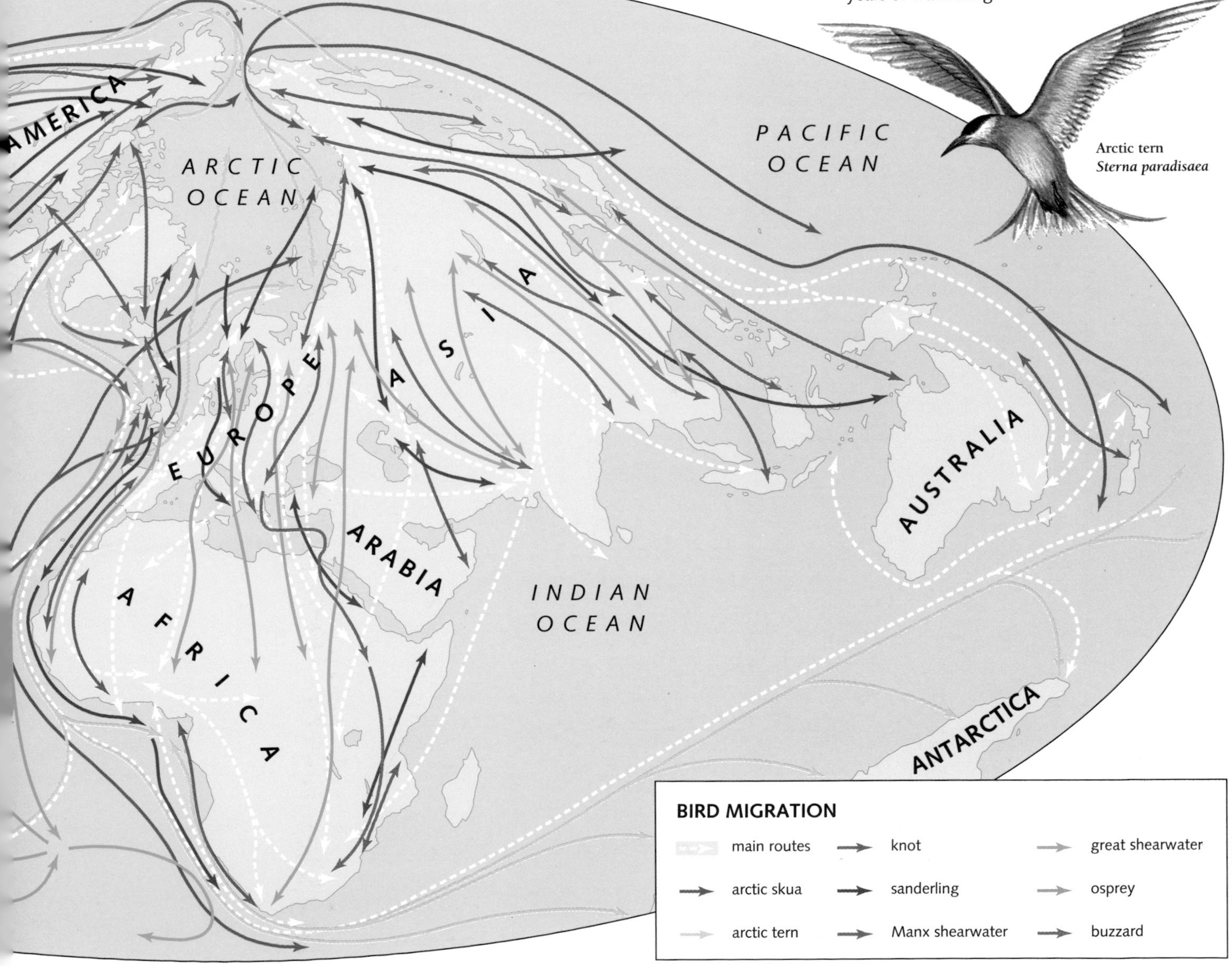

PACIFIC OCEAN

AMERICA

ARCTIC OCEAN

EUROPE

ASIA

ARABIA

AFRICA

INDIAN OCEAN

AUSTRALIA

ANTARCTICA

Arctic tern
Sterna paradisaea

BIRD MIGRATION

main routes	knot	great shearwater
arctic skua	sanderling	osprey
arctic tern	Manx shearwater	buzzard

Mammal Migration
The caribou of the Canadian Arctic

Land mammals are so diverse that it is not surprising to find a huge range of migratory behaviour. Those with an annual migration pattern generally have separate feeding ranges in winter and summer or in the wet and dry seasons. Their behaviour is dictated by local conditions rather than species characteristics, since other herds of the same animal stay in one place all year, if food is sufficient.

The herds of caribou that occupy the 'barrenground' region west of Hudson Bay are one of nature's most spectacular migrants. They live on the tundra in the summer, feeding off grass and lichen. At the onset of winter, the freezing and thawing of

snow on the open tundra creates layers of ice which the caribou cannot break with their hooves. They therefore turn south and west in the search for food.

In the forests that edge the tundra, the snow stays powdery. The caribou herds range through these forests in winter, scraping away the soft snow to get at the vegetation underneath. Between February and April the caribou start to move north again, led by pregnant females returning to the calving grounds. By the beginning of May large herds gather at the tree line between the forest and tundra, then walk about 500 kilometres within three or four weeks. The calving grounds are found in remote,

Encircling the Arctic
The caribou of North America and the reindeer found on the northern edge of Europe and Asia (*far right*) are subspecies of *Rangifer tarandus* and inhabit coniferous forest and tundra. Populations restricted to one or another of these habitats have short seasonal migrations; those in transition zones in-between, like the barren-ground caribou, travel much further.

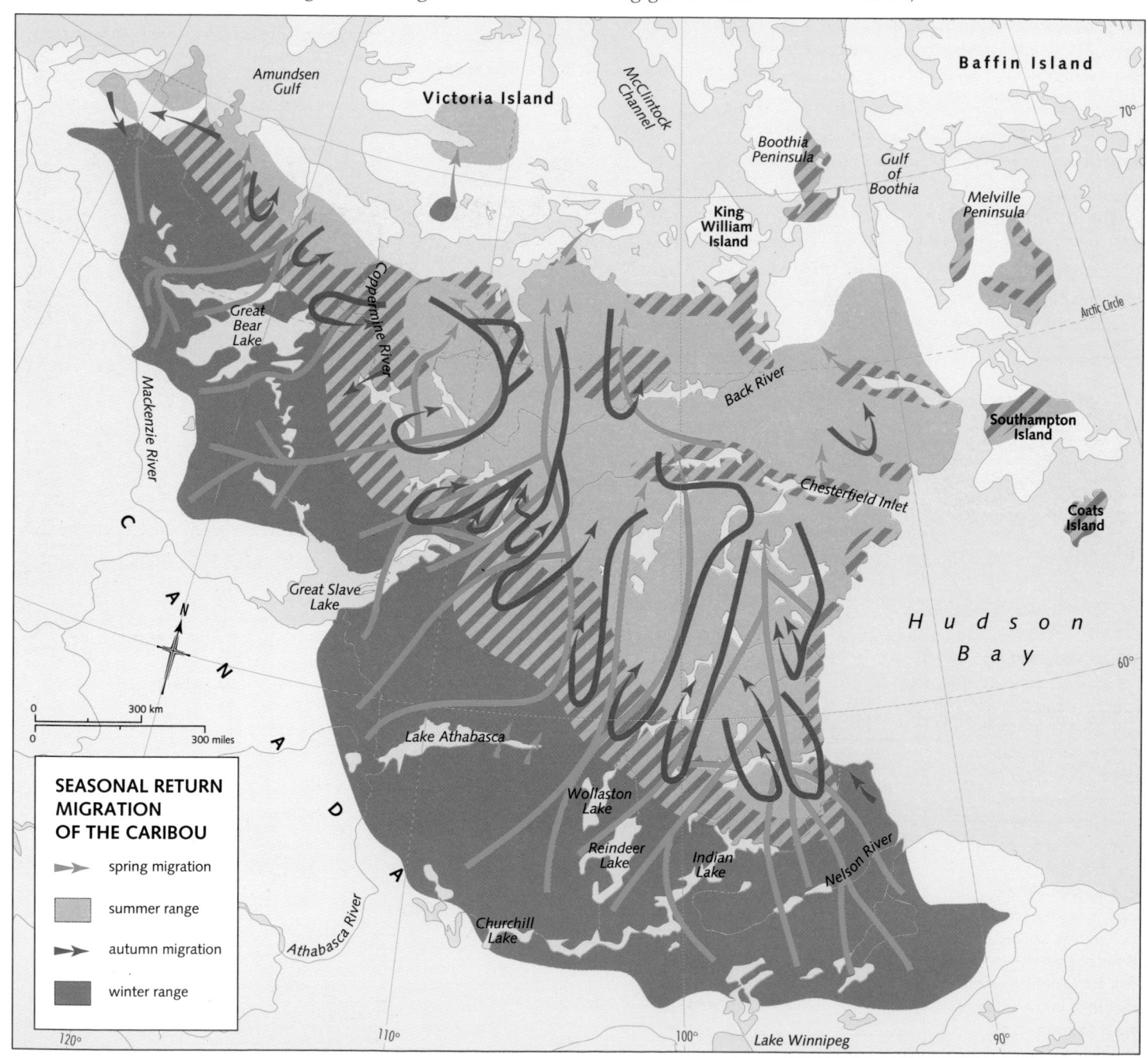

SEASONAL RETURN MIGRATION OF THE CARIBOU

→ spring migration

□ summer range

→ autumn migration

■ winter range

inhospitable places in order to be safe from predators. Almost immediately after calving, the animals move a further 200 kilometres to summer feeding grounds. In July the herds disperse and in August they travel south in groups of two or three. They travel up to 65 kilometres a day in autumn and arrive back at the tree line by the end of September.

The great spring migration

Caribou can live in small groups of 100 or so, but in spring the caribou of northeast Canada gather in herds of up to 100,000 (*see photograph*) in order to migrate to calving grounds and summer ranges. As they stretch out, these herds may reach over a distance of 300 kilometres, taking weeks to pass. But these vast herds are dwarfed by earlier migrations of grazing animals. Before human intervention decimated their numbers, for example, the springboks of southern Africa would set out to find new pastures every three or four years, usually in response to drought. In 1896 a herd of about one million, estimated to be 220 kilometres long and 20 kilometres wide, was seen.

Caribou
Rangifer tarandus

Tracking a single group

In 1956 researchers began to track a group of caribou to see whether or not they had an annual migration pattern. They discovered that the herd did follow the same general pattern but never returned to the same place during the two-and-a-half-year study. This supports the view that, like other grazing animals, caribou migrate for food and a safe place for calving rather than being driven by instinct to return to the same places every year.

ARCTIC WANDERERS
range of caribou and reindeer

CARIBOU MIGRATION

major caribou herds

migration

forest

tundra

northern limit of trees

Marine Migration
Environments for breeding and feeding

Many species of marine animals migrate over large distances. Some, like the Pacific salmon, make one spectacular return journey in their lives; others, such as baleen whales, have an annual cycle. The evolution of migratory habits in marine animals is as puzzling as it is in land animals and birds. The underlying reasons are feeding and reproduction, but how these are balanced out varies greatly from one species to another.

Baleen whales live in polar regions in summer, where their main food sources are abundant. The extension of pack ice in winter drives them towards the tropics, where food is much more scarce, so that they have to live off their summer fat. The waters of the tropics are, however, ideal for calving. The young survive much better in warmer seas, feeding off their mothers rather than krill. It is easy to envisage the gradual development of this migratory habit. The ancestors of today's whales presumably went further afield each winter until they found the warm seas they needed. Their routes follow natural currents.

The evolutionary mechanism that causes the Pacific salmon to leave its ocean existence and journey up fast-rushing rivers and streams, up waterfalls and through rapids, to reach its birthplace, spawn and then die, is more perplexing. This dramatic life-cycle shows that evolution brings about wonders of behaviour as well as of form.

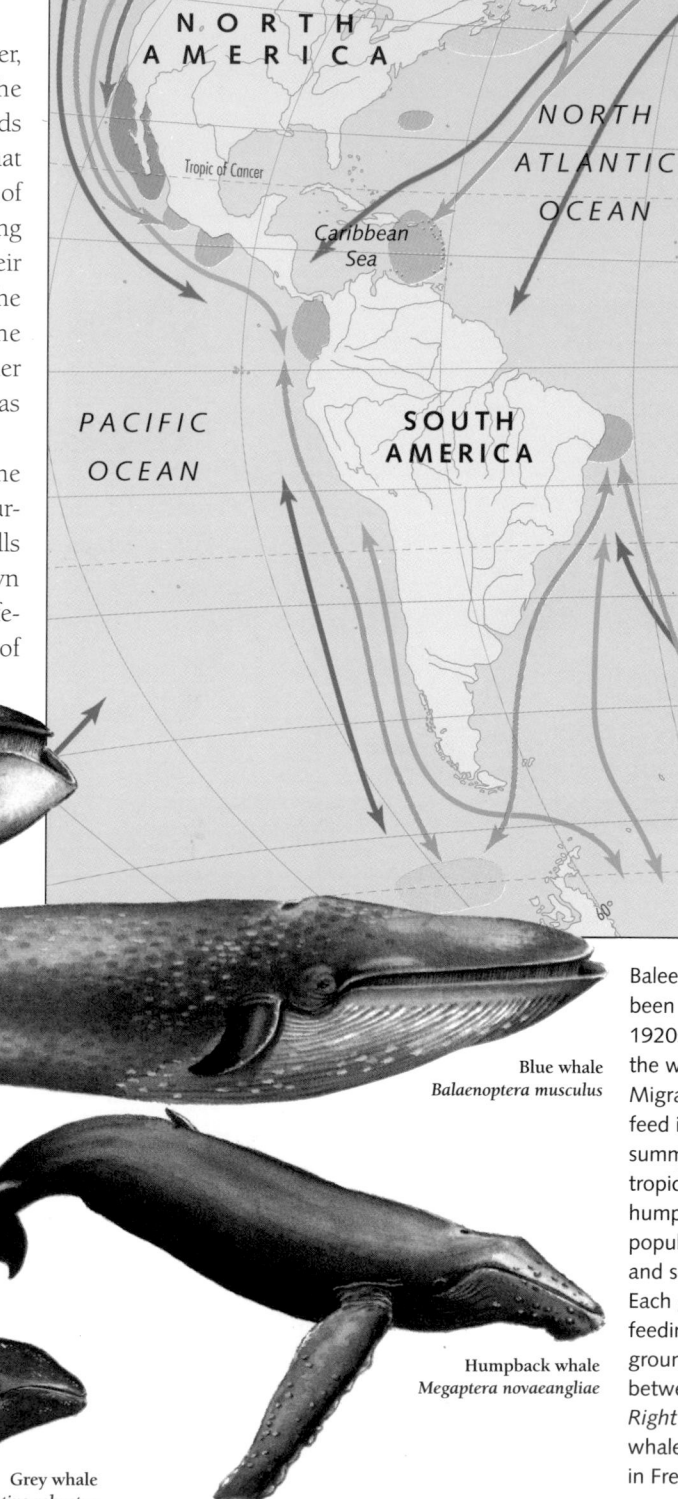

Bowhead whale
Balaena mysticetus

Blue whale
Balaenoptera musculus

Baleen whales (*right*) are named after the horny plates across their mouth cavities, which are used to strain the tiny krill and plankton that are their principal food source. Baleen whales migrate further than other whales in search of food and breeding places.

Grey whale
Eschrichtius robustus

Humpback whale
Megaptera novaeangliae

Baleen whale migration has been studied since the 1920s using tags lodged in the whales' outer blubber. Migrating whales generally feed in polar latitudes in summer and breed in the tropics in winter. There are humpback whale populations in both northern and southern hemispheres. Each group has its own feeding and breeding grounds, and its own routes between the two.
Right: a school of humpback whales cooperative feeding in Frederick Sound, Alaska.

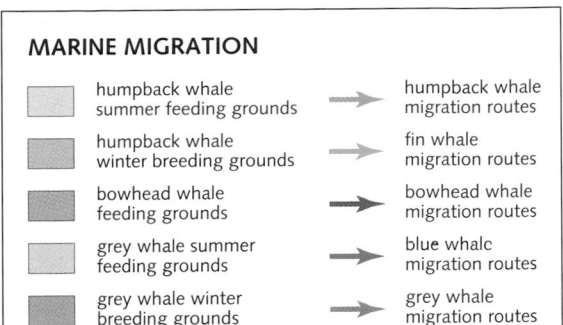

MARINE MIGRATION

humpback whale summer feeding grounds		humpback whale migration routes	
humpback whale winter breeding grounds		fin whale migration routes	
bowhead whale feeding grounds		bowhead whale migration routes	
grey whale summer feeding grounds		blue whale migration routes	
grey whale winter breeding grounds		grey whale migration routes	

Green turtle
Chelonia mydas

The green turtle swims 2,500 kilometres from Ascension Island to feed on the South American coast, returning to breed every three years. One theory is that this migration pattern has evolved in parallel with the opening of the South Atlantic. As Africa and South America separated, the islands near the latter were pushed outwards and sank, becoming seamounts. New islands continually arose at the mid-Atlantic ridge, so the turtles had to go further and further to reach an island on which to breed.

However, the timespan of vertebrate species is usually around 10 million years – far shorter than the 200 million years that the South Atlantic took to open. This behaviour must therefore have been passed on to the present species via several generations of speciation. But only a small population of green turtles follows the route. The chances of this behaviour being retained by this group challenges our understanding of the mechanisms of evolution.

ancient continents		volcanic islands
turtle migration		seamounts (former islands now below sea level)
plate spread direction		

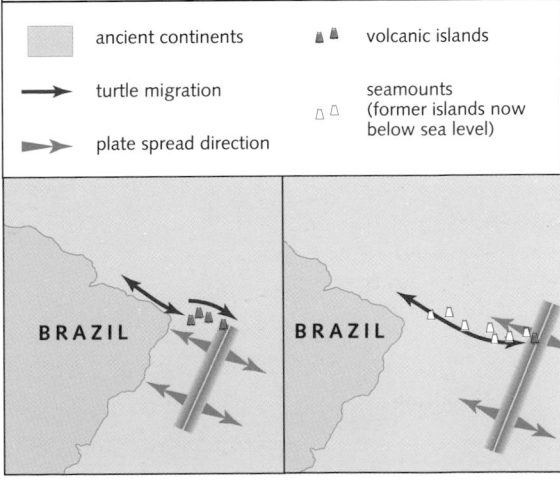

Section IX: Islands and Endemics

Islands provide laboratories for the study of evolution, increasing our understanding of the processes at work in all environments. The origins of islands can often be dated, and the processes of species arrival and establishment may be charted in detail.

Islands have fascinated naturalists and evolutionary biologists for centuries. Not only do they often contain unique plants and animals, but they can also be seen as small-scale worlds that biologists can study in detail. The continuing fascination of islands has fuelled major advances in evolutionary theory, and theories based on the study of islands affect how we regard the history of life on Earth.

The first sentences that Darwin wrote in the *Origin* are:

> When on board H.M.S. 'Beagle', as naturalist, I was much struck
> with certain facts in the distribution of the inhabitants of South
> America...These facts seemed to me to throw some light on the
> origin of species--that mystery of mysteries.

Darwin had observed similarities between the living and the fossil mammals of South America and postulated a link in time between the two: in other words, evolutionary change. He also observed that the islands he visited around the continent – the Falklands, Chiloé and the Galapagos Islands – all had faunas that seemed to be subsets of South American fauna, proving a link in space. These must have arrived by chance dispersal events. There were therefore both links extending back to the fossils and links among the species of islands and the neighbouring mainland.

Biogeographers had hitherto explained present distributions as the static results of creation. Some stuck firmly to the idea of a single creation, others permitted multiple creations over time, and yet others saw the possibility of many dispersal events after the initial creation(s). No one could explain what Darwin and others had seen before: that contiguous species looked similar, and that islands usually contained plants and animals that looked like those on neighbouring mainlands but were not identical. Darwin pointed out that the creationists could not explain the fact that oceanic islands contained so few species, and that certain groups of animals, such as terrestrial mammals, salamanders and true freshwater fish, are generally absent. He emphasized his third, critical piece of evidence for evolution from island biogeography:

> Why should the species which are supposed to have been creat-
> ed in the Galapagos archipelago, and nowhere else, bear so plain
> a stamp of affinity to those created in America?

Darwin's explanation – dispersal and evolution – now seems so obvious that it is difficult to understand how intelligent naturalists could have clung to purely creationist models for so long.

Evolution and ecology

Islands are characterized by low species diversities. This has ecological consequences, and also provides the basis for certain current models in evolutionary ecology. The low diversity means that most plants and animals on islands occupy wider, or different, niches than their mainland relatives. Galapagos finches, for instance, eat seeds as their relatives elsewhere do, but they have also broadened their dietary range to include cactus, fruit and insects in the absence of the birds and mammals that occupy those niches on the South American mainland.

Islands often lack large predatory animals, mainly because of the difficulties that these face in crossing wide expanses of ocean. This was so frequent on Pacific and Indian ocean islands that sailors, with their cats and dogs, could soon wipe out whole species, which had lost all powers of defence and escape. The absence of predators led to the evolution of large flightless birds on many islands: the kiwis and moas of New Zealand and the dodos of Mauritius. Moas and dodos became victims to the first predators they encountered – humans.

The balance of floras is also different on islands. Many trees have large seeds, which are not well adapted to flotation in seawater or transportation through the air, so islands often lack the kinds of trees found on nearby mainlands. Other plants may become tree-like instead. Plants related to the small tarweeds of North America have colonized the Hawaiian Islands, evolving into species of the *Daubutia* genus of shrubs and trees that can reach a height of eight metres. Their relatives the Hawaiian silverswords (*Argyroxiphium*) occupy many habitats, from boglands and rainforests to bare hillsides composed of volcanic lava – a much broader range than any related mainland group.

Island plants and animals may be smaller or larger than normal. Dramatic dwarfing was seen on some Mediterranean islands such as Malta, the dwarf elephants and giant dormice of which are now sadly extinct. Changes in size are a result of the impoverished floras and faunas. Species present on the mainland are absent on islands, and others evolve to fill vacant niches: the moas of New Zealand and the giant tortoises of the Galapagos Islands took the role of medium- and large-sized mammalian herbivores that would normally be present.

Islands have provided the foundation for a major set of theories in evolutionary ecology. Robert MacArthur and Edward Wilson published their *Theory of Island Biogeography* in 1967 and spawned a new discipline in ecology. They observed that when a new island emerges it is devoid of life. Seeds and stray animals arrive piecemeal. Over time, some of these unwilling migrants become established as breeding populations and the formerly bare island gradually becomes green. New species continue to arrive, but after a while many will fail to become established because competitors are already there. Other new arrivals may establish a toehold and insinuate (find a new niche to occupy) or cause an established species to become extinct.

Islands of the South Pacific [1]

The origins of island flora

The southern Pacific is a vast area of open ocean dotted with tiny islands. These are all volcanic islands, formed by geological disturbances in the complex of plates that make up the Pacific seafloor. Volcanic islands are relatively short-lived. They erupt, building up enough lava to reach above the surface of the sea, and are then rapidly eroded by the action of waves and water. The whole process typically takes less than a few million years – a very short time in geological terms. But in that time volcanic islands can build up a complex ecosystem of plants and animals (*see pages 160–161*). The origin of the flora and fauna of these remote specks of land tells us a great deal about how plants and animals move around, the circumstances under which they produce new species, and the influence of size of territory, ocean currents and winds on the distribution of flora and fauna – not just on islands but in all environments.

We know that volcanic islands are too young for land plants to have evolved there independently. In any case, the plant species that are found on islands are often the same as those seen on mainland sites, or are closely related to them. The plants must therefore have been transported to the islands, presumably as seeds or spores from another land mass.

On the islands of the southwest Pacific, the main influences on the content of the flora are the land masses lying to the west. Because of their own geological history as separated continents, Australia and Asia have entirely different floras. This makes it relatively easy for us to trace the origin of the different plant genera that are found on the outlying islands. Overall, Asian or 'Old World' plants account for a higher percentage of the island flora, but both Australian and Asian floras steadily lose their influence as distance from the islands increases. The flora of the furthest islands contains less than 1.5 per cent Australian genera and less than 15 per cent Asian genera. On these easterly archipelagoes, the influence of American flora is a great deal stronger. Endemic genera (that is, genera that are unique to that particular place) and endemic species within an existing genera also evolve once a population has been established.

Plants are brought to islands in a variety of ways, by birds or by flotation on air and ocean currents (*see pages 154–155*). Ocean currents and wind direction do have an influence on all of these methods, but it seems that the greatest influence is distance from the source.

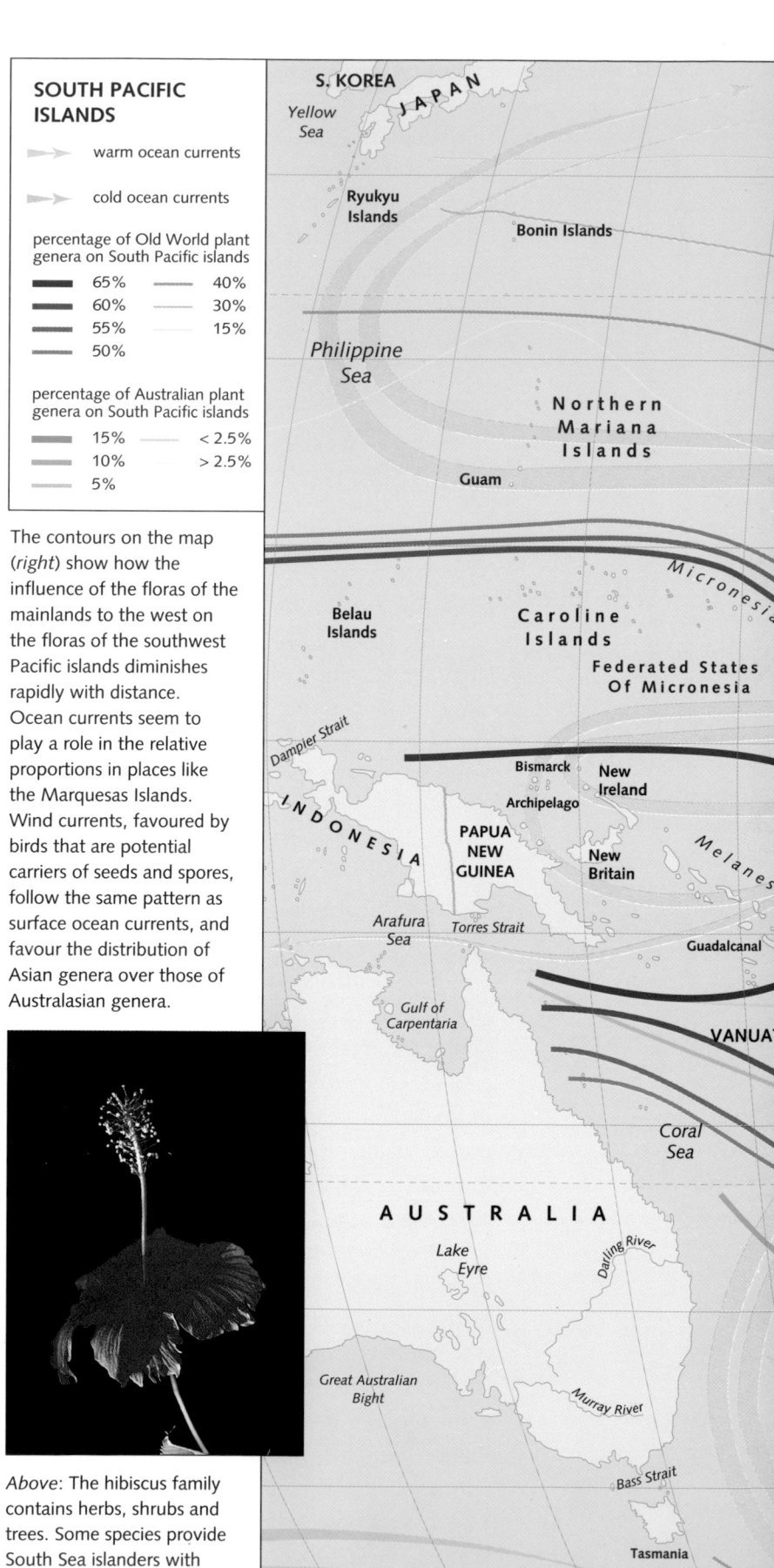

SOUTH PACIFIC ISLANDS

→ warm ocean currents

→ cold ocean currents

percentage of Old World plant genera on South Pacific islands

65%	40%
60%	30%
55%	15%
50%	

percentage of Australian plant genera on South Pacific islands

15%	< 2.5%
10%	> 2.5%
5%	

The contours on the map (*right*) show how the influence of the floras of the mainlands to the west on the floras of the southwest Pacific islands diminishes rapidly with distance. Ocean currents seem to play a role in the relative proportions in places like the Marquesas Islands. Wind currents, favoured by birds that are potential carriers of seeds and spores, follow the same pattern as surface ocean currents, and favour the distribution of Asian genera over those of Australasian genera.

Above: The hibiscus family contains herbs, shrubs and trees. Some species provide South Sea islanders with timber and fibrous bark.

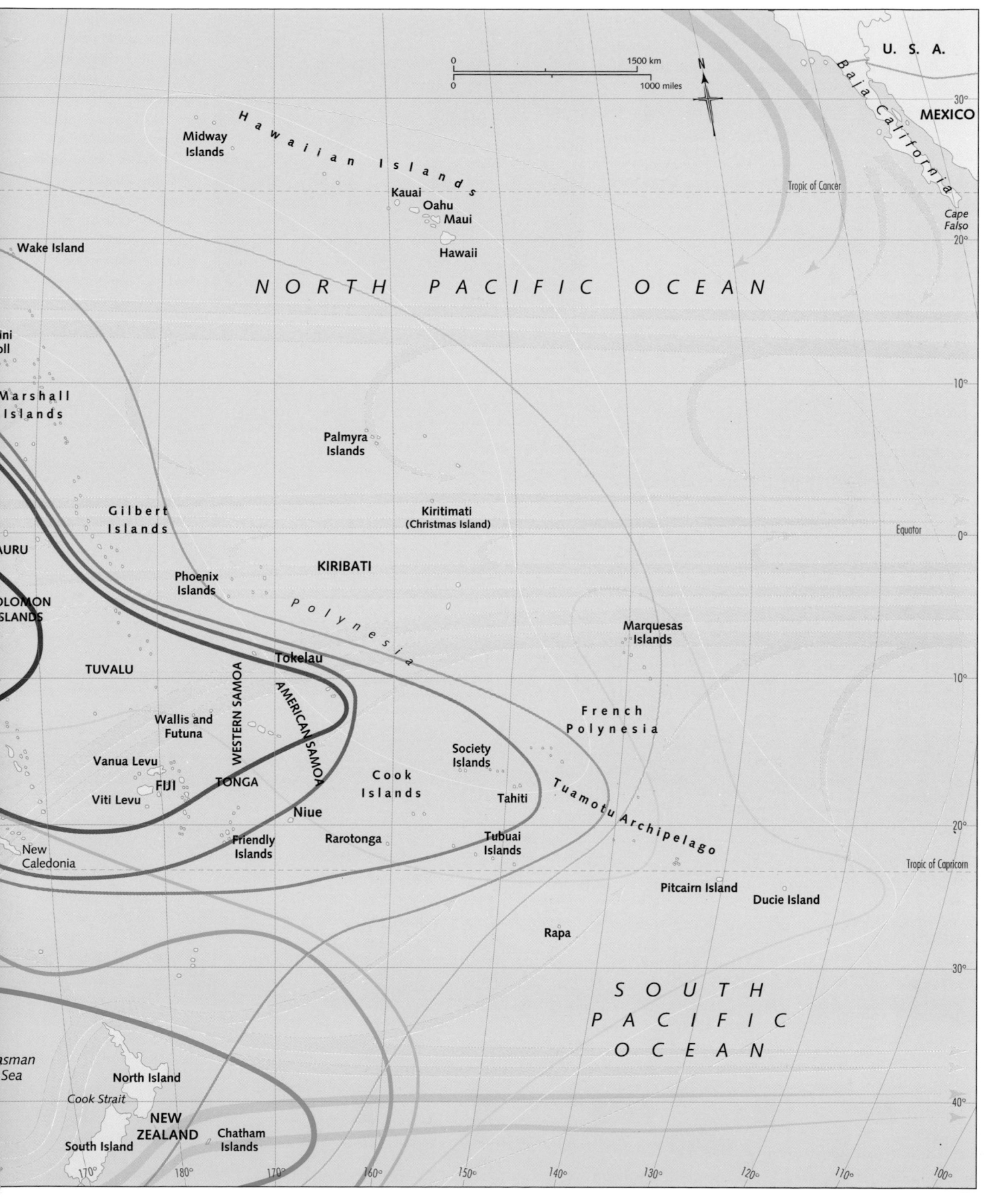

0   1500 km
0   1000 miles

N

U. S. A.

Baja California

MEXICO

30°

Tropic of Cancer

Cape Falso

20°

Hawaiian Islands

Midway
Islands

Kauai
Oahu
Maui

Hawaii

Wake Island

N O R T H   P A C I F I C   O C E A N

10°

ini
oll

**Marshall
Islands**

Palmyra
Islands

10°

**Gilbert
Islands**

Kiritimati
(Christmas Island)

Equator   0°

AURU

KIRIBATI

OLOMON
SLANDS

Phoenix
Islands

Polynesia

Marquesas
Islands

TUVALU

Tokelau

10°

Wallis and
Futuna

WESTERN SAMOA

AMERICAN SAMOA

French
Polynesia

Vanua Levu

Society
Islands

Cook
Islands

FIJI

TONGA

Viti Levu

Tahiti

Tuamotu Archipelago

Niue

New
Caledonia

Friendly
Islands

Rarotonga

Tubuai
Islands

20°

Tropic of Capricorn

Pitcairn Island

Ducie Island

Rapa

30°

S O U T H

P A C I F I C

O C E A N

asman
Sea

North Island

Cook Strait

**NEW
ZEALAND**

Chatham
Islands

South Island

40°

170°   180°   170°   160°   150°   140°   130°   120°   110°   100°

Islands of the South Pacific [2]

Dispersal and endemic species

Animals, plants and birds make their way to oceanic islands by a variety of methods. Seeds or breeding pairs arrive and, with luck, establish a self-generating population. Species can then diversify to occupy different ecological roles from those that they occupied on the mainland. Whichever species gets there first has a distinct advantage, for by diversifying into other roles it effectively excludes would-be competitors.

The development of new and endemic species depends on a number of factors. Two of the most important are the island's size and its distance from the source of originators and potential competitors.

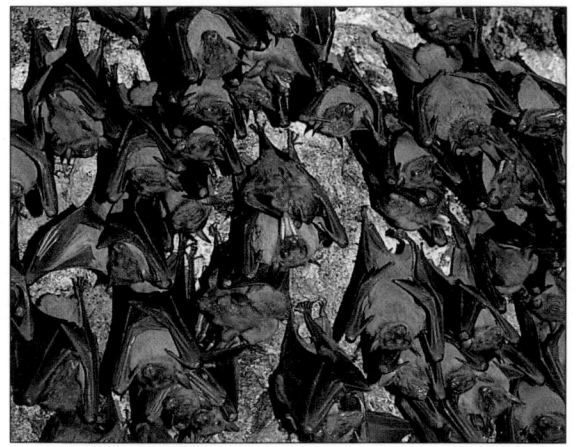

Left: Fruit bats are the only naturally occurring mammals on many small South Pacific islands. *Below*: The number of species of pisonia trees, fruit bats and amphibians on South Pacific islands declines with the distance from the source (the mainlands of Asia and Australia). The total number of endemic genera of conifer and flowering plants follows the same pattern.

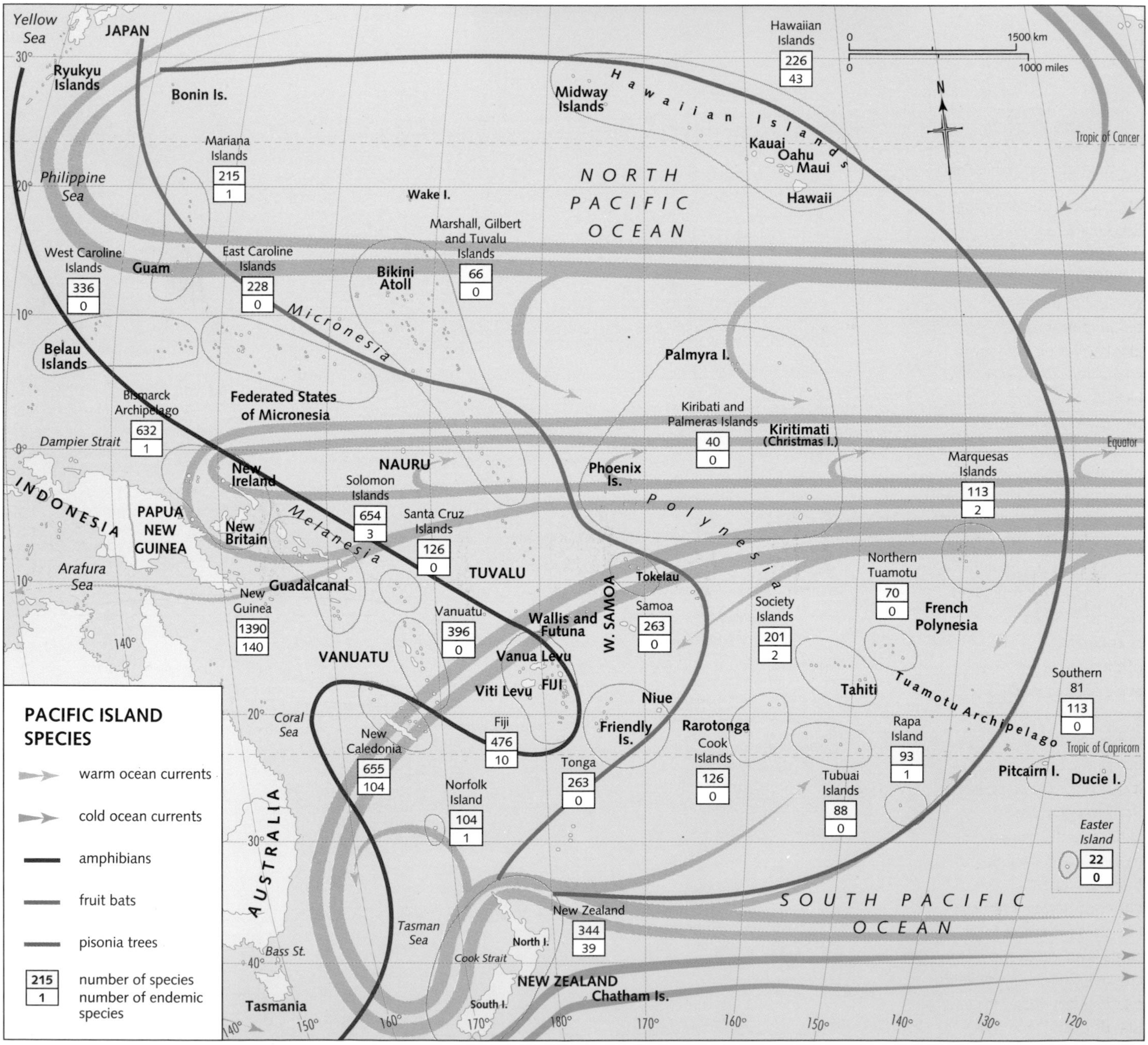

PACIFIC ISLAND SPECIES
warm ocean currents
cold ocean currents
amphibians
fruit bats
pisonia trees
215 / 1 number of species / number of endemic species
Yellow Sea
JAPAN
Ryukyu Islands
Bonin Is.
Philippine Sea
Mariana Islands 215 / 1
Wake I.
Hawaiian Islands 226 / 43
Midway Islands
Hawaiian Islands
Kauai
Oahu
Maui
Hawaii
NORTH PACIFIC OCEAN
0 1500 km
0 1000 miles
N
Tropic of Cancer
West Caroline Islands 336 / 0
Guam
East Caroline Islands 228 / 0
Marshall, Gilbert and Tuvalu Islands 66 / 0
Bikini Atoll
Micronesia
Belau Islands
Palmyra I.
Bismarck Archipelago 632 / 1
Federated States of Micronesia
Kiribati and Palmeras Islands 40 / 0
Kiritimati (Christmas I.)
Dampier Strait
Equator
INDONESIA
NAURU
Phoenix Is.
Marquesas Islands 113 / 2
New Ireland
Solomon Islands 654 / 3
Polynesia
PAPUA NEW GUINEA
New Britain
Melanesia
Santa Cruz Islands 126 / 0
Arafura Sea
TUVALU
Northern Tuamotu 70 / 0
Guadalcanal
Tokelau
New Guinea 1390 / 140
Vanuatu 396 / 0
Wallis and Futuna
W. SAMOA
Samoa 263 / 0
Society Islands 201 / 2
French Polynesia
VANUATU
Vanua Levu
Viti Levu
FIJI
Niue
Tahiti
Tuamotu Archipelago
Southern 81 113 / 0
Coral Sea
New Caledonia 655 / 104
Fiji 476 / 10
Friendly Is.
Rarotonga
Cook Islands 126 / 0
Rapa Island 93 / 1
Tropic of Capricorn
Tonga 263 / 0
Pitcairn I.
Ducie I.
Norfolk Island 104 / 1
Tubuai Islands 88 / 0
AUSTRALIA
Easter Island 22 / 0
New Zealand 344 / 39
SOUTH PACIFIC OCEAN
Tasman Sea
North I.
Bass St.
Cook Strait
NEW ZEALAND
Chatham Is.
Tasmania
South I.

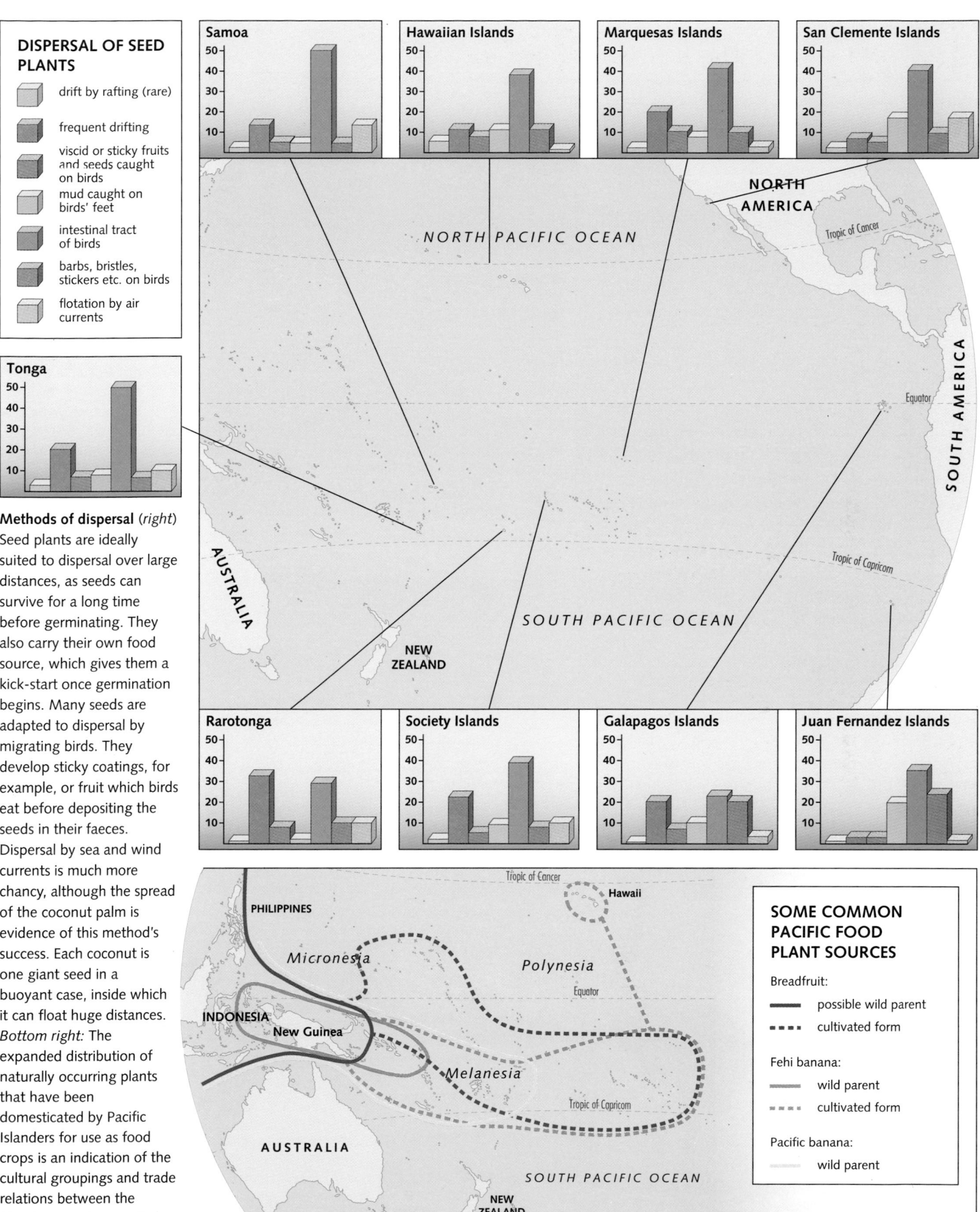

DISPERSAL OF SEED PLANTS

- drift by rafting (rare)
- frequent drifting
- viscid or sticky fruits and seeds caught on birds
- mud caught on birds' feet
- intestinal tract of birds
- barbs, bristles, stickers etc. on birds
- flotation by air currents

Samoa

Hawaiian Islands

Marquesas Islands

San Clemente Islands

Tonga

Methods of dispersal (*right*) Seed plants are ideally suited to dispersal over large distances, as seeds can survive for a long time before germinating. They also carry their own food source, which gives them a kick-start once germination begins. Many seeds are adapted to dispersal by migrating birds. They develop sticky coatings, for example, or fruit which birds eat before depositing the seeds in their faeces. Dispersal by sea and wind currents is much more chancy, although the spread of the coconut palm is evidence of this method's success. Each coconut is one giant seed in a buoyant case, inside which it can float huge distances. *Bottom right:* The expanded distribution of naturally occurring plants that have been domesticated by Pacific Islanders for use as food crops is an indication of the cultural groupings and trade relations between the islands over a long period.

Rarotonga

Society Islands

Galapagos Islands

Juan Fernandez Islands

SOME COMMON PACIFIC FOOD PLANT SOURCES

Breadfruit:

— possible wild parent

- - - - cultivated form

Fehi banana:

— wild parent

- - - - cultivated form

Pacific banana:

— wild parent

New Zealand

The migration of flora and fauna from the southern continents

The isolation of New Zealand makes it a particularly interesting place for evolutionary biologists. Geologically it is both a piece of continent and an oceanic island. Most oceanic islands are of recent origin in geological terms, but New Zealand has existed for several hundred million years, and although at some time in its history it was in proximity to the supercontinent Gondwanaland, it has been out on its own for at least 60 million years.

New Zealand lies 1,600 kilometres (1,000 miles) from the nearest land mass, Australia. The native mammal population of Australia is made up almost entirely of marsupials and monotremes, which have diversified into niches occupied by placental mammals on northern continents (*see pages 96–97*). In New Zealand the only indigenous mammals are bats. This is presumably because New Zealand was separate from Australia when marsupial mammals arrived there, and only the bats could fly across. In the absence of mammals, flightless birds have been able to thrive. These probably arrived from Australia rather than evolving from flying forebears that had colonized the islands. Some have grown to large sizes and have adapted to fill the roles taken by mammals on other continents and islands. The tallest species of moa bird (*Dinornis*), which was hunted to extinction in the late 18th century, reached a height of over 3 metres (10 feet).

The changing geography of the southern continents has brought the flora and fauna of different land masses into the New Zealand sphere at different times. This is reflected in the fossil record of the islands for the past 150 million years (*right*). We know more about marine animals because they are most easily preserved as fossils, but land plants and animals are present in sufficient numbers to show the pattern of the migration of life to New Zealand over the past 150 million years.

Moas

There were about 20 species of these extinct kiwi-like birds in New Zealand, all flightless. Some species were very large, and all were endowed with a long neck and powerful limbs. The moas were all hunted to extinction by early Maoris when they first arrived on the islands. The last bird was killed in the 1800s.

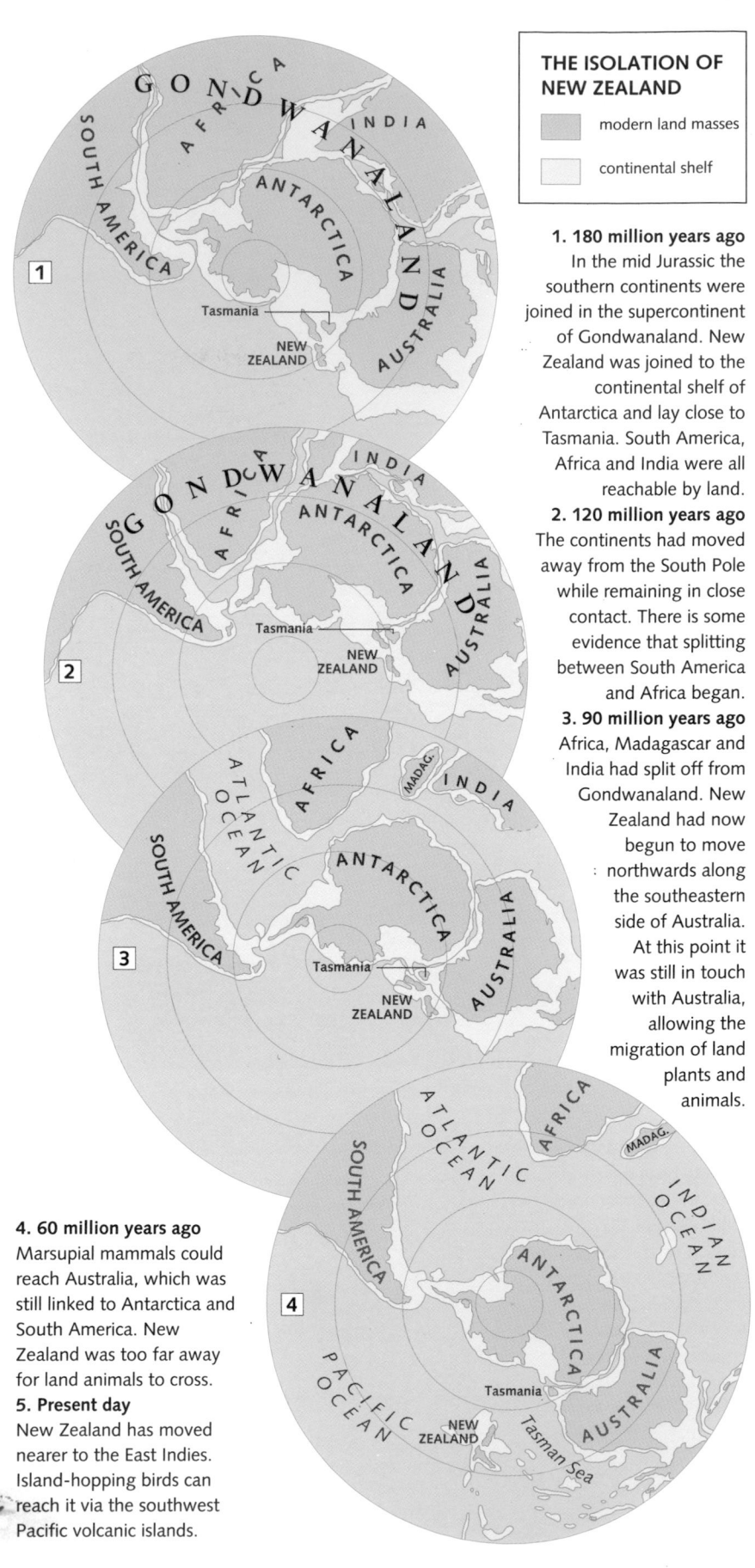

THE ISOLATION OF NEW ZEALAND

modern land masses

continental shelf

1. 180 million years ago
In the mid Jurassic the southern continents were joined in the supercontinent of Gondwanaland. New Zealand was joined to the continental shelf of Antarctica and lay close to Tasmania. South America, Africa and India were all reachable by land.

2. 120 million years ago
The continents had moved away from the South Pole while remaining in close contact. There is some evidence that splitting between South America and Africa began.

3. 90 million years ago
Africa, Madagascar and India had split off from Gondwanaland. New Zealand had now begun to move northwards along the southeastern side of Australia. At this point it was still in touch with Australia, allowing the migration of land plants and animals.

4. 60 million years ago
Marsupial mammals could reach Australia, which was still linked to Antarctica and South America. New Zealand was too far away for land animals to cross.

5. Present day
New Zealand has moved nearer to the East Indies. Island-hopping birds can reach it via the southwest Pacific volcanic islands.

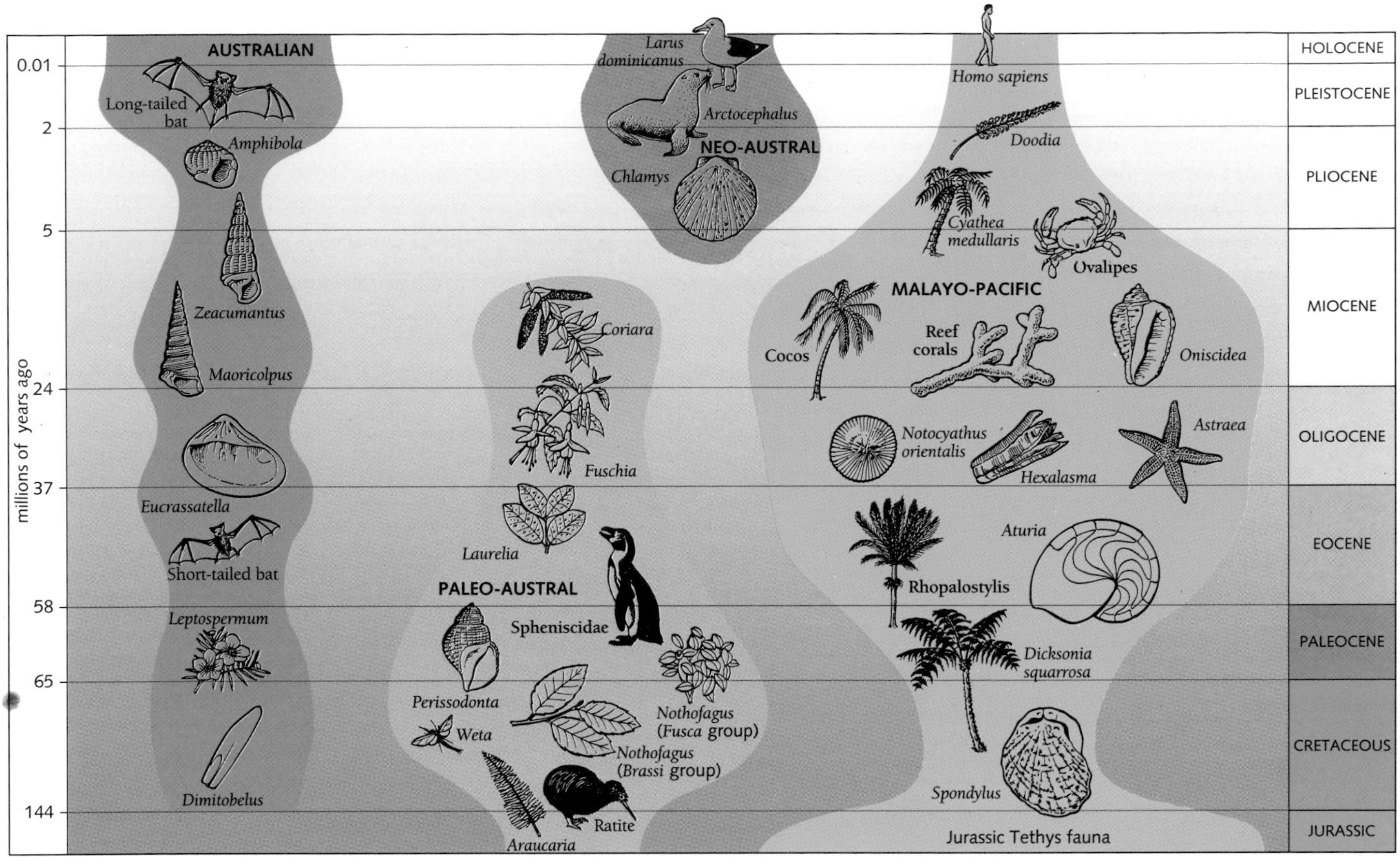

AUSTRALIAN
Long-tailed bat
Amphibola
Zeacumantus
Maoricolpus
Eucrassatella
Short-tailed bat
Leptospermum
Dimitobelus
Larus dominicanus
Arctocephalus
NEO-AUSTRAL
Chlamys
Coriara
Fuschia
Laurelia
PALEO-AUSTRAL
Spheniscidae
Perissodonta
Weta
Nothofagus (Fusca group)
Nothofagus (Brassi group)
Ratite
Araucaria
Homo sapiens
Doodia
Cyathea medullaris
Ovalipes
MALAYO-PACIFIC
Cocos
Reef corals
Oniscidea
Notocyathus orientalis
Hexalasma
Astraea
Aturia
Rhopalostylis
Dicksonia squarrosa
Spondylus
Jurassic Tethys fauna
millions of years ago
0.01
2
5
24
37
58
65
144
HOLOCENE
PLEISTOCENE
PLIOCENE
MIOCENE
OLIGOCENE
EOCENE
PALEOCENE
CRETACEOUS
JURASSIC

Above: The timing of the migrations of plants and animals into new territories has been, and still is, influenced by continental movement. Nowhere is this better shown than in New Zealand, which is home to a wide variety of flora and fauna from all over the southern hemisphere.

New Zealand's fossil record

Flora and fauna from the Australian mainland have been a consistent element in the New Zealand fossil record since the Cretaceous period. Paleo-Austral elements originated in South America, Antarctica and the seas around the southern supercontinent of Gondwanaland. These were a dominant influence around the end of the Cretaceous period 65 million years ago, but came to an end as the continents moved further apart and Antarctica became covered in ice. Malayo-Pacific elements, particularly marine animals, grew as New Zealand moved northwards. Recently, plants originating in Southeast Asia have made their way to the islands. Humans arrived from the Pacific by boat. More recently still, changing ocean currents have carried marine animals and birds from South America and the southern ocean (the Neo-Austral group) to New Zealand's shores.

AFRICA
ATLANTIC OCEAN
INDIAN OCEAN
SOUTHERN OCEAN
SOUTH AMERICA
5
ANTARCTICA
AUSTRALIA
Tasmania
Tasman Sea
NEW ZEALAND
PACIFIC OCEAN

Right: The pattern of ocean currents in the South Seas has altered since the Miocene, bringing more marine fauna and flora from Antarctica and South America.

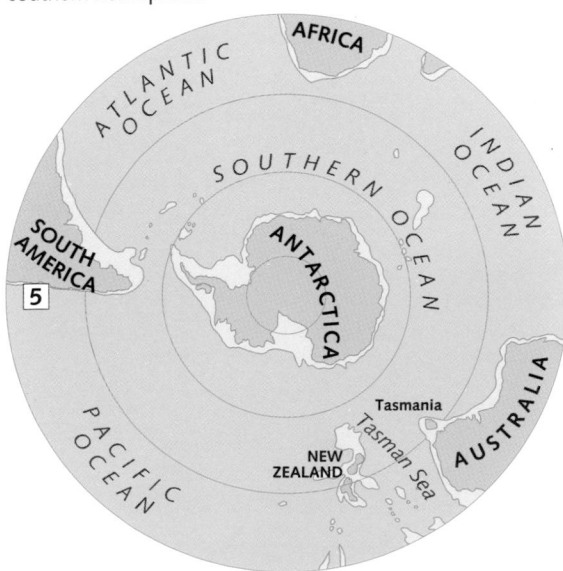

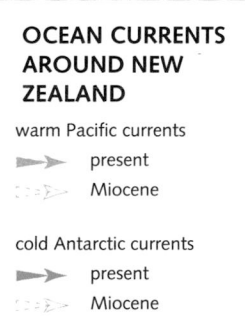

OCEAN CURRENTS AROUND NEW ZEALAND
warm Pacific currents
present
Miocene
cold Antarctic currents
present
Miocene
AUSTRALIA
PACIFIC OCEAN
Westward Drift
NEW ZEALAND
SOUTHERN OCEAN
ANTARCTICA

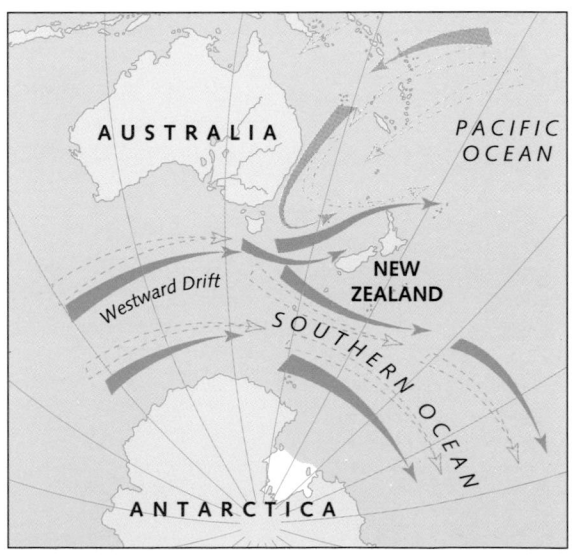

Island Birds

Hummingbirds of the West Indies and white-eyes of the East Indies

Birds that inhabit chains of islands are of special interest in the study of evolution. Individual birds can move from one island to another, but the colonization of an island is likely to be fiercely resisted by any species that occupies the same ecological ground. The range of birds occupying a small island is therefore bounded much more exactly than in mainland areas, where species territories often grade into one another. Species often inhabit more than one island, but it is rare to find a species occupying half an island and sharing territory with another species that has similar ecological requirements.

The hummingbirds of the West Indies are closely related, yet they belong to ten different genera. Unlike the finches of the Galapagos (*see pages 18–19*), they have not evolved from a common ancestor on the islands but seem instead to have descended from separate forebears, who colonized the islands independently of one another.

Where more than one species appears on an island, these species are always of different genera. It seems possible for hummingbirds of different families to live together, but not those of the same family. The explanation for this is that the latter occupy the same habitats and therefore fend off would-be colonizers from their own family. A generic ancestor may have colonized several islands, which in turn evolved separate species that no longer have any contact with one another. The only exception to this rule of one genera per island is the two species of *Anthracothorax* on Puerto Rico, one of which lives in the uplands and the other in the low-lying part.

There is also only one large and one small hummingbird per island in the West Indies, apart from on Jamaica, Hispaniola and Puerto Rico, which each have two species of large hummingbird. The greater variation in the habitats on these larger islands allows similar species to occupy slightly different ecological niches. Apart from on Puerto Rico, the two species are not from the same genera, so allopatric speciation (the evolution of one species from another by geographical separation) has not taken place. The species must have colonized the islands separately.

The Zosteropidae family or white-eyes has successfully colonized the islands of the Malay archipelago and the southwestern Pacific. Species distribution on each island depends on its size, but there are marked exceptions – seven species on the small island of Flores, for example. White-eyes have adapted to differing environments within each island and throughout the region, allowing for a great diversity of species. This remarkable family has also evolved different sized species that often share the same habitat. ●

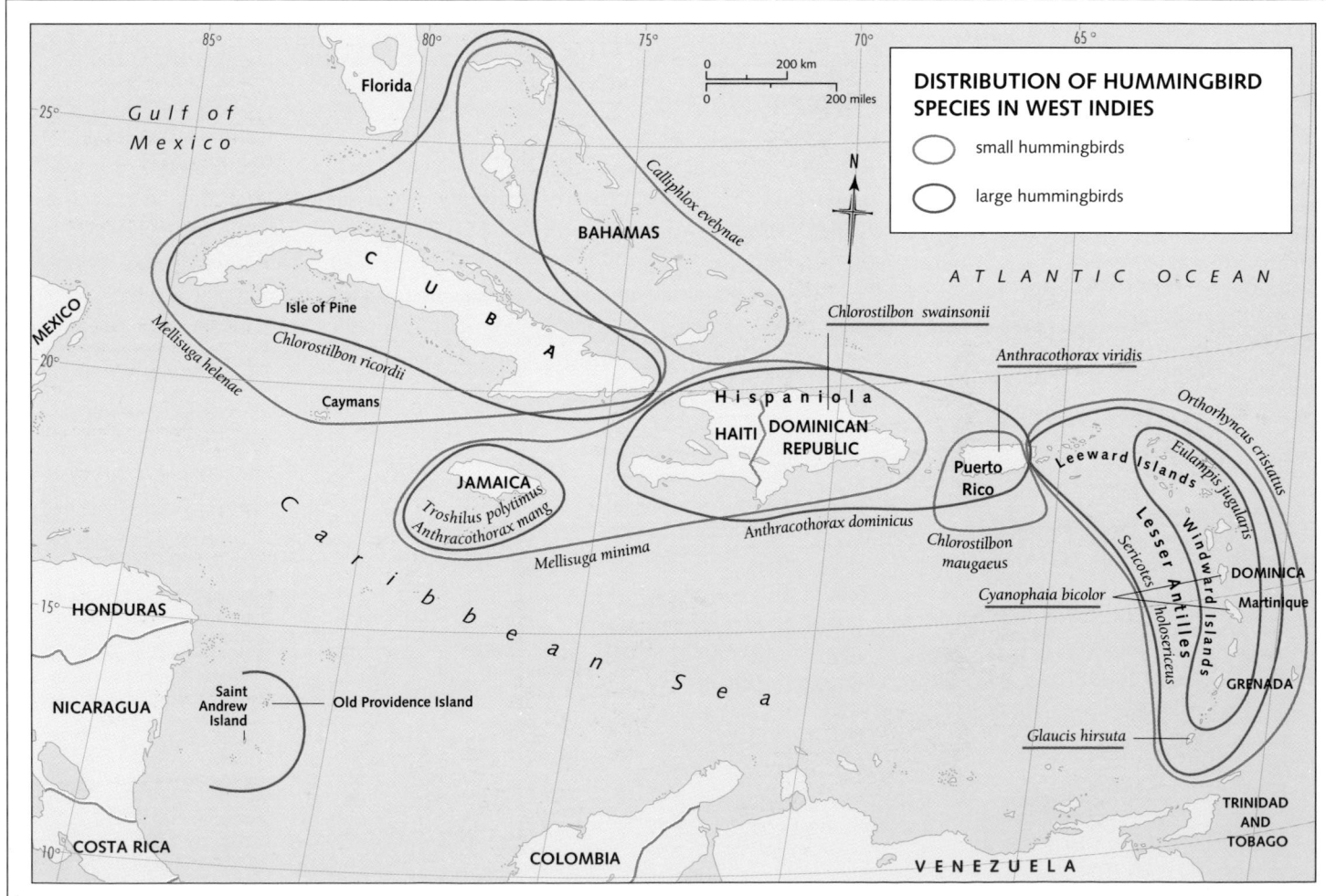

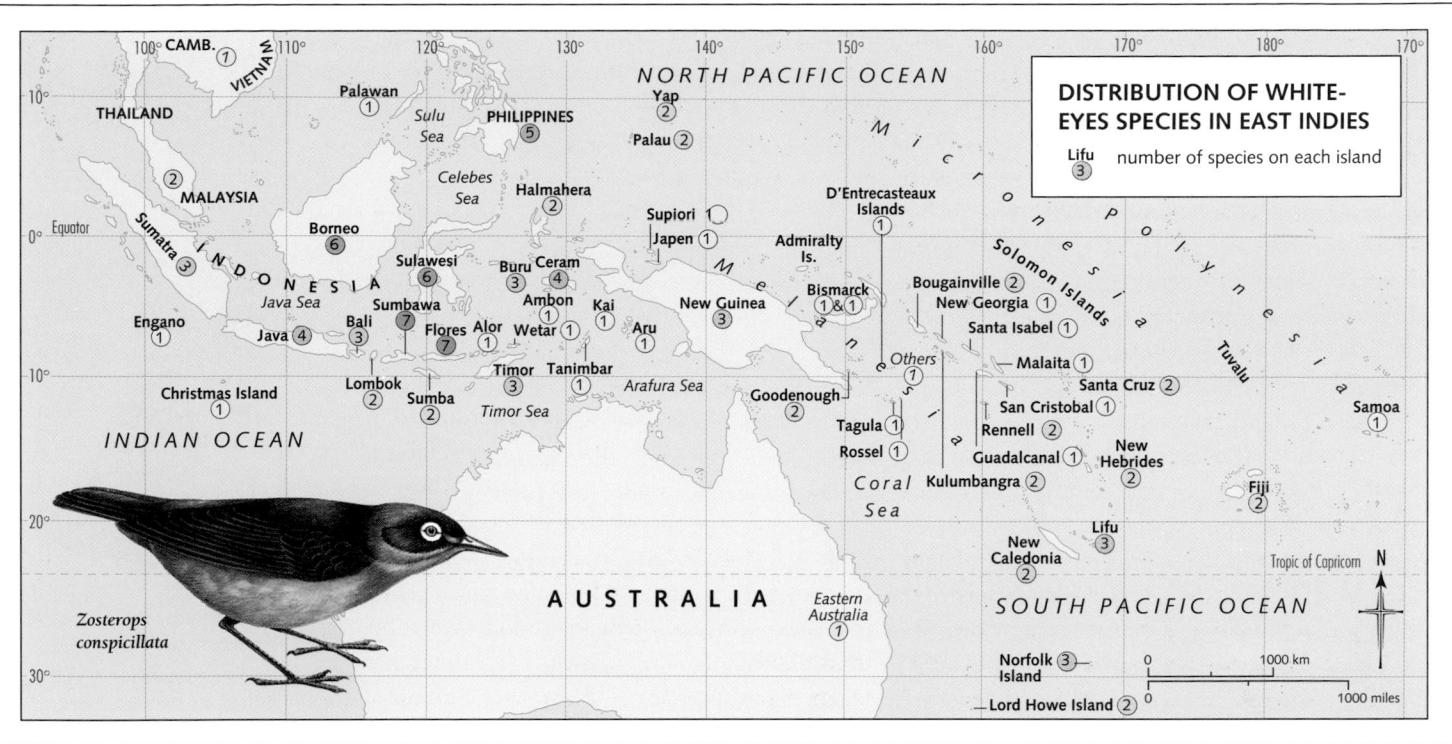

DISTRIBUTION OF WHITE-EYES SPECIES IN EAST INDIES

Lifu ③ number of species on each island

Zosterops conspicillata

Calliplox evelynae

Trochilus polytimus

Mellisuga helenae

Orthorhyncus cristatus

Chlorostilbon ricordii

Sericotes holosericeus

Anthracothorax mango

Cyanophaia bicolor

Glaucis hirsuta

Eulampis jugularis

Bahamas
Large: *Chlorostilbon ricordii*
Small: *Calliplox evelynae*

Cuba
Large: *Chlorostilbon ricordii*
Small: *Mellisuga helenae*

Jamaica
Large: *Anthracothorax mango;
Trochilus polytimus*
Small: *Mellisuga minima*

Hispaniola
Large: *Chlorostilbon swainsonii;
Anthracothorax dominicus*
Small: *Mellisuga minima*

Puerto Rico
Large: *Anthracothorax dominicus;
Anthracothorax viridis*
Small: *Chlorostilbon maugaeus*

Virgin Islands
Large: *Sericotes holosericeus*
Small: *Orthorhyncus cristatus*

Lesser Antilles (except Grenada)
Large: *Sericotes holosericeus;
Eulampis jugularis*
Small: *Orthorhyncus cristatus*

Also on Dominica & Martinique
Cyanophaia bicolor (large)

Grenada
Large: *Sericotes holosericeus; Glaucis hirsuta*
Small: *Orthorhyncus cristatus*

Saint Andrew and Old Providence
Anthracothorax provostii (large)

Coral Islands
From volcanoes to ecosystems

The Earth's crust seems solid and unchanging, but we know from earthquakes and volcanic eruptions that it is a mobile, dynamic entity with a history of movement and change. The plates that make up the crust are moving relative to one another, producing friction and disturbance along their boundaries. The plates that make up the seafloors are thin and young in comparison to the continents. Where two oceanic plates are in collision, seafloor eruptions occur. A mountain of lava can result, and if this is large enough to reach the surface of the sea, a volcanic island is formed. These islands are grouped around the active plate boundaries of the Earth's crust, or over 'hot spots' where volcanic activity occurs above a hot region in the layers below the crust.

A volcanic island begins life as hot lava. Erosion by wind, rain and waves begins immediately, forming a ring of debris around the island. Over time this is pounded into sand by waves and tides. Meanwhile, marine plants are washed into the intertidal zone, attracting sea birds. These bring seeds from land plants, some of which will germinate in the sandy conditions. These pioneer plants help to form a layer of soil which enables other plants to become established. An increase in plants will bring more birds bearing seeds. Insects also arrive via birds or driftwood. The vegetation of the island establishes zones (*right*), with beach grasses and palms on the outer fringes and dense forest in the higher interior.

Wave action will make the original cone shape of the volcano into a stepped form, with a shallow shelf around the edge of the island. If conditions are right, reef-forming corals will begin to grow on the outer edge of the shelf. These corals are colonial animals which combine by fusing their outer walls. New corals build on the remains of previous generations to form massive structures. The reefs are wonderfully diverse environments, providing habitats for a huge variety of shellfish, plants and fish. The steep outer edge of the reef is particularly rich in nutrients, as water pushes up from the deeper parts of the ocean, which are a favourite site for plankton. Between the reefs and the islands a shallow lagoon is formed, which develops its own independent ecosystem.

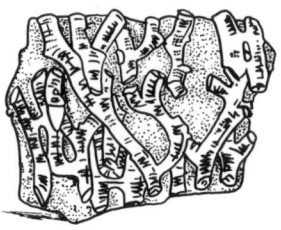

Corals
Corals can appear individually or in tubular (*top*) or colony form. Coral polyps (*bottom*) are encased in hard skeletons consisting of lime extracted from the seawater and have tentacles to draw in nutrients.

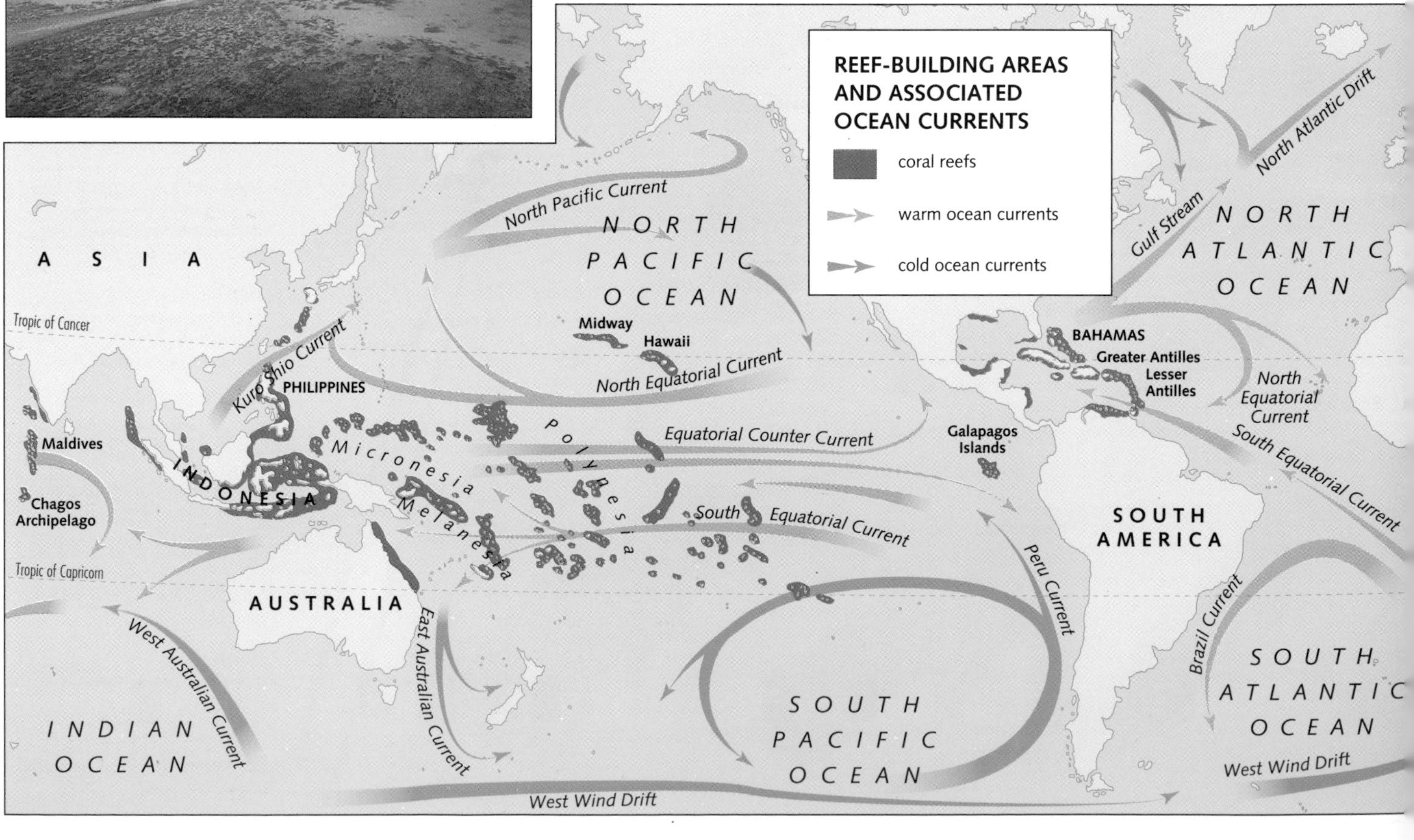

REEF-BUILDING AREAS AND ASSOCIATED OCEAN CURRENTS

- coral reefs
- → warm ocean currents
- → cold ocean currents

ASIA

Tropic of Cancer

North Pacific Current

NORTH PACIFIC OCEAN

Midway
Hawaii

Kuro Shio Current

PHILIPPINES

North Equatorial Current

Maldives

Micronesia

Equatorial Counter Current

Chagos Archipelago

INDONESIA

Melanesia

Polynesia

South Equatorial Current

Tropic of Capricorn

AUSTRALIA

West Australian Current

East Australian Current

INDIAN OCEAN

SOUTH PACIFIC OCEAN

West Wind Drift

North Atlantic Drift

Gulf Stream

NORTH ATLANTIC OCEAN

BAHAMAS
Greater Antilles
Lesser Antilles

North Equatorial Current

Galapagos Islands

South Equatorial Current

SOUTH AMERICA

Peru Current

Brazil Current

SOUTH ATLANTIC OCEAN

West Wind Drift

ZONES OF LIFE ON A BARRIER REEF

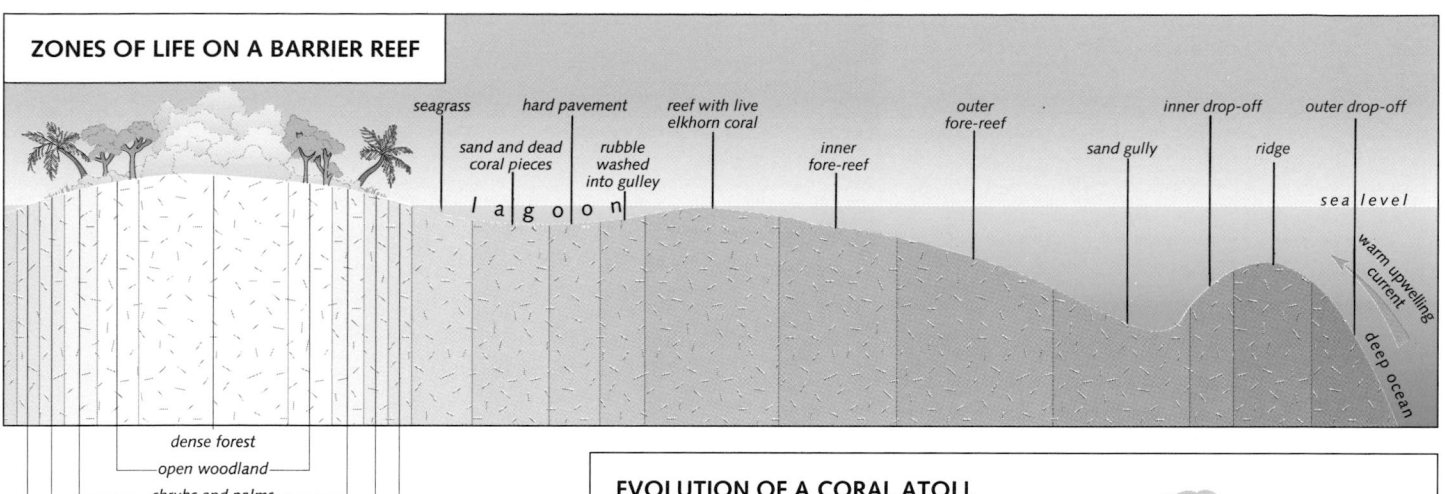

seagrass

sand and dead coral pieces

hard pavement

rubble washed into gulley

reef with live elkhorn coral

inner fore-reef

outer fore-reef

inner drop-off

outer drop-off

sand gully

ridge

sea level

warm upwelling current

deep ocean

l a g o o n

dense forest
open woodland
shrubs and palms
seaweed, grasses
beach, intertidal plants and animals

FIJI ISLANDS

🪸 coral

Yasawa Group

Vanua Levu

Taveuni

Lau Group

Viti Levu

Suva

Lakebu

Kadavu

ASIA

UROPE

ARABIA

Red Sea

FRICA

Seychelles

MADAGASCAR

Benguela Current

INDIAN OCEAN

Left: Corals need a minimum temperature of 20°C and are therefore largely limited to tropical waters. Warm ocean currents allow the Great Barrier Reef to extend south of the Tropic of Capricorn. The most northerly reefs are in the Red Sea. Cold ocean currents prevent reef growth on the west side of Africa and South America. Eroded shelves around volcanic islands are ideal foundations for coral reefs, as are the offshore shelves of some continents. The Great Barrier Reef sits on a shelf that was part of Australia, before faulting and rising sea levels resulted in flooding by warm shallow seas.

Far left: A new island begins to form in the Indian Ocean.

EVOLUTION OF A CORAL ATOLL

A volcanic island is formed by eruptions of a particular type of lava on the seafloor, which rapidly build up into the classic volcanic cone. These cones are often thousands of metres high, and only the tips protrude above sea level.

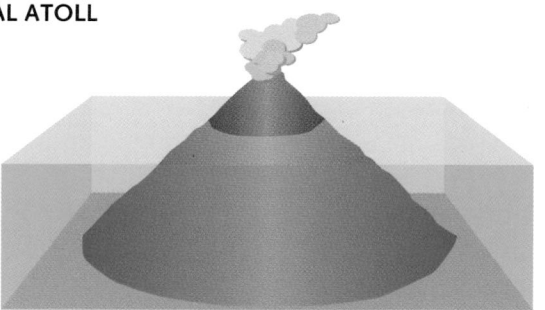

Upwelling currents around the island's margins encourage the growth of reef-forming corals. The reef follows the shape of the volcanic cone, and is therefore often roughly circular.

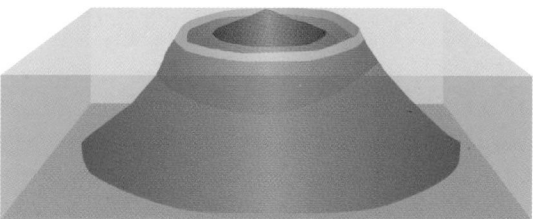

The island is eroded by wind, water and waves. The cone sinks in time, through rising sea levels or subsidence. If this is gradual, the reef will grow upwards, and layers of living coral will form on its dead portion.

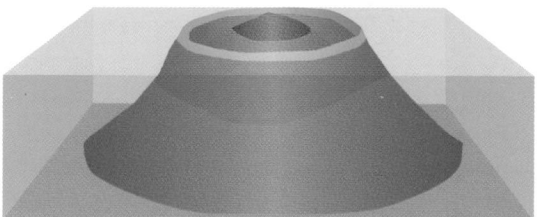

When the volcano sinks below sea level, an atoll may remain. The cone's natural subsidence will eventually overtake the coral's ability to keep building, though there are reefs of great depths in parts of the tropics and in the fossil record.

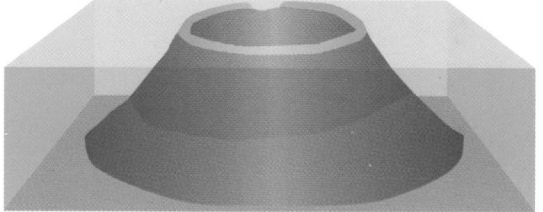

Volcanic islands, coral atolls and evolution
On his voyage aboard the HMS *Beagle* in the 1830s, Darwin noticed that volcanic islands display a variety of forms but that these fit into a consistent pattern. Each of the forms is a stage in the evolution of an island, which eventually leads to the creation of an atoll.

Birds of New Guinea
Habitats and territories

There are many different types of kingfisher on New Guinea and its surrounding islands. One of the groups, which is known as the paradise kingfishers (comprising *Tanysiptera galatea* and *Tanysiptera hydrocharis*), shows a pattern of distribution that can help us to gain an understanding of the way in which the evolution of species and subspecies works.

On the main island of New Guinea, the paradise kingfisher group is represented by three subspecies (*Tanysiptera galatea meyeri*, *Tanysiptera galatea minor* and *Tanysiptera galatea galatea*). Each of these subspecies has a wide distribution and each is virtually indistinguishable from the others in appearance. The explanation for their apparent similarity is that they are all in some contact with one another genetically, even though they occupy opposite ends of the island and opposite sides of the central dividing mountain chain. This partial isolation has caused some evolutionary differentiation, but odd individuals from the three subspecies must wander into neighbouring territories, and mating must then take place. This contact maintains a flow of genetic material or gene flow.

However, the paradise kingfishers that are found on the islands surrounding New Guinea are quite different, and most have now been recognized as distinct species – that is, they are no longer able to breed with other members of the paradise kingfisher family. In each case, the species occupy tiny areas on islands that are not far from the main island. The distance is evidently enough, however, to have prevented individuals from straying back to the mainland and encroaching upon a neighbour's territory. Gene flow between the populations therefore ceased, and the small isolated populations that were located around the edges of the main zone diverged rapidly from their parent stock. This is a classic example of geographic, or allopatric, speciation in which populations living on the periphery of the main group become isolated, diverge rapidly and go on to form a separate species (*see pages 54–59*).

One island form, *Tanysiptera hydrocharis*, evolved in isolation on the Aru Islands. These were once joined to a patch of land that is now part of the main island. *Tanysiptera hydrocharis* is still present on that patch of the main island, and it does not interbreed with the mainland subspecies in the area. This phenomenon shows that *Tanysiptera hydrocharis* has diverged sufficiently from the stock that inhabits the main island to have become a separate species.

A biogeographic relationship seen also in the southern Rockies is demonstrated by the distribution of particular bird species in New Guinea. On the island of New Guinea there is a large central mountain range, with a series of much smaller ranges along the north coast and on the northwest peninsula. A very large number of New Guinea bird species, almost 200 in all, are restricted to high altitudes. The mountainous areas therefore act as a chain of islands for these birds.

The numbers on the map indicate the number of such species that were recorded at particular locations. The relationship between the number of species and the area of the montane range is clearly shown. Unlike the Rocky Mountain mammals (*see pages 108–109*), these birds can travel from one montane zone to another with ease.

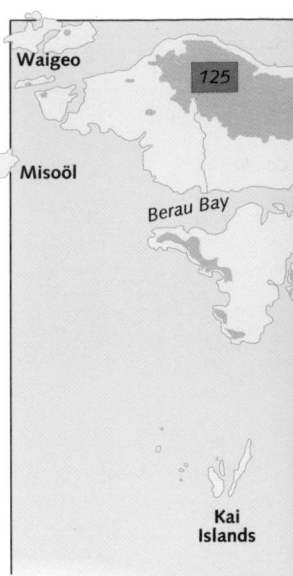

Common paradise kingfisher
Tanysiptera galatea

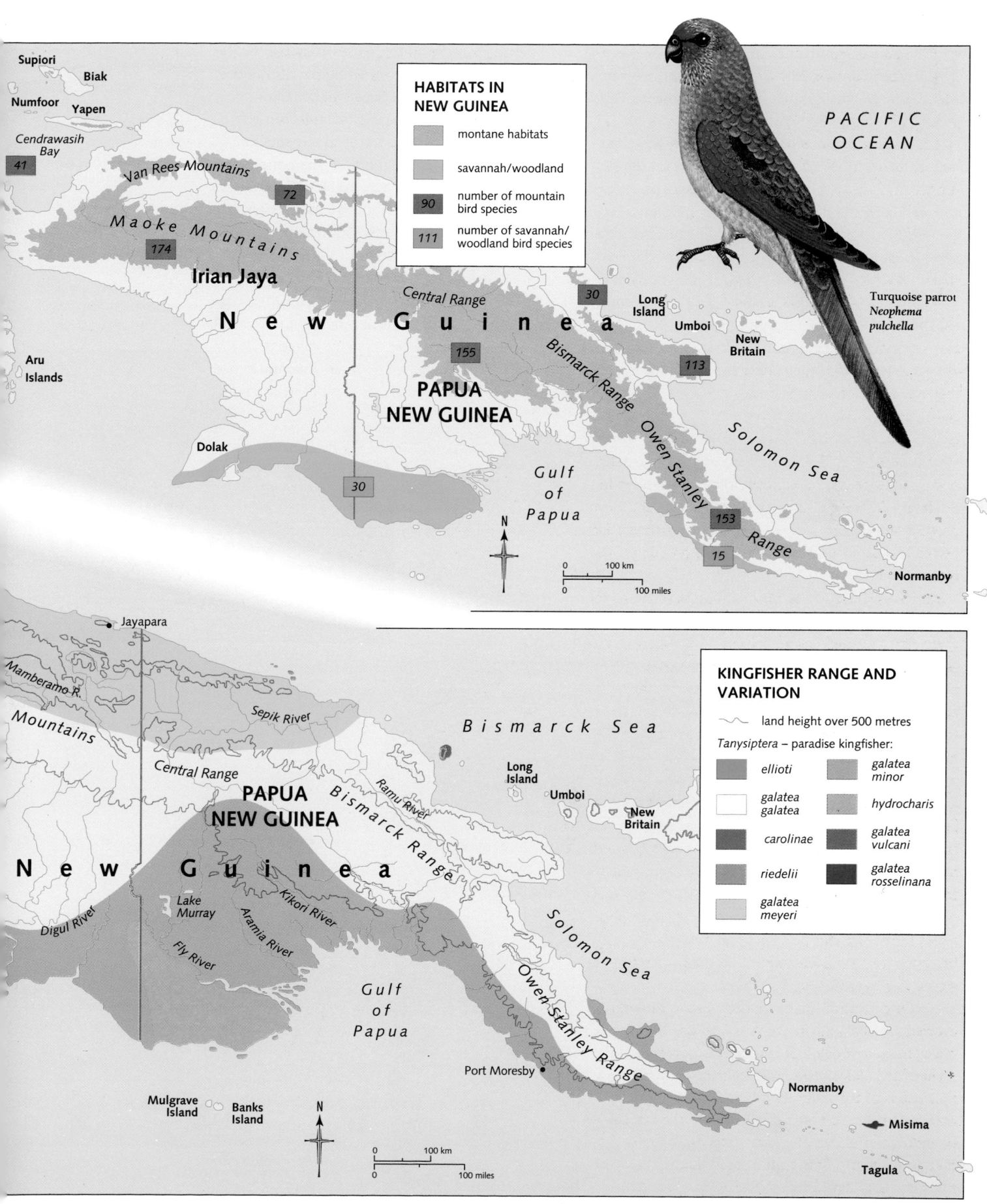

HABITATS IN NEW GUINEA

- montane habitats
- savannah/woodland
- **90** number of mountain bird species
- *111* number of savannah/woodland bird species

Cendrawasih Bay 41

Supiori

Biak

Numfoor

Yapen

Van Rees Mountains 72

Maoke Mountains 174

Irian Jaya

N e w G u i n e a

Aru Islands

Central Range

30

Long Island

Umboi

New Britain

155

PAPUA NEW GUINEA

Bismarck Range

Owen Stanley

113

Solomon Sea

Dolak

30

Gulf of Papua

153

Range

15

Normanby

PACIFIC OCEAN

Turquoise parrot
Neophema pulchella

N
0 100 km
0 100 miles

Jayapara

Mamberamo R.

Mountains

Central Range

Sepik River

PAPUA NEW GUINEA

Bismarck Range

Ramu River

Bismarck Sea

Long Island

Umboi

New Britain

N e w G u i n e a

Digul River

Lake Murray

Kikori River

Aramia River

Fly River

Gulf of Papua

Owen Stanley Range

Solomon Sea

Port Moresby

Mulgrave Island

Banks Island

Normanby

Misima

Tagula

N
0 100 km
0 100 miles

KINGFISHER RANGE AND VARIATION

- land height over 500 metres

Tanysiptera – paradise kingfisher:

- *ellioti*
- *galatea galatea*
- *carolinae*
- *riedelii*
- *galatea meyeri*
- *galatea minor*
- *hydrocharis*
- *galatea vulcani*
- *galatea rosselinana*

Canary Islands
Floral affinities

Four island groups off Africa's west coast – the Canary Islands, Madeira, the Azores and the Cape Verde Islands – contain one of the world's most interesting floras. Plant groups on these islands have strong relations or affinities with groups all over the world. What is most peculiar is that these groups often occur on the islands and in places thousands of kilometres away, but nowhere in between.

The four groups, plus a portion of the northwest African coast, form a biogeographic zone known as Macaronesia. The flora of the region has been studied in relation to plant groups elsewhere in order to gain an insight into disjunct distributions. The

The inadequacies of traditional dispersal methods in accounting for the diversity of affinity of Macaronesian flora is emphasized by the map opposite. All five plant families are found on the Canary Islands and Cape Verde, and some on the Azores. But particular groups are also found in South Africa, East Africa, East Asia, Australasia and South America. The only credible explanation for the huge distances between these distributions is a combination of continental movement and a historically widespread range that is now much reduced. In Macaronesia we see relicts of floras that once appeared in South America, Africa, Europe, India and Australia.

The bougainvillea (*below*) grows throughout the tropics and is thought to have originated in South America. It grows naturally on the Canary Islands but is also grown for ornament. Palms, also seen in this picture, are the classic tropical and subtropical plant. Outside the latitudes 40°N and 30°S the bud gets killed by frost and the plant dies. Specially bred palms are grown for ornament outside the range.

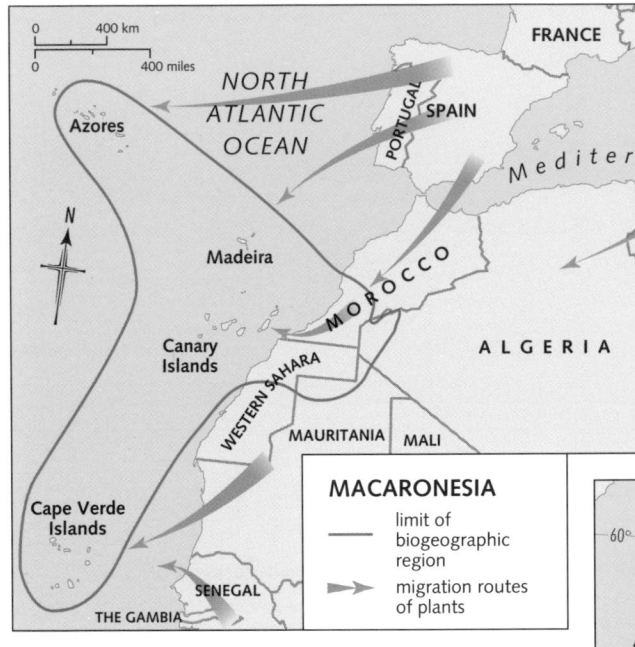

MACARONESIA

—— limit of biogeographic region

➤ migration routes of plants

answers seem to lie more in the region's history than in present biological conditions.

The endemic plants of Macaronesia have affinities to groups from at least eight regions. Groups with affinities to Mediterranean flora form the largest group – not surprisingly given the proximity of the region and similarity of climate. But how did the plants reach the islands? Possible migration routes are marked on the map above, but understanding the processes is more complicated. The map on the right shows the distributions of four families. As well as Macaronesia, they occur in biogeographic regions around the Mediterranean, the European Atlantic coastal provinces, the Sahara and East Africa. They must have reached the islands either by seed dispersal or via land bridges. Dispersal is the favoured route for species present in other areas, but for those endemic to Macaronesia the geographical history of the region is the place to look for explanations.

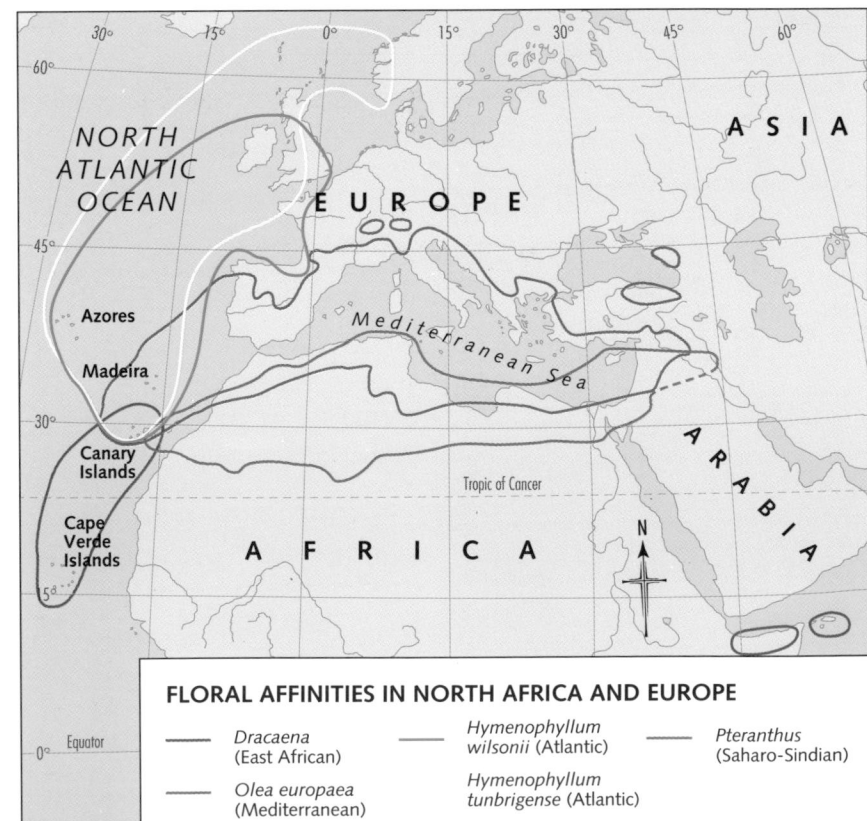

FLORAL AFFINITIES IN NORTH AFRICA AND EUROPE

—— *Dracaena* (East African)

—— *Olea europaea* (Mediterranean)

—— *Hymenophyllum wilsonii* (Atlantic)

Hymenophyllum tunbrigense (Atlantic)

—— *Pteranthus* (Saharo-Sindian)

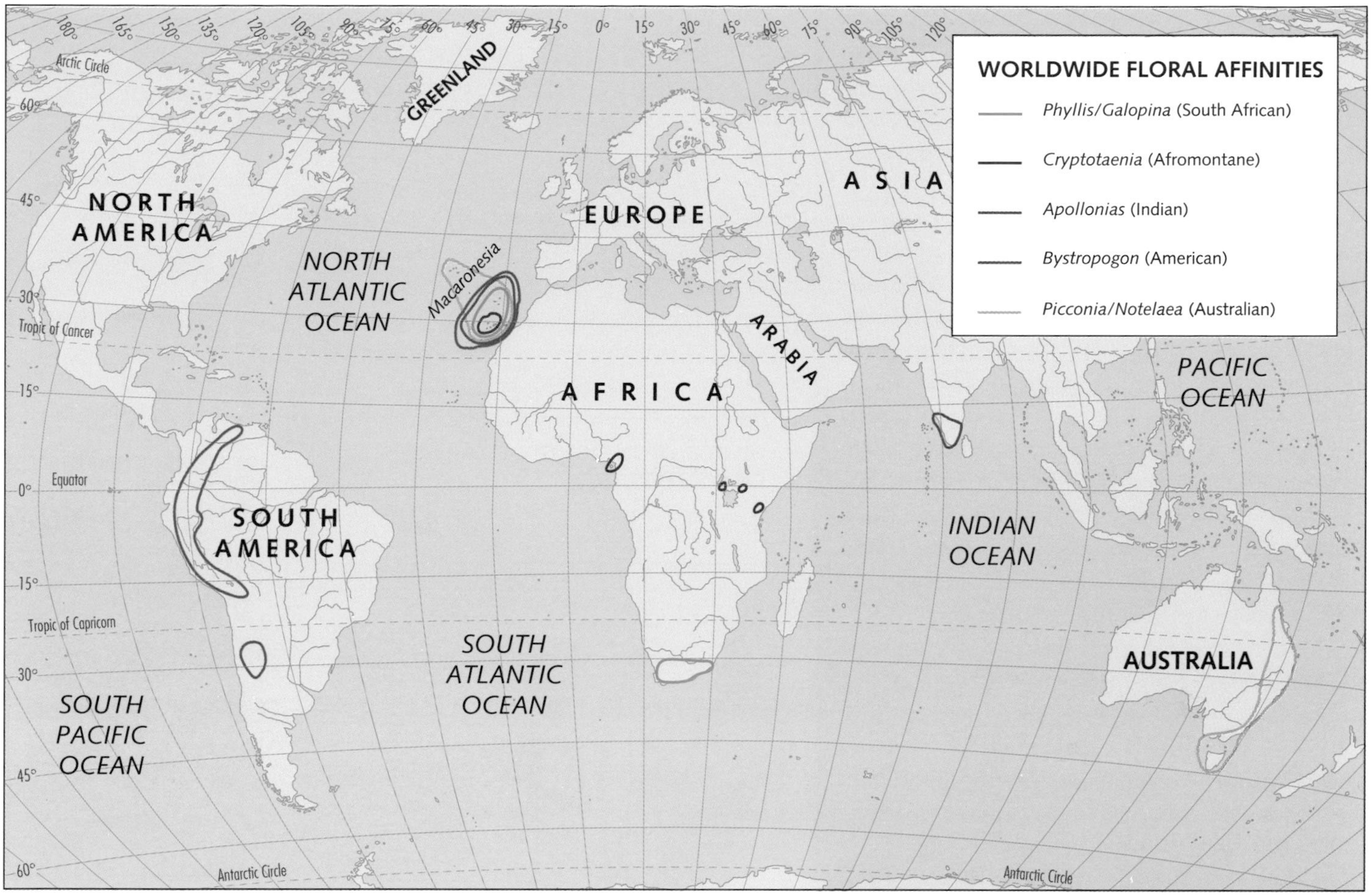

WORLDWIDE FLORAL AFFINITIES

——	*Phyllis/Galopina* (South African)
——	*Cryptotaenia* (Afromontane)
——	*Apollonias* (Indian)
——	*Bystropogon* (American)
——	*Picconia/Notelaea* (Australian)

The ancient laurel forests (*right*) are a natural treasure of the Canary Islands and Madeira. Most are now under strict protection. Climatic conditions and perhaps the geographical isolation have enabled many species that have died out elsewhere to survive.

Flowering plants first appeared in the Cretaceous. Macaronesia was on the edge of the newly forming Central Atlantic. Africa was attached to South America, Europe, North America, India, Australasia and Antarctica. The southern continents were joined as Gondwanaland. As the continents split up, populations became separate branches. Many plant distributions have since been reduced by changes in climate, leaving relicts, as seen in the map above.

As well as their affinities with plants from the Mediterranean and other parts of the world, plants on the islands of Macaronesia have complex internal relationships. One group, *Dendrosonchus,* has 17 species on Macaronesia, at least one of which occurs on each group of islands and on the mainland. The origin and evolutionary history of the group has been studied, and this has enabled possible migration routes between the islands to be mapped out (*left*). The two oldest members of the group occur in the most ancient volcanic areas of Tenerife and Gran Canaria, so it is presumed that species moved out over those islands and then migrated to the other island groups.

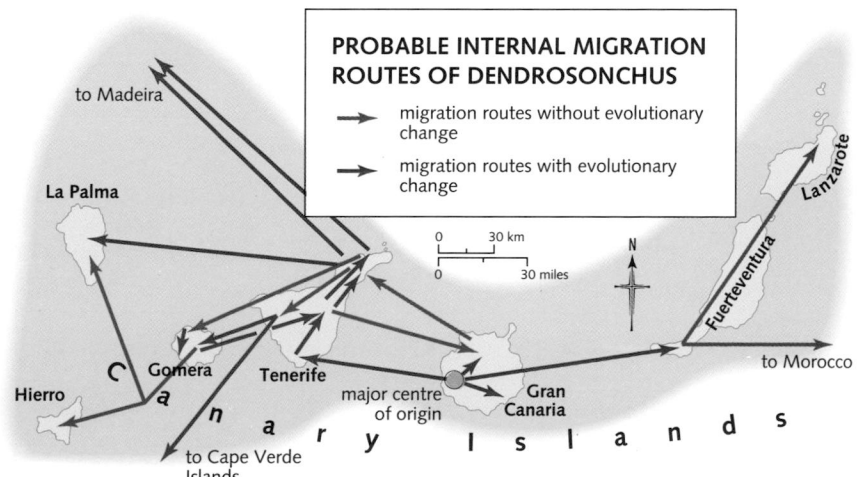

PROBABLE INTERNAL MIGRATION ROUTES OF DENDROSONCHUS

→ migration routes without evolutionary change

→ migration routes with evolutionary change

0   30 km
0   30 miles

N

to Madeira

La Palma

Gomera

Tenerife

Hierro

to Cape Verde Islands

major centre of origin

Gran Canaria

Fuerteventura

Lanzarote

to Morocco

C a n a r y   I s l a n d s

The Flora of the Aegean Islands
Land bridges in the Mediterranean

The presence of plants on any island leads us to ask how they got there. There are a number of possibilities, and it is certain that the total flora of any ecosystem, whether it be an island or part of larger land mass, is made up of plants that have arrived by different means.

When a species on an island matches a species on the nearest mainland, it is reasonable to presume that the former has been transported across to the island, probably in the form of seeds carried by birds, wind or water currents. Where the island plant shows a variation from the mainland species but is still part of the same species, then this type of dispersal might still apply, although transportation might be more difficult because of distance or other factors. Contact between the main population and the island would need to be infrequent, allowing the island population to evolve different characteristics without losing its ability to breed with its parent group.

Where an island species is related to, but separate from, those on the mainland, a different mechanism is likely to have operated. The plants must have arrived on the island by some method that is no longer available. The key to this process lies in finding out what has happened to the physical geography of the island since the plant first arrived.

The Aegean Islands, which lie in the eastern Mediterranean, are particularly interesting in this context, since at various times during the lifetime of much of their flora they have been connected to the mainland – but in different configurations. The Mediterranean is known from the salt layers on the seabed to have flooded and dried out as many as 40 times around 7–6 million years ago. As the sea level dropped, land bridges formed between groups of islands and between the islands and the mainlands of what are now Greece and Turkey. The distributions of the flora should reflect these historical land bridges.

On the opposite page, the distributions of three species from the genus *Procopiana* and Greek representatives of the *Crepis neglecta* group are shown, together with the 200 metre isobath, to demonstrate the configuration of land bridges in the Aegean region in the past. In these groups of relict flora, species distributions follow the old land masses. The map above, however, shows species of *Ptilostemon gnaphaloides*, whose distribution bears no relation to the ancient geography of the region. Another explanation is needed, demonstrating nature's ability to confound simple solutions, no matter how elegant.

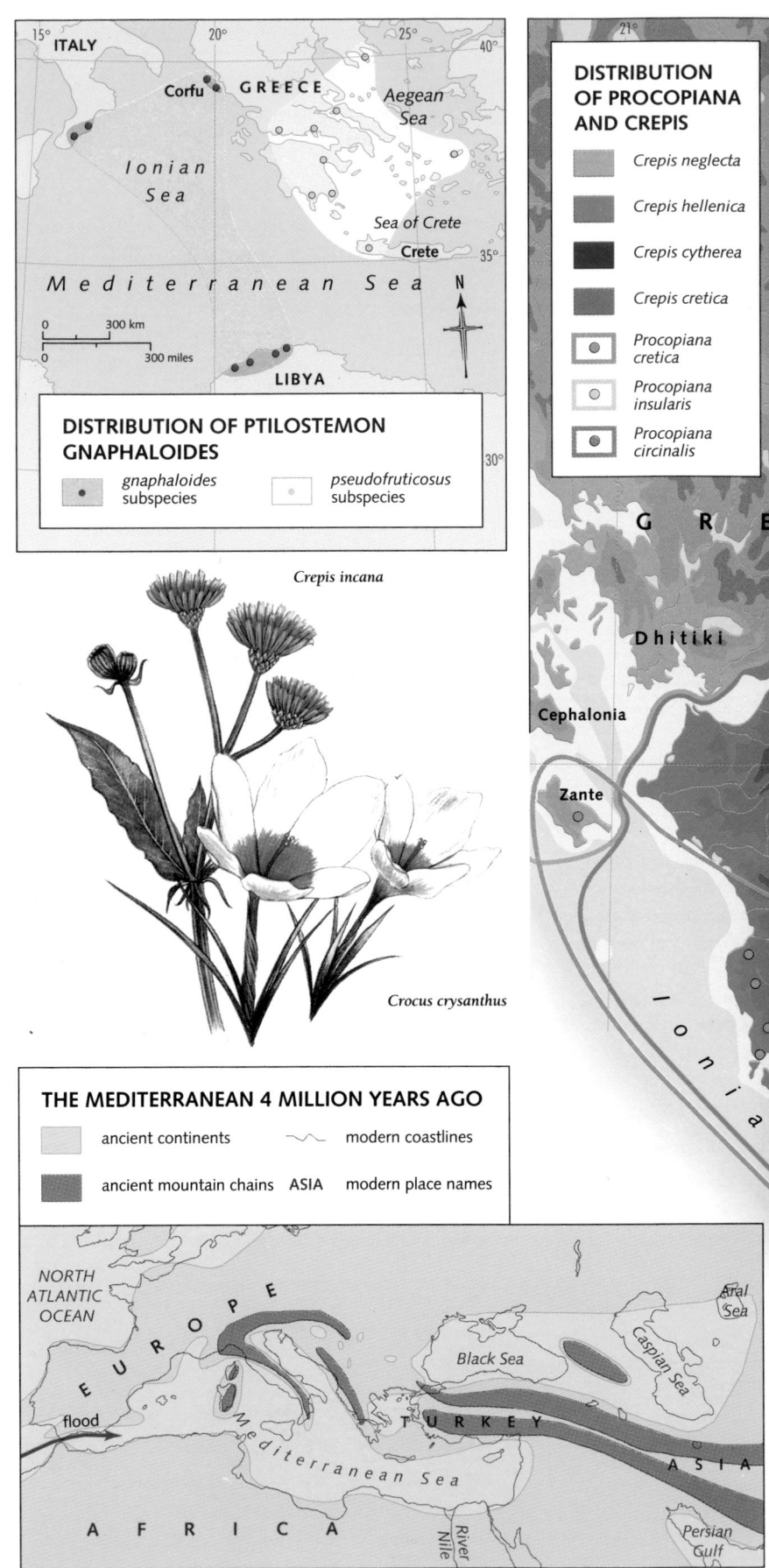

DISTRIBUTION OF PTILOSTEMON GNAPHALOIDES

- *gnaphaloides* subspecies
- *pseudofruticosus* subspecies

Crepis incana

Crocus crysanthus

DISTRIBUTION OF PROCOPIANA AND CREPIS

- *Crepis neglecta*
- *Crepis hellenica*
- *Crepis cytherea*
- *Crepis cretica*
- *Procopiana cretica*
- *Procopiana insularis*
- *Procopiana circinalis*

THE MEDITERRANEAN 4 MILLION YEARS AGO

- ancient continents
- ancient mountain chains
- modern coastlines
- ASIA modern place names

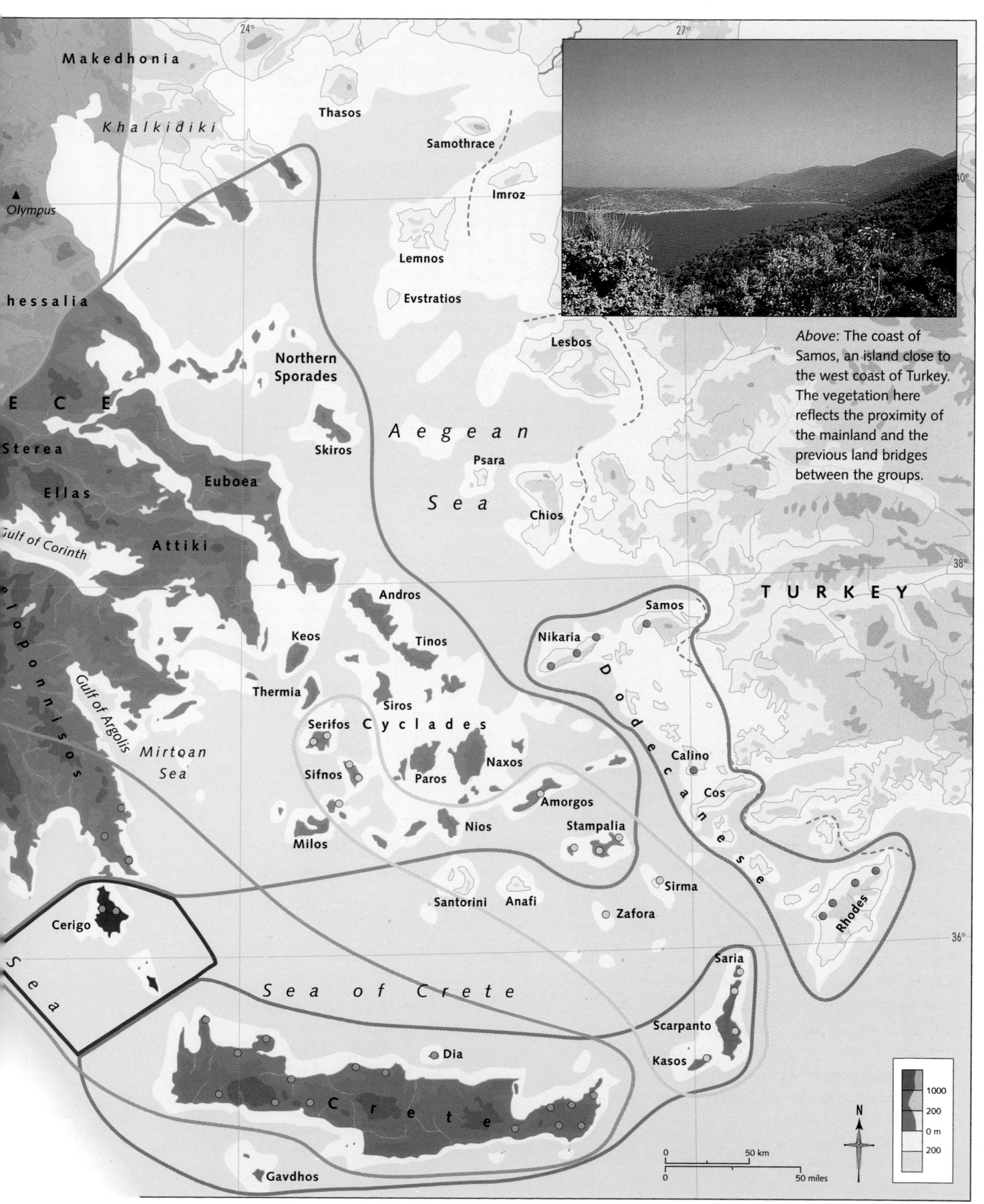

Above: The coast of Samos, an island close to the west coast of Turkey. The vegetation here reflects the proximity of the mainland and the previous land bridges between the groups.

Makedhonia

Khalkidiki

Thasos

Samothrace

Imroz

Olympus

Lemnos

hessalia

Evstratios

Lesbos

E C E

Northern
Sporades

Aegean

Sterea

Skiros

Psara

Sea

Ellas

Euboea

Chios

Gulf of Corinth

Attiki

TURKEY

Andros

Samos

Peloponnisos

Keos

Tinos

Nikaria

Gulf of Argolis

Thermia

Siros

Dodecanese

Mirtoan
Sea

Serifos

Cyclades

Calino

Sifnos

Paros

Naxos

Cos

Milos

Nios

Amorgos

Stampalia

Sirma

Santorini

Anafi

Zafora

Rhodes

Cerigo

Saria

Sea

Sea of Crete

Scarpanto

Dia

Kasos

C r e t e

Gavdhos

	1000
	200
	0 m
	200

0   50 km

0   50 miles

N

Section X: Human Intervention

Human beings have had a major impact on some aspects of evolution, although we tend to exaggerate our own long-term importance. Evolutionary ecologists attempt to predict the effects of major modifications to landscapes by using their knowledge of modern ecology and evolution in the past.

I t is difficult to disentangle science from emotion in attempting to understand the impact of human activity on global diversity, and biology has become linked to a variety of political agendas that are growing in importance. There is no doubt that *Homo sapiens* has had a greater effect on the Earth than any other species. Some commentators suggest that the planet is losing one species every month, while others claim that the total is nearer to one species every day, or even every hour. Are we living through the sixth mass extinction right now?

Pollution is an enormous problem. A city of 1 million inhabitants produces 500,000 tonnes of sewage, 2,000 tonnes of garbage and 1,000 tonnes of air pollutants each year. Controlled treatment of these waste products can vastly diminish their effects, but considerable change to the atmosphere and to natural waters and soils is inevitable. The burning of coal and oil produces carbon dioxide, sulphur, nitrogen and dust, which react with other gases to produce ozone and sulphuric and nitric acids. These substances have caused pollution in cities and widespread human fatalities in the past 200 years. Acid rain kills plants, and ozone breaks down the protective layers in the upper atmosphere, allowing excess ultraviolet light to reach the Earth's surface. This may increase the incidence of skin cancer.

Many major rivers and lakes are now so polluted by domestic waste and industrial effluent that almost no life can survive. Excess organic matter consumes oxygen in the water as it decays, and fish and other animals die in the anaerobic conditions. Fertilizer use also disturbs the natural balance, and lead, zinc and mercury poisoning caused by runoff from metal-mining tips and processing works have resulted in horrific human and animal fatalities. Acid rain affects lakes, and many waters have suffered a massive decline in fish stocks. Soils are polluted by metals and by artificial fertilizers and pesticides that are known to kill wildlife and humans yet are still widely in use in many countries. Pesticides persist in agricultural food produce and natural food chains for decades.

The spread of cities, roads and farmland is eating away natural habitats, and in many countries rainforest is being removed at a high rate, either for the commercial production of hardwood timber or to provide living space for subsistence farmers. Attempts are being made to control pollution and habitat destruction in many countries by legislation on motor vehicles, factory output and agricultural pesticide use. The felling of trees in the rainforest for timber could also be stopped, but growing human populations need food, and there is no obvious solution to the problem of the spread of subsistence farmers. The best that can be expected is a balance between human and wildlife needs. Such approaches are being pioneered in several parts of the world. Large nature reserves are being established in which plants and animals can live unmolested but humans may practise low-impact farming and controlled tourism. At the base of many of these approaches is evolutionary ecology theory.

Evolutionary ecology

Ecologists MacArthur and Wilson developed a quantified approach to geographical ecology by focusing on islands, which may provide analogues for the understanding of larger and more complex ecosystems. They worked from the premise of a new island in the process of being colonized by chance migrants, and found that the rate of successful colonizations declines over time, while the rate of local extinction increases as more species become established. If a new species becomes established, it must displace a pre-existing species. At a certain point, the rate of colonization and the rate of extinction cross over, marking the island's equilibrium diversity (the number of species that can be accommodated).

MacArthur and Wilson suggested that small islands support small numbers of species and large islands support more (the 'species-area effect'), and that the number of species depends on distance from the nearest mainland. Ecologists seized on this as a seminal proposal that could be used as a model for all kinds of ecological situations. The programme was extended to practical applications. Conservationists studied interaction patterns among plants and animals in fully natural systems, for instance, and used their observations of the natural ranges of species to predict the minimum suitable size for a game preserve.

These ideas and models depend on the assumption of a fixed carrying capacity, both for islands and the Earth as a whole. A lively debate has been running since the 1970s as to the validity of this idea. Certainly, say the critics, there is evidence for the species-area effect, but experiments on the colonization of islands suggest that species numbers continue to rise the longer the island exists. The 'carrying capacity' is therefore flexible, and depends as much on time to permit evolution and specialization as on island size. Large continent-scale migrations such as the Great American Interchange do not in any way fit the expectations of an equilibrium diversity pattern. The same is true of the long-term history of the diversity of life: there is evidence for temporary limits to diversity, but not for any cap on the total number of species that the Earth can sustain in the long term. Does this mean that the modelling of game preserve sizes on the equilibrium diversity pattern is pointless? Or does it imply that many species survive much better than we think by modifying their habits and insinuating into existing habitats, thereby raising their overall diversity?

From Wild Plants to Staple Crops
The origins of the world's major food plants

Plants are the source of nutrition for all life forms on Earth. They are the only organisms that are capable of converting the energy of the sun into usable food. Almost all animals, whether they are herbivores, omnivores or carnivores, ultimately depend on plants for their survival. The domestication of certain wild plants for use as a reliable food source was a turning point in human history. Indeed, the invention of agriculture has, quite literally, changed the face of the planet, as well as paving the way for the development of highly sophisticated human societies.

The plants that are used in agriculture have themselves been changed over the centuries by selective breeding, but agriculture has also changed much of the vegetation found on the Earth's surface. Some wild plants and animals have adapted to these changes, others have not. This tension between the needs of agriculture and the preservation of natural habitats will continue.

Human beings existed for several hundred thousand years as hunter-gatherers. For food they killed animals and caught fish, picked wild fruit, dug up roots and munched green leaves. Then, about 10,000 years ago, a crucial step in the development of human civilization took place. Human beings realized that they could make plants reproduce and grow simply by pushing seeds, bulbs or tubers into the ground. Agriculture was born, and since that time human survival has depended on the ability to cultivate plants successfully.

It is no coincidence that the earliest human civilizations developed in those regions where a large variety of potential food plants grew in the wild. These areas were in subtropical or temperate zones, and usually contained hills or mountain ranges which provided a wide range of habitats for wild plants. It seems likely that the first plants to be cultivated for food were tubers and bulb plants. These are much easier to grow than seed plants, and they require much less knowledge of the life-cycle of the plant. To grow a potato, for example, one simply puts a tuber that has been found in the ground back into the soil. At the end of the next growing season it will have produced a crop of potatoes. Taking a seed from the stem of a grass plant and planting it is a much more sophisticated action. For this reason archaeologists and botanists are fairly sure that tubers came before cereals.

The first regions of the world where agriculture was practised were therefore probably the highland

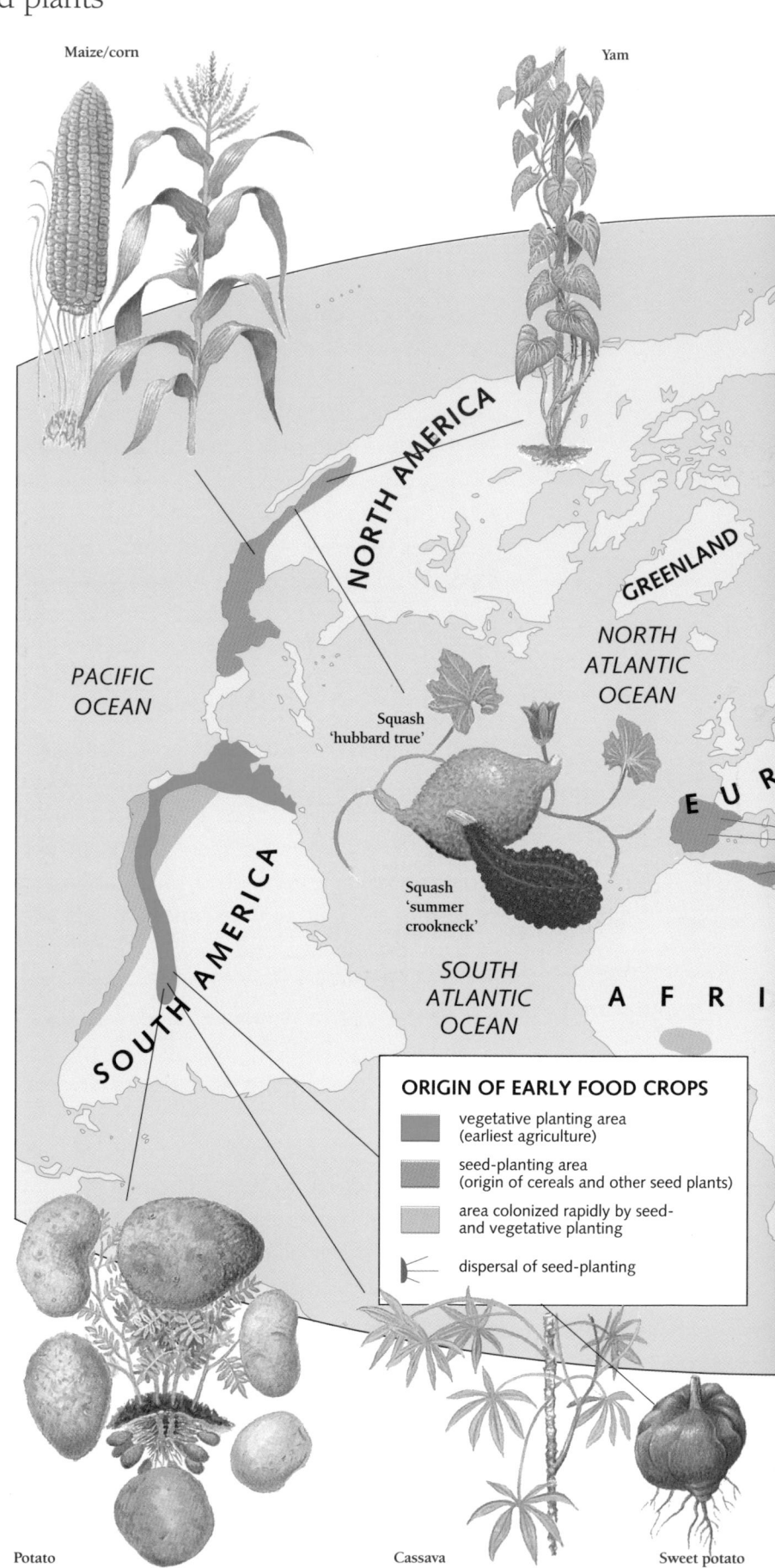

Maize/corn

Yam

NORTH AMERICA

GREENLAND

NORTH ATLANTIC OCEAN

PACIFIC OCEAN

Squash 'hubbard true'

EUR

Squash 'summer crookneck'

SOUTH AMERICA

SOUTH ATLANTIC OCEAN

A F R I

ORIGIN OF EARLY FOOD CROPS

- vegetative planting area (earliest agriculture)
- seed-planting area (origin of cereals and other seed plants)
- area colonized rapidly by seed- and vegetative planting
- dispersal of seed-planting

Potato

Cassava

Sweet potato

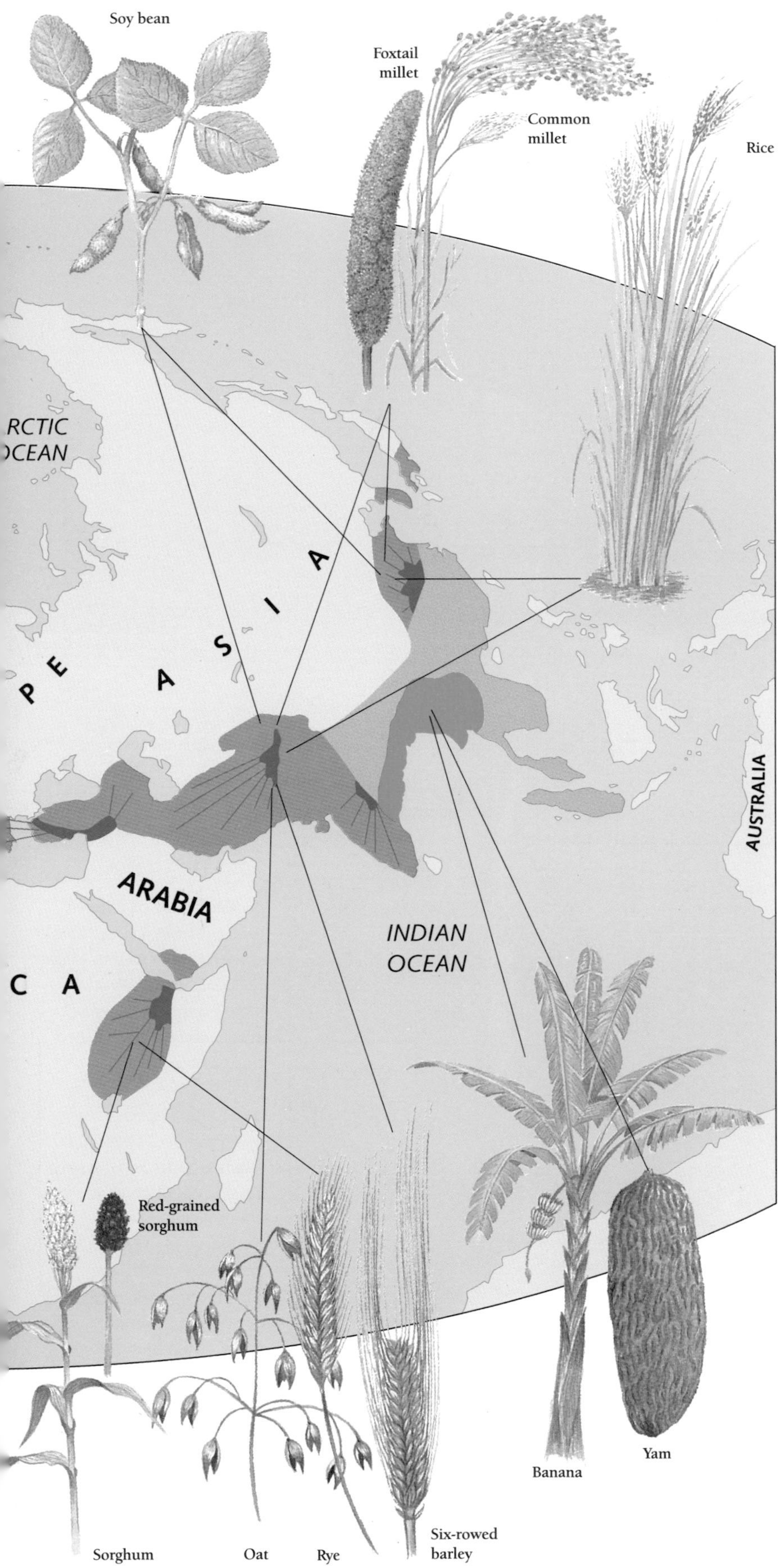

Soy bean

Foxtail millet

Common millet

Rice

ARCTIC OCEAN

E ASIA PE

ARABIA

INDIAN OCEAN

AUSTRALIA

CA

Red-grained sorghum

Sorghum

Oat

Rye

Six-rowed barley

Banana

Yam

areas of northwest South America, where potatoes, sweet potatoes and cassava originated, and Southeast Asia, which is the home of the yam and the banana. These areas formed strategic crossroads of human communication, so early agricultural techniques spread rapidly. The development of seed-planting took off later in areas in which there were suitable wild grasses – the Ethiopian Highlands, Northern China, Mexico and Guatemala. Grain crops have several advantages over tuber crops like potatoes. They can be dried and stored against non-growing seasons, and because they are light they can be much more easily transported.

The earliest cereal crop plants would have been wild plants that humans beings brought into

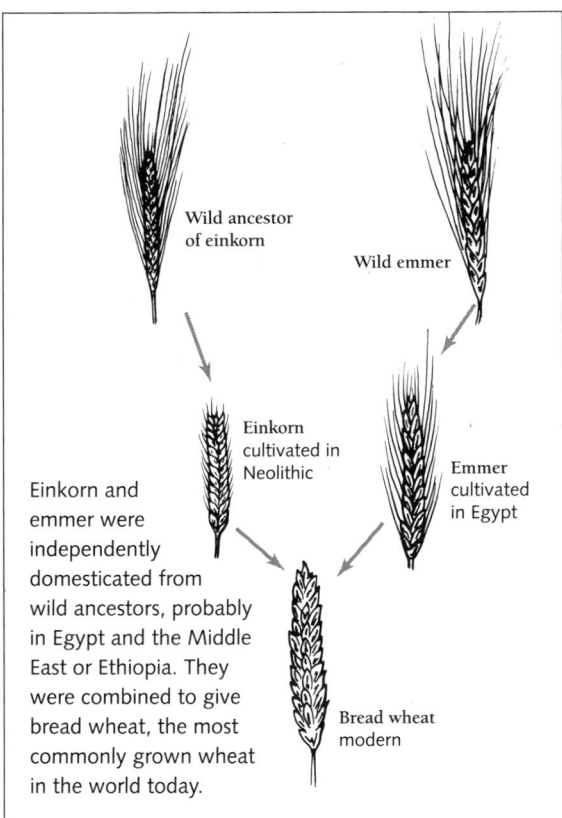

Wild ancestor of einkorn

Wild emmer

Einkorn cultivated in Neolithic

Emmer cultivated in Egypt

Einkorn and emmer were independently domesticated from wild ancestors, probably in Egypt and the Middle East or Ethiopia. They were combined to give bread wheat, the most commonly grown wheat in the world today.

Bread wheat modern

cultivation. These are known as 'habitation weeds', and would have included wild varieties of wheat, barley, soy and maize. Other plants would have appeared alongside these as they were cultivated. These 'weeds of cultivation' would have included rye, oats, mustard and rape. As in tropical gardens today, plants would have been grown together, and this would have created hotbeds of accidental hybridization. As well as crossbreeding with one other, cultivated plants continued to cross with their wild equivalents, giving rise to hundreds of different plant varieties.

Invaders From Around the World
Exotic animals in the British Isles

The native fauna of the British Isles (the islands of Britain and Ireland) is impoverished in comparison to the mainland of the European continent. Offshore islands often have fewer species than their 'parent' land mass, but in this case there is a particular reason for Britain's meagre diversity of vertebrate animals.

As recently as 18,000 years ago, ice sheets extended as far south as the Thames Valley in Britain and the Shannon in Ireland. When these ice sheets withdrew, the animals that had been driven south began to migrate north in order to reclaim their former territory. As the ice melted, however, the sea levels rose, cutting Britain and Ireland off from the rest of Europe and creating a barrier to further migrations.

The number of species that have been imported into these islands by humans is partly a reflection of Britain's traditions as a seafaring and trading nation and of its long history of interest in natural science. By the end of the 19th century, the British Empire extended around the globe, and British explorers, traders and naturalists visited every continent, bringing back living and dead specimens of every kind of plant and animal life. Some of these were imported for serious study, but many animals were displayed in zoos and wildlife parks, while others were farmed for fur, for meat or for sport. In addition, there were those, such as the black rat, that came by human agency but without human intent.

Of all these exotic species, a small but significant proportion has become 'established'. By this we mean that there are populations that exist in the wild and perpetuate themselves without human intervention. Some are ideally suited to the host environment, while others have found peculiar habitats. Cichlid fish from East Africa, for instance, are thought to live in the warm water that flows into rivers from industrial cooling systems .

All species introduced into a new environment have an effect on native populations. Some of these effects are easy to measure – the grey squirrel has pushed the formerly ubiquitious red squirrel into the forests of northern Britain – while others are unknown. Some may seem innocuous to humans but are deeply damaging to other species. The vertebrate fauna of the British Isles has undoubtedly been enlarged by exotic species, but whether it has been enriched remains a difficult judgement to make.

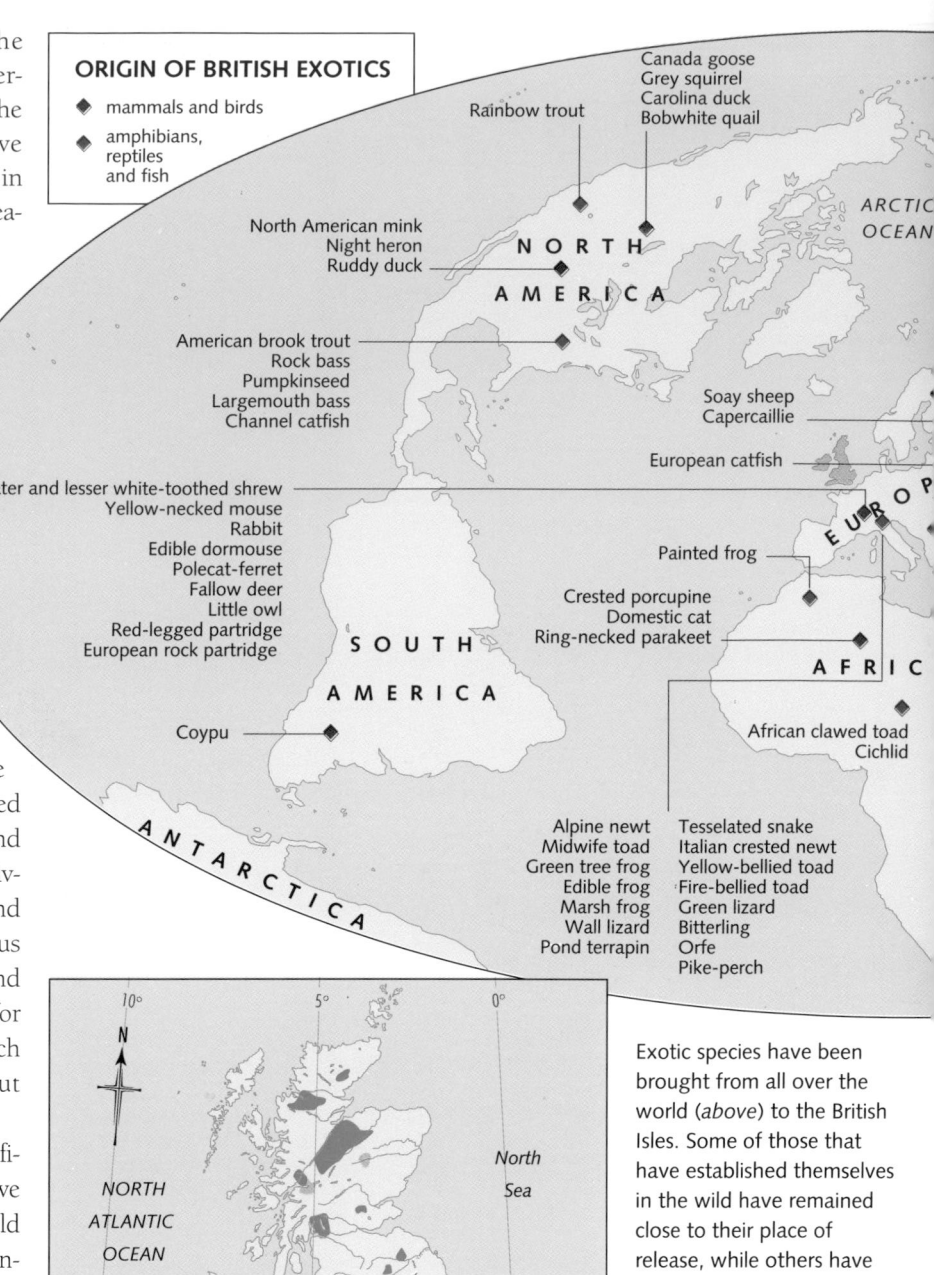

ORIGIN OF BRITISH EXOTICS

◆ mammals and birds

◆ amphibians, reptiles and fish

Rainbow trout

Canada goose
Grey squirrel
Carolina duck
Bobwhite quail

North American mink
Night heron
Ruddy duck

ARCTIC OCEAN

NORTH AMERICA

American brook trout
Rock bass
Pumpkinseed
Largemouth bass
Channel catfish

Soay sheep
Capercaillie

European catfish

Greater and lesser white-toothed shrew
Yellow-necked mouse
Rabbit
Edible dormouse
Polecat-ferret
Fallow deer
Little owl
Red-legged partridge
European rock partridge

EUROPE

Painted frog

Crested porcupine
Domestic cat
Ring-necked parakeet

SOUTH AMERICA

AFRICA

Coypu

African clawed toad
Cichlid

Alpine newt
Midwife toad
Green tree frog
Edible frog
Marsh frog
Wall lizard
Pond terrapin

Tesselated snake
Italian crested newt
Yellow-bellied toad
Fire-bellied toad
Green lizard
Bitterling
Orfe
Pike-perch

ANTARCTICA

10°   5°   0°

N

NORTH ATLANTIC OCEAN

North Sea

55°

Irish Sea

0   100 km
0   100 miles

English Channel

50°

fallow deer

little owl

wild goat

Exotic species have been brought from all over the world (*above*) to the British Isles. Some of those that have established themselves in the wild have remained close to their place of release, while others have spread far and wide. Large mammals are restricted by habitat, particularly in such a heavily populated and intensively farmed environment. The fallow deer (*left*), which was introduced 3,000 years ago, became widespread when much of the islands were forested. The capercaillie (*right*) is the only example of the successful reintroduction of a vertebrate into the wild in Britain. Having disappeared in 1785, birds were imported and released in Scotland in 1837.

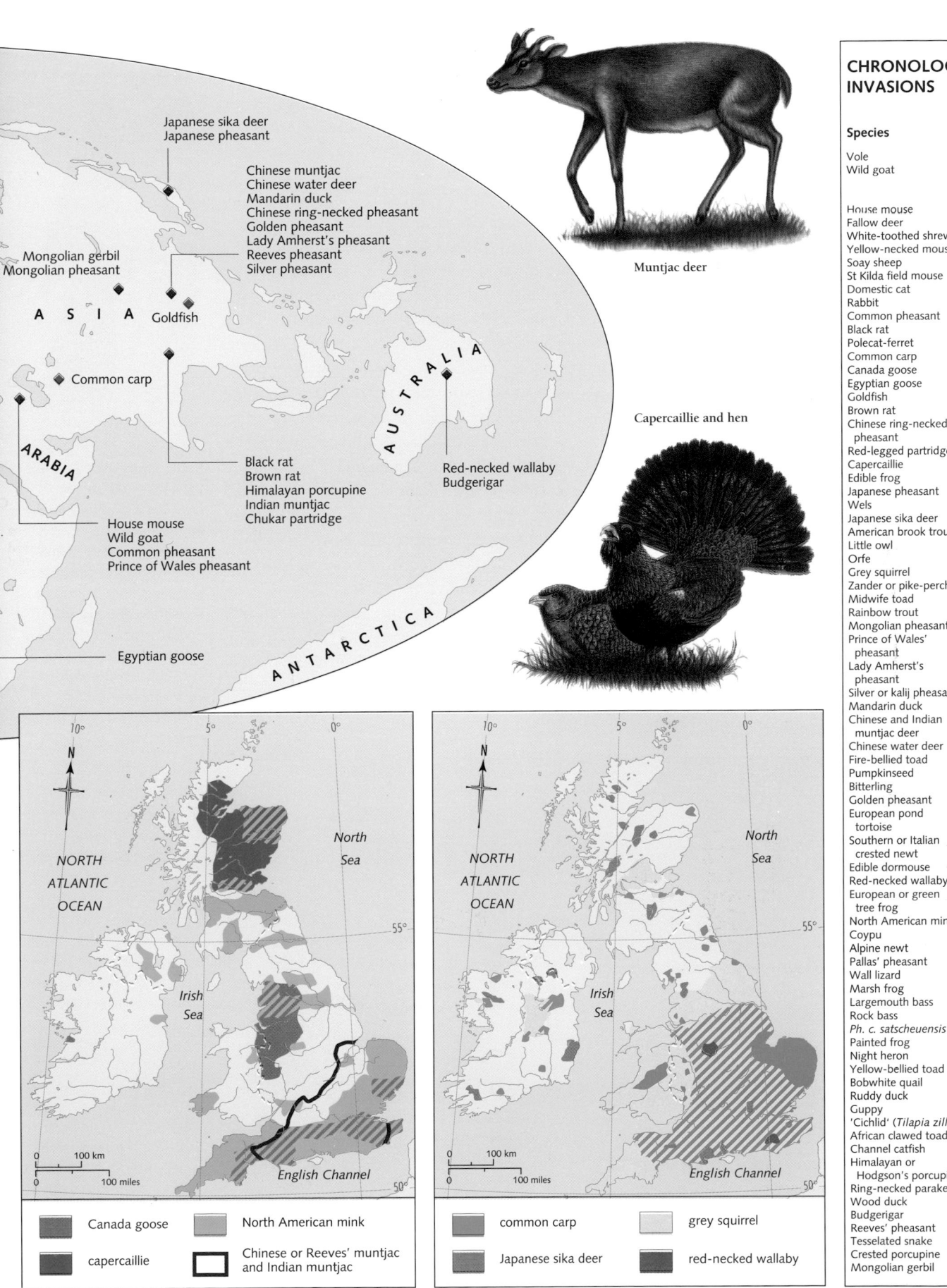

Muntjac deer

Capercaillie and hen

Map labels

Japanese sika deer
Japanese pheasant

Chinese muntjac
Chinese water deer
Mandarin duck
Chinese ring-necked pheasant
Golden pheasant
Lady Amherst's pheasant
Reeves pheasant
Silver pheasant

Mongolian gerbil
Mongolian pheasant

A S I A

Goldfish

Common carp

A R A B I A

Black rat
Brown rat
Himalayan porcupine
Indian muntjac
Chukar partridge

House mouse
Wild goat
Common pheasant
Prince of Wales pheasant

A U S T R A L I A

Red-necked wallaby
Budgerigar

A N T A R C T I C A

Egyptian goose

Lower left map

10° 5° 0°

N

NORTH
ATLANTIC
OCEAN

North
Sea

55°

Irish
Sea

0 100 km
0 100 miles

English Channel

50°

Legend	
Canada goose	North American mink
capercaillie	Chinese or Reeves' muntjac and Indian muntjac

Lower middle map

10° 5° 0°

N

NORTH
ATLANTIC
OCEAN

North
Sea

55°

Irish
Sea

0 100 km
0 100 miles

English Channel

50°

Legend	
common carp	grey squirrel
Japanese sika deer	red-necked wallaby

CHRONOLOGY OF INVASIONS

Species	Date
Vole	? c. 2000 BC
Wild goat	? late Stone/ early Bronze Age
House mouse	c. 1000 BC
Fallow deer	c. 1000 BC
White-toothed shrew	? Pre-hist.
Yellow-necked mouse	? Pre-hist.
Soay sheep	? Pre-hist.
St Kilda field mouse	? c. AD 900
Domestic cat	bef. M. Ages
Rabbit	bef. 1176
Common pheasant	bef. 1177
Black rat	? 12th c.
Polecat-ferret	late 13th c.
Common carp	before 1596
Canada goose	17th c.
Egyptian goose	17th c.
Goldfish	? c. 1691
Brown rat	1728-29
Chinese ring-necked pheasant	by 1768
Red-legged partridge	c. 1770
Capercaillie	1837-38
Edible frog	1837
Japanese pheasant	c. 1840
Wels	1853
Japanese sika deer	1860
American brook trout	1868
Little owl	1874
Orfe	1874
Grey squirrel	1876
Zander or pike-perch	1878
Midwife toad	1878
Rainbow trout	1884
Mongolian pheasant	c. 1898
Prince of Wales' pheasant	c. 1898
Lady Amherst's pheasant	late 19th c.
Silver or kalij pheasant	late 19th c.
Mandarin duck	from c. 1900
Chinese and Indian muntjac deer	c. 1900
Chinese water deer	c. 1900
Fire-bellied toad	since c. 1900
Pumpkinseed	early 20th c.
Bitterling	early 20th c.
Golden pheasant	20th c.
European pond tortoise	20th c.
Southern or Italian crested newt	20th c.
Edible dormouse	1902
Red-necked wallaby	c. 1908
European or green tree frog	1900-10
North American mink	1929
Coypu	1929
Alpine newt	1920s-30s
Pallas' pheasant	bef. 1930
Wall lizard	1932
Marsh frog	1934-35
Largemouth bass	1934 or 35
Rock bass	bef. 1937
Ph. c. satscheuensis	1942
Painted frog	1947
Night heron	1950
Yellow-bellied toad	c. 1954
Bobwhite quail	1956
Ruddy duck	1957
Guppy	1963
'Cichlid' (Tilapia zillii)	1963
African clawed toad	1967
Channel catfish	1969
Himalayan or Hodgson's porcupine	1969
Ring-necked parakeet	1969
Wood duck	late 1960s
Budgerigar	1969
Reeves' pheasant	1970
Tesselated snake	c. 1970
Crested porcupine	1972
Mongolian gerbil	1973

Exploring the Southern Ocean
European domestication of distant lands

The French explorer Kerguelen discovered the group of Southern Ocean islands that bears his name in 1772. On his return to Paris, he described the islands as being like southern France. Four years later Captain James Cook spent Christmas anchored in a harbour in the Kerguelen Islands, which he described as desolate and remote. Cook's description was the more accurate of the two. But what is desolate to one species is an opportunity for many others. The effect of the visits of sailors to these islands is an early example of the consequences of the introduction of exotic species.

In their voyages, first exploratory, then for trade, the crews of European sailing ships valued these tiny islands as places to rest and find shelter, and as a source of fresh food and water. The animal life of the southern islands was sparse, however, and sometimes non-existent. The sailors therefore decided to introduce their own domestic animals to the islands. A small breeding population would be left on an island on an outward journey. Then, either on the return leg or on a future voyage, the sailors would call in to harvest a supply of fresh meat. Sailing ships also brought animals to these islands accidentally – principally rats, mice and cats.

This method is a testament to the farsightedness of the early global travellers and to the accuracy of their navigation skills. But what was good for European travellers was potentially harmful to the flora and fauna that was specially adapted to life on these islands. As there was very little cataloguing of this flora and fauna before the release of the European animals, it is impossible to be certain of the precise effect that they had, but a comparison of inaccessible grassy areas with those grazed by the herbivores (mainly cattle and sheep) shows that much of the flora has been destroyed.

On most of these islands there are few trees and the islands are often too small to have supported a population of predatory mammals. Birds are therefore able to nest on the ground in relative safety. Flightless species can also develop. In these circumstances, the introduction of predatory mammals such as cats and rats can have a devastating effect. This has been seen in more recent introductions, where the existing populations had been observed and recorded by naturalists.

In 1894, a single domestic cat was introduced to Stephen Island, which lies off the north coast of New Zealand, by the keeper of the newly built lighthouse. Within a short time it had exterminated the entire population of the flightless wren, *Xenicus lyalli*, which had been endemic to the island. Cats, rats and pigs introduced to Mauritius in the 17th and 18th centuries decimated the eggs and juveniles of the flightless dodo (*Raphus solitarius*), while humans hunted the adults. Most of the native reptiles disappeared in the same way.

Laughing kookaburra *Dacelo gigas*

Below: Cook's three voyages to the southern hemisphere between 1768 and 1780 provided a map for future European seafarers. Settlements followed, but the introduction of animals and crops from 'home' was disastrous, due to a lack of understanding of the native environment. The islands of the Southern Ocean were inhospitable to settlers, but were used as staging posts on long ocean voyages.

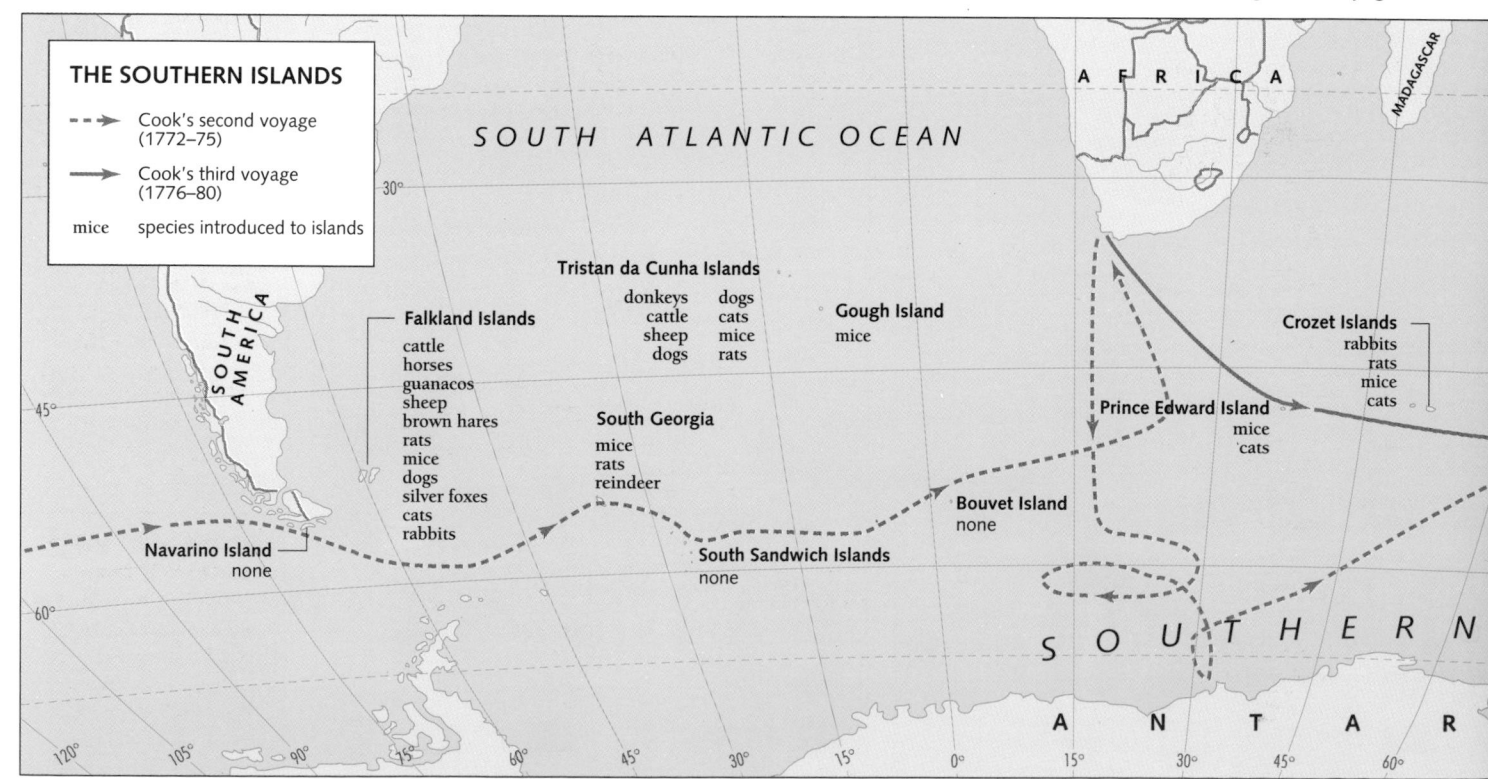

THE SOUTHERN ISLANDS

- - -> Cook's second voyage (1772–75)

—> Cook's third voyage (1776–80)

mice   species introduced to islands

SOUTH ATLANTIC OCEAN

AFRICA

MADAGASCAR

30°

SOUTH AMERICA

45°

60°

120° 105° 90° 75° 60° 45° 30° 15° 0° 15° 30° 45° 60°

Tristan da Cunha Islands
donkeys dogs
cattle cats
sheep mice
dogs rats

Gough Island
mice

Crozet Islands
rabbits
rats
mice
cats

Falkland Islands
cattle
horses
guanacos
sheep
brown hares
rats
mice
dogs
silver foxes
cats
rabbits

South Georgia
mice
rats
reindeer

Prince Edward Island
mice
cats

Bouvet Island
none

Navarino Island
none

South Sandwich Islands
none

SOUTHERN

ANTAR

ORIGIN OF SPECIES INTRODUCED TO NEW ZEALAND

AUSTRALIA

Mammals:
opossum, wallaby

Birds:
Cape Barren goose, black swan, brown quail, eastern rosella, white cockatoo, laughing kookaburra, white-backed magpie, black-backed magpie

POLYNESIA

Mammals:
native dog, Maori rat

ASIA

Mammals:
thar, Axis deer, Sambar deer, Japanese deer

Birds:
chukor, laceneck dove, Indian myna, pea fowl

AMERICA

Mammals:
wapiti, Virginia deer, mule deer, moose

Birds:
Canada goose, Californian quail, Virginian quail

BRITAIN

Mammals:
hedgehog, stoat, ferret, weasel, European dog, cat, black rat, brown rat, mouse, rabbit, hare, wild cattle, wild sheep, wild goat, red deer, fallow deer, wild pig, wild horse

Birds:
English mallard, English pheasant, skylark, song thrush, blackbird, hedge sparrow, rook, starling, house sparrow, chaffinch, redpoll, goldfinch, greenfinch, yellow bunting

EUROPE

Mammals:
chamois

Birds:
small brown owl

Chamois
Rupicapra rupicapra

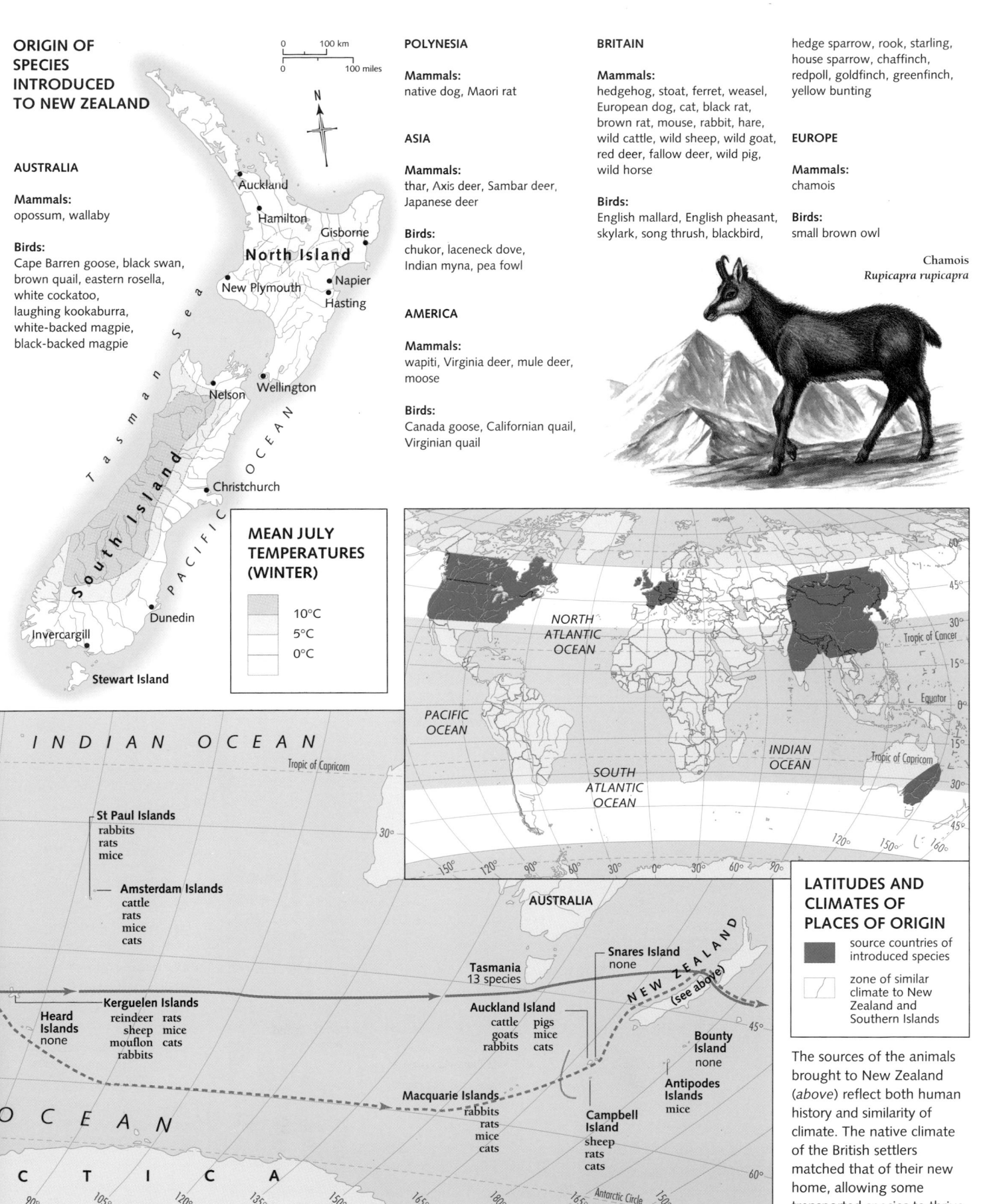

MEAN JULY TEMPERATURES (WINTER)

- 10°C
- 5°C
- 0°C

North Island

- Auckland
- Hamilton
- Gisborne
- New Plymouth
- Napier
- Hasting

South Island

- Nelson
- Wellington
- Christchurch
- Dunedin
- Invercargill
- Stewart Island

Tasman Sea
PACIFIC OCEAN

LATITUDES AND CLIMATES OF PLACES OF ORIGIN

- source countries of introduced species
- zone of similar climate to New Zealand and Southern Islands

The sources of the animals brought to New Zealand (*above*) reflect both human history and similarity of climate. The native climate of the British settlers matched that of their new home, allowing some transported species to thrive.

INDIAN OCEAN
Tropic of Capricorn

St Paul Islands
rabbits
rats
mice

Amsterdam Islands
cattle
rats
mice
cats

Heard Islands
none

Kerguelen Islands
reindeer   rats
sheep   mice
mouflon   cats
rabbits

Tasmania
13 species

AUSTRALIA

Auckland Island
cattle   pigs
goats   mice
rabbits   cats

Macquarie Islands
rabbits
rats
mice
cats

Campbell Island
sheep
rats
cats

Snares Island
none

NEW ZEALAND
(see above)

Bounty Island
none

Antipodes Islands
mice

Antarctic Circle

NORTH ATLANTIC OCEAN
PACIFIC OCEAN
SOUTH ATLANTIC OCEAN
INDIAN OCEAN
Tropic of Cancer
Equator
Tropic of Capricorn

The American Muskrat in Europe
Success and failure of an accidental invasion

The American muskrat (*Ondatra zibethica*) first arrived in mainland Europe in 1905. Its initial introduction was made for commercial purposes, as were the many subsequent introductions – the muskrat is highly valued for its fur. There are now millions of muskrats inhabiting the river systems of central and northeastern Europe, Siberia and Kazakhstan. Muskrats were also introduced to the British Isles in the early part of this century, but on these islands a deliberate policy of extermination was carried out, so that by 1935 the muskrat had been totally eradicated from England, Scotland and Ireland. The deliberate encouragement in eastern Europe and deliberate eradication in Britain and Ireland are at opposite ends of the spectrum of human intervention in species distribution.

The continent of North America lies within the Nearctic biogeographic region (*see pages 134–135*). In common with all other Nearctic mammals, muskrats do not occur naturally outside the boundaries of the region. The Palearctic biogeographic region, which encompasses Europe and Asia north of the Himalayas, has a similar climate to the Nearctic region. When species from one region are introduced to the other, they stand a good chance of becoming established there.

The invasion of an English river system

The American muskrat lives in and around rivers and other watercourses. It can spread rapidly up- and downstream, as shown in the map (*far right*) of the main rivers in the English county of Shropshire in the 1930s. The muskrat was eradicated from the areas around each of its points of introduction in the British Isles by the use of riverside traps. One thousand were caught in the rivers Forth and Fearn in Scotland alone. More recently, English waterways were invaded by the coypu (*Myocastor coypus*) from South America. This thrived in the aquatic environment of the Norfolk Broads before being eradicated in the early 1980s.

Palearctic muskrat distribution (*below*)
The range of the North American muskrat in the Palearctic region seems to be continually expanding. Climatic similarities between the Palearctic and the Nearctic region, which includes North America, has facilitated its expansion there. In Siberia, where human intervention is minimal, the distribution and numbers have grown particularly rapidly.

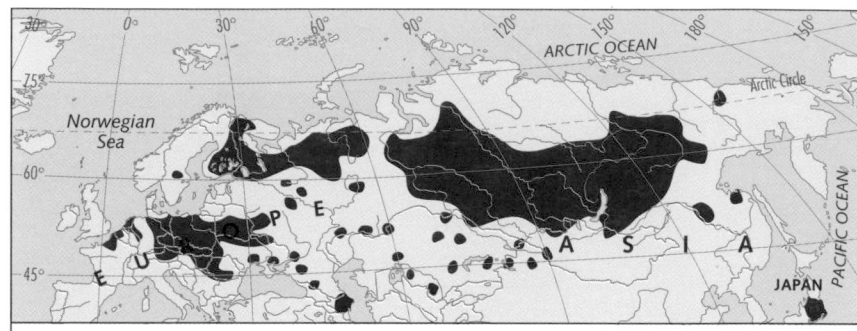

NORTH AMERICAN MUSKRAT DISTRIBUTION IN EUROPE AND ASIA

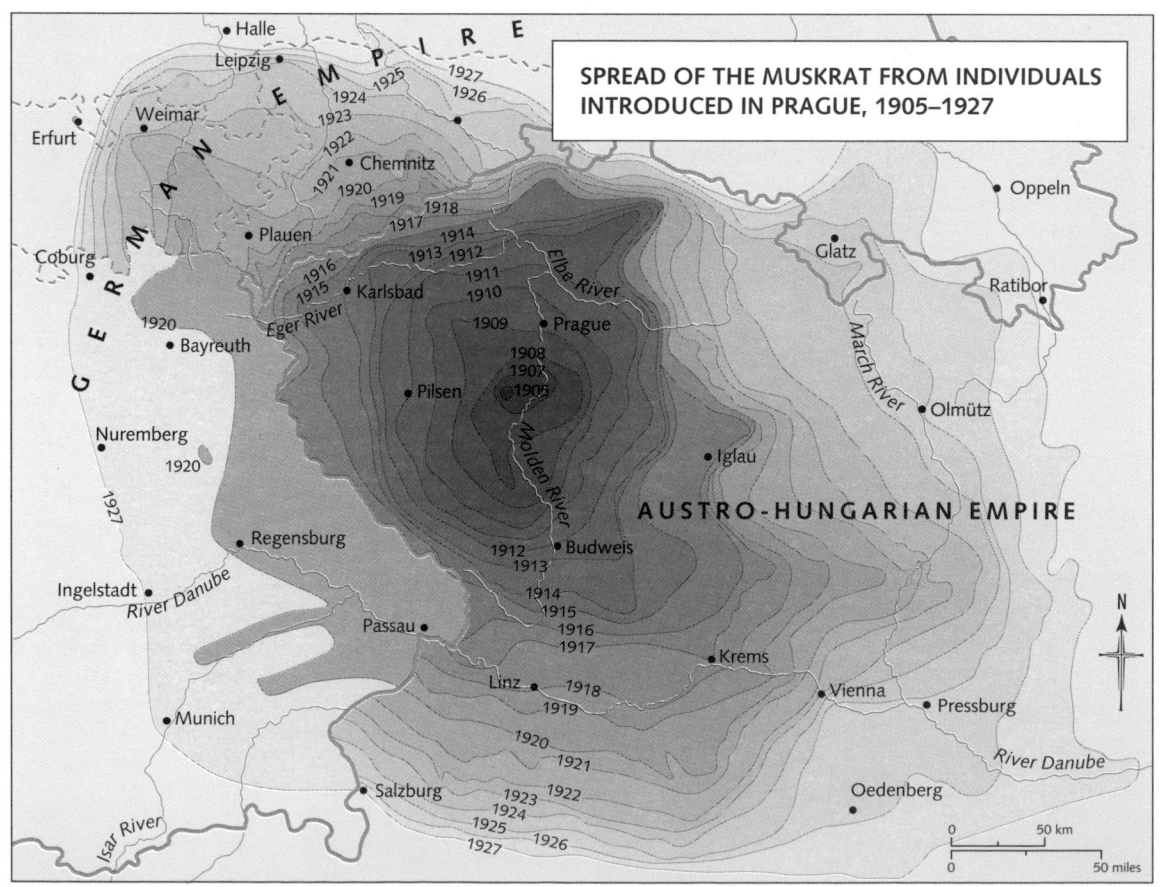

SPREAD OF THE MUSKRAT FROM INDIVIDUALS INTRODUCED IN PRAGUE, 1905–1927

The spread of the muskrat in Central Europe (*left*)
The first incidence of the North American muskrat in mainland Europe was the introduction of five individuals that were kept by a landowner near Prague in the Austro-Hungarian empire (now in the Czech Republic) in 1905. These animals were allowed to breed in the wild and rapidly established a viable population. Within ten years of their introduction, the area of muskrat habitation stretched in a rough circle extending 150 kilometres (90 miles) from Prague. Within 12 years it had spread to 250 kilometres (150 miles). American muskrats are now found over a vast area of the Palearctic biogeographic region (*above*), which includes Europe and northern Asia.

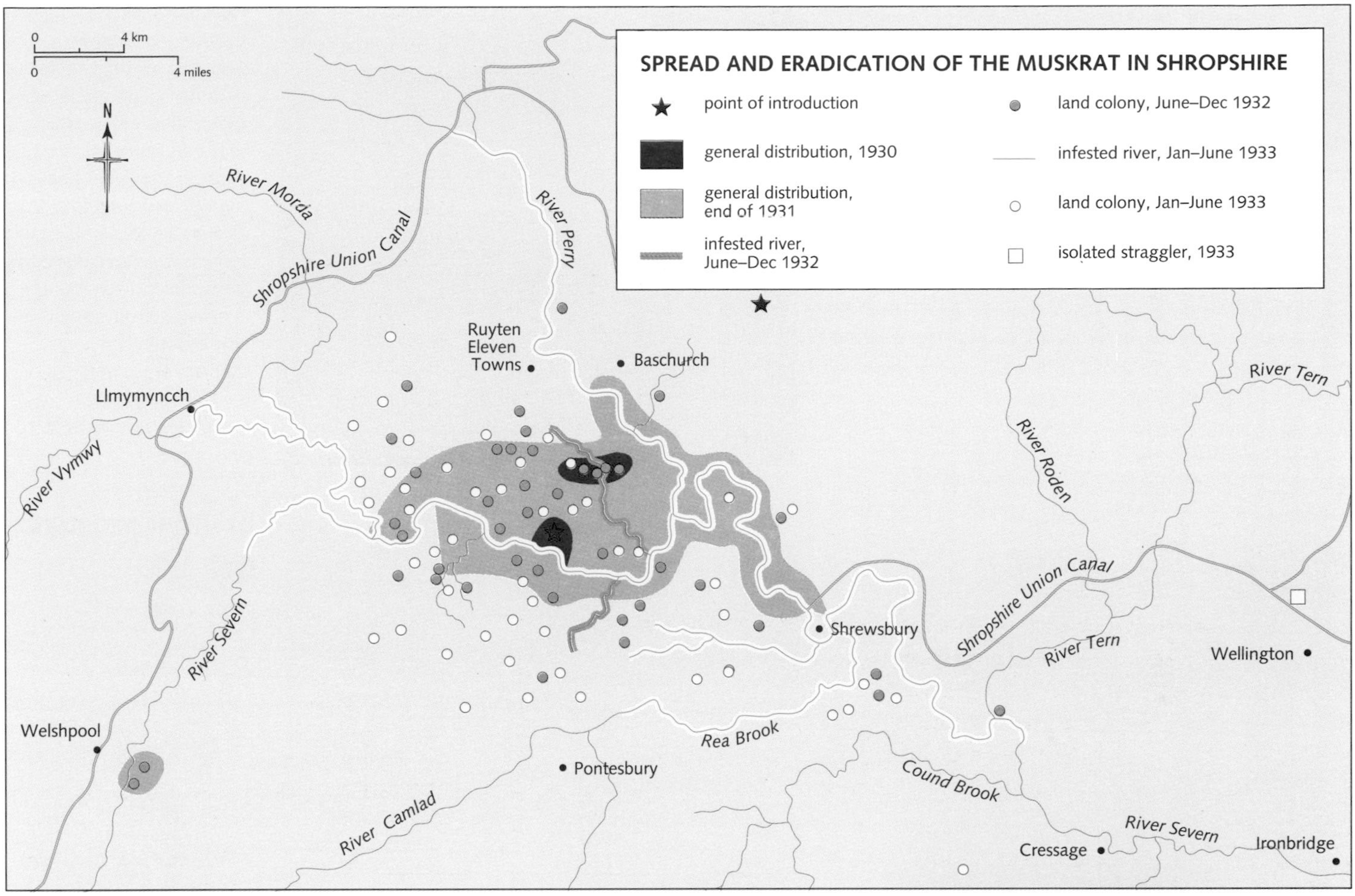

SPREAD AND ERADICATION OF THE MUSKRAT IN SHROPSHIRE

★ point of introduction
● land colony, June–Dec 1932
■ general distribution, 1930
— infested river, Jan–June 1933
▨ general distribution, end of 1931
○ land colony, Jan–June 1933
▰ infested river, June–Dec 1932
□ isolated straggler, 1933

River Morda
Shropshire Union Canal
River Perry
Ruyten Eleven Towns
● Baschurch
River Tern
Llmymyncch
River Roden
River Vyrnwy
River Severn
● Shrewsbury
Shropshire Union Canal
River Tern
Wellington ●
Welshpool ●
Rea Brook
● Pontesbury
Cound Brook
River Camlad
River Severn
Cressage ●
Ironbridge ●

0 4 km
0 4 miles
N

Encouraging dispersal

Because of the commercial value of their fur (known as musquash), North American muskrats have been deliberately introduced into several European countries. In Finland, over 200 separate introductions have occurred since 1922. Numerous colonies were also established in the river systems of northern Russia, Siberia and Kazakhstan by the Soviet government. Introductions are usually made to fur farms, but escapes from these are common, and wild breeding populations expand rapidly.

The capture of muskrats in the whole of the European and west Asian region has reached a level of around 240,000 animals per year. The alteration to the aquatic environment made by the presence of millions of these foreign invaders is felt by competing mammals and waterfowl. Invading species may be reduced or eradicated by artificial means, like the removal of the muskrat from the British Isles. Also, natural conditions often allow a rapid initial increase in population that is followed by a gradual decline, until some sort of equilibrium is reached.

Ondatra zibethica (right) North American muskrats build their nests close to waterways, where their ability to swim protects them from land-based predators. This adherence to aquatic habitats makes the muskrat easier to trap, but on the other hand the waterways act as effective channels for the rapid spread of these mammals.

The Rainforests of Southeast Asia
Destruction and conservation

From historical evidence and scientific surveys it has been estimated that in 1979 the area of tropical rainforest stood at about 56 per cent of its prehistoric area (the coverage before humans appeared). In the 1970s about 1 per cent of the rainforest area was lost every year. In the 1980s the rate of destruction increased, and by the end of the decade almost 2 per cent was lost annually. At this rate, half of the world's remaining rainforest will be gone in 30 years' time. Using mathematical models based on previous experience of habitat destruction,

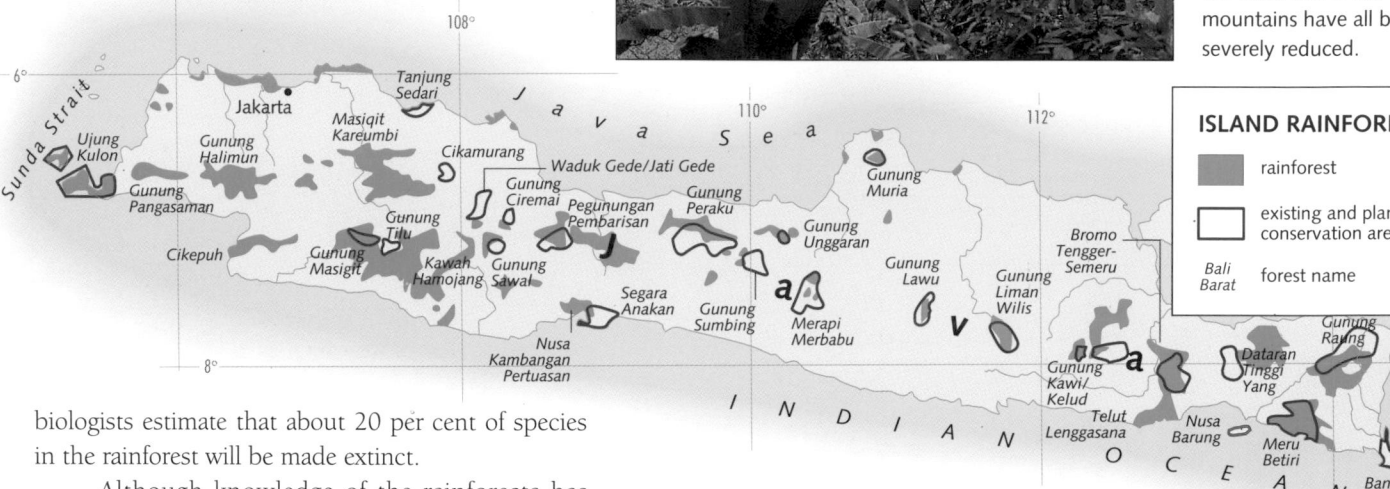

Rainforests still cover large areas of Southeast Asia, though increasingly only in areas unsuited to farming, such as this forest in the highlands of Penang in Malaysia (*left*). Java, Bali, Lombok and Sumbawa (*below*) are among the most densely populated islands in the world and many of the remaining forest pockets are now conservation zones. Different types of forest on the coast and in the mountains have all been severely reduced.

ISLAND RAINFORESTS

- rainforest
- existing and planned conservation areas
- *Bali Barat* forest name

biologists estimate that about 20 per cent of species in the rainforest will be made extinct.

Although knowledge of the rainforests has increased dramatically in the past 20 years, we know very little about the vast numbers of species of plants and small animals that inhabit them. Many species have never been recorded, so we are unaware even of how much we will be destroying. Whole groups of organisms will disappear, depriving us of the chance to learn more about life on our planet and to wonder at its continual variety.

Plants that we do know about, because of their commercial value, are the timber-producing trees. Many are under threat of extinction, and some have undoubtedly already been lost. Particular species are protected by law in most tropical countries, but breaches continue and the demand for tropical timber seems endless. Better equipment has led to increased destruction, while the concentrated use of a few species is unnecessary. A more intelligent use of a wider variety of species would benefit both the loggers and the forests. The selective logging of valuable timbers need not be destructive to the forest ecosystem, but commercial logging is usually the first step in the process of forest clearance. The roads built to bring out timber allow in settlers eager to clear the land for agriculture.

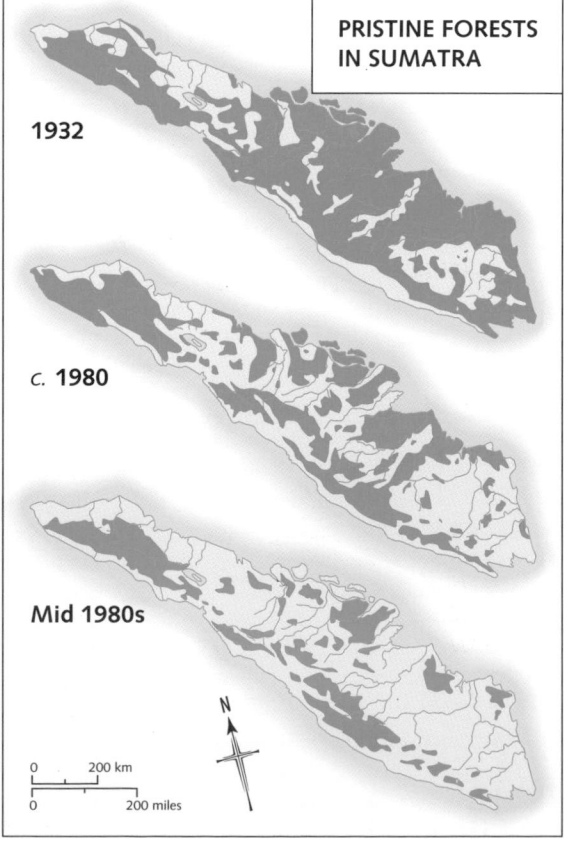

PRISTINE FORESTS IN SUMATRA

1932

c. 1980

Mid 1980s

0   200 km
0   200 miles

The island of Sumatra (*left*) is thought to be losing its natural vegetation faster than any other part of Indonesia. It has a relatively high population density and large areas have been cleared for agriculture and industrial plantations. Coastal swamp forests on the northeastern side have been drained for farmland. On the southern lowlands there were once great stands of ironwood, *Eusideroxylon zwageri*. This exceptionally hard wood, which is of great commercial value, has now been almost totally destroyed. Today around half of the original forest cover on Sumatra remains. In 1975, 42 per cent of the island was covered in forest, but that figure is much lower now.

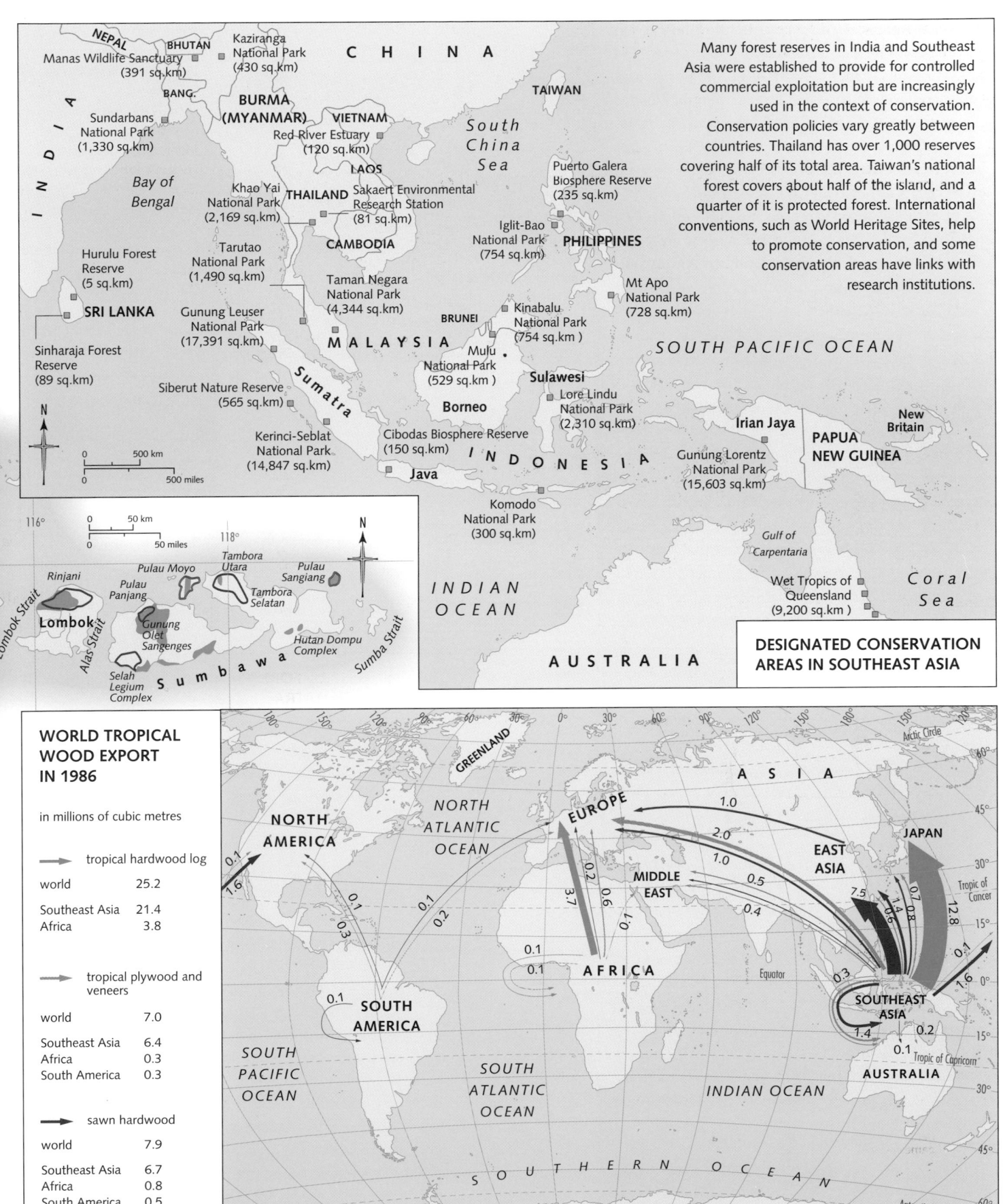

NEPAL
BHUTAN
Kaziranga
National Park
(430 sq.km)
Manas Wildlife Sanctuary
(391 sq.km)

C H I N A

BANG.

BURMA
(MYANMAR)
VIETNAM
Sundarbans
National Park
(1,330 sq.km)
Red River Estuary
(120 sq.km)

LAOS

TAIWAN

South
China
Sea

I N D I A

Bay of
Bengal

Khao Yai
National Park
(2,169 sq.km)
THAILAND
Sakaert Environmental
Research Station
(81 sq.km)

Puerto Galera
Biosphere Reserve
(235 sq.km)

Hurulu Forest
Reserve
(5 sq.km)
Tarutao
National Park
(1,490 sq.km)
CAMBODIA

Iglit-Bao
National Park
(754 sq.km)
PHILIPPINES

SRI LANKA
Taman Negara
National Park
(4,344 sq.km)
Gunung Leuser
National Park
(17,391 sq.km)
BRUNEI
Kinabalu
National Park
(754 sq.km)
Mt Apo
National Park
(728 sq.km)

SOUTH PACIFIC OCEAN

Sinharaja Forest
Reserve
(89 sq.km)
M A L A Y S I A
Mulu
National Park
(529 sq.km)

N

Siberut Nature Reserve
(565 sq.km)
Sumatra
Borneo
Sulawesi
Lore Lindu
National Park
(2,310 sq.km)

New
Britain

Irian Jaya
PAPUA
NEW GUINEA

0   500 km
0   500 miles
Kerinci-Seblat
National Park
(14,847 sq.km)
Cibodas Biosphere Reserve
(150 sq.km)
I N D O N E S I A
Java
Gunung Lorentz
National Park
(15,603 sq.km)

Komodo
National Park
(300 sq.km)

Gulf of
Carpentaria

116°
0   50 km
0   50 miles
118°
N

Wet Tropics of
Queensland
(9,200 sq.km)
Coral
Sea

Rinjani
Pulau Moyo
Tambora
Utara
Pulau
Sangiang
I N D I A N
O C E A N
Pulau
Panjang
Tambora
Selatan
Lombok
Gunung
Olet
Sangenges
Hutan Dompu
Complex
Lombok Strait
Alas Strait
Selah
Legium
Complex
S u m b a w a
Sumba Strait
A U S T R A L I A

Many forest reserves in India and Southeast
Asia were established to provide for controlled
commercial exploitation but are increasingly
used in the context of conservation.
Conservation policies vary greatly between
countries. Thailand has over 1,000 reserves
covering half of its total area. Taiwan's national
forest covers about half of the island, and a
quarter of it is protected forest. International
conventions, such as World Heritage Sites, help
to promote conservation, and some
conservation areas have links with
research institutions.

**DESIGNATED CONSERVATION
AREAS IN SOUTHEAST ASIA**

WORLD TROPICAL WOOD EXPORT IN 1986

in millions of cubic metres

→ tropical hardwood log

world	25.2
Southeast Asia	21.4
Africa	3.8

→ tropical plywood and veneers

world	7.0
Southeast Asia	6.4
Africa	0.3
South America	0.3

→ sawn hardwood

world	7.9
Southeast Asia	6.7
Africa	0.8
South America	0.5

The Serengeti Park
An artificial environment?

In many parts of the world, conflict has arisen between the economic and social needs of humans and the ecological needs of plants and animals. The growth in the human population has been accelerating, particularly in the past 50 years, with no obvious slowdown in sight. This rise in population has happened in parallel with an intensification of agriculture and an extension of the amount of land under cultivation.

The establishment of conservation areas, or game reserves, is a deliberate attempt to prevent agriculture and hunting from intruding on the 'natural' world. These conservation areas are, of course, artificial and are another example of human intervention, albeit benign. The most successful, such as the famous Serengeti Park in Tanzania, have been carefully tailored to suit the needs of their inhabitants.

The continent of Africa has the greatest concentration of large mammals in the world. As a simple rule of thumb, bigger animals need more space. Large herbivores, such as elephants and giraffes, need huge areas of land over which to range. Herds tend to feed and move on, and it is essential that vegetation has time to recover before the herd returns. Grazing animals such as wildebeests and antelopes have established seasonal migration patterns in some

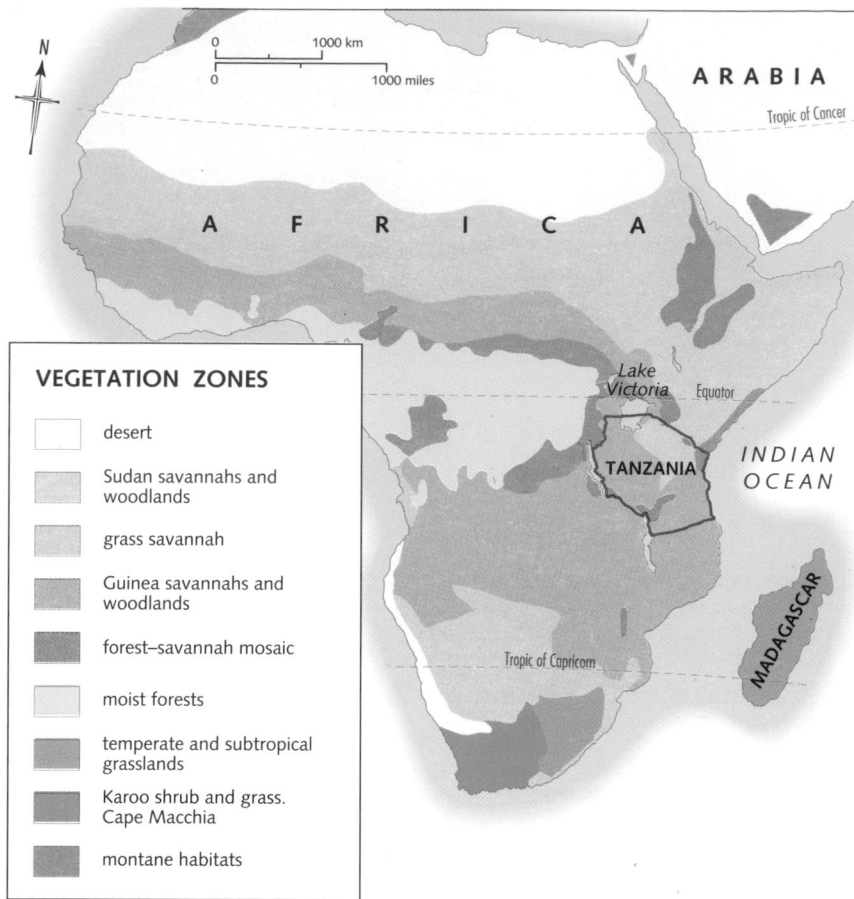

VEGETATION ZONES

- desert
- Sudan savannahs and woodlands
- grass savannah
- Guinea savannahs and woodlands
- forest–savannah mosaic
- moist forests
- temperate and subtropical grasslands
- Karoo shrub and grass. Cape Macchia
- montane habitats

TANZANIAN NATIONAL PARKS AND GAME RESERVES

- —— national border
- —— national parks and game reserves
- —— roads
- • towns/cities

regions that involve journeys over thousands of kilometres of country. The big carnivores that feed off the nomadic herds need the same space.

A healthy breeding population of animals depends upon a degree of genetic diversity. Inbreeding promotes the persistence of potentially harmful recessive genes, so animals must be able to breed with individuals from outside their family group. In the case of large mammals such as lions, where family groups occupy large territories, a vast area is needed in order to maintain genetic diversity. This means that a conservation area that contains large animals must be considerable in size in order to be sustainable.

The Serengeti Park covers an area of 14,500 square kilometres (5,600 square miles) in northern Tanzania. Much of the area is open grassland with savannah, woodland and scrub. There are rivers, swamps and lakes, and occasional high rock outcrops standing above the plain. The park is home to the most spectacular concentration of animal wildlife in the world. Here, the grazing animals of the African plains thrive without the dual pressures of agriculture and hunting, which have reduced their numbers elsewhere on the continent.

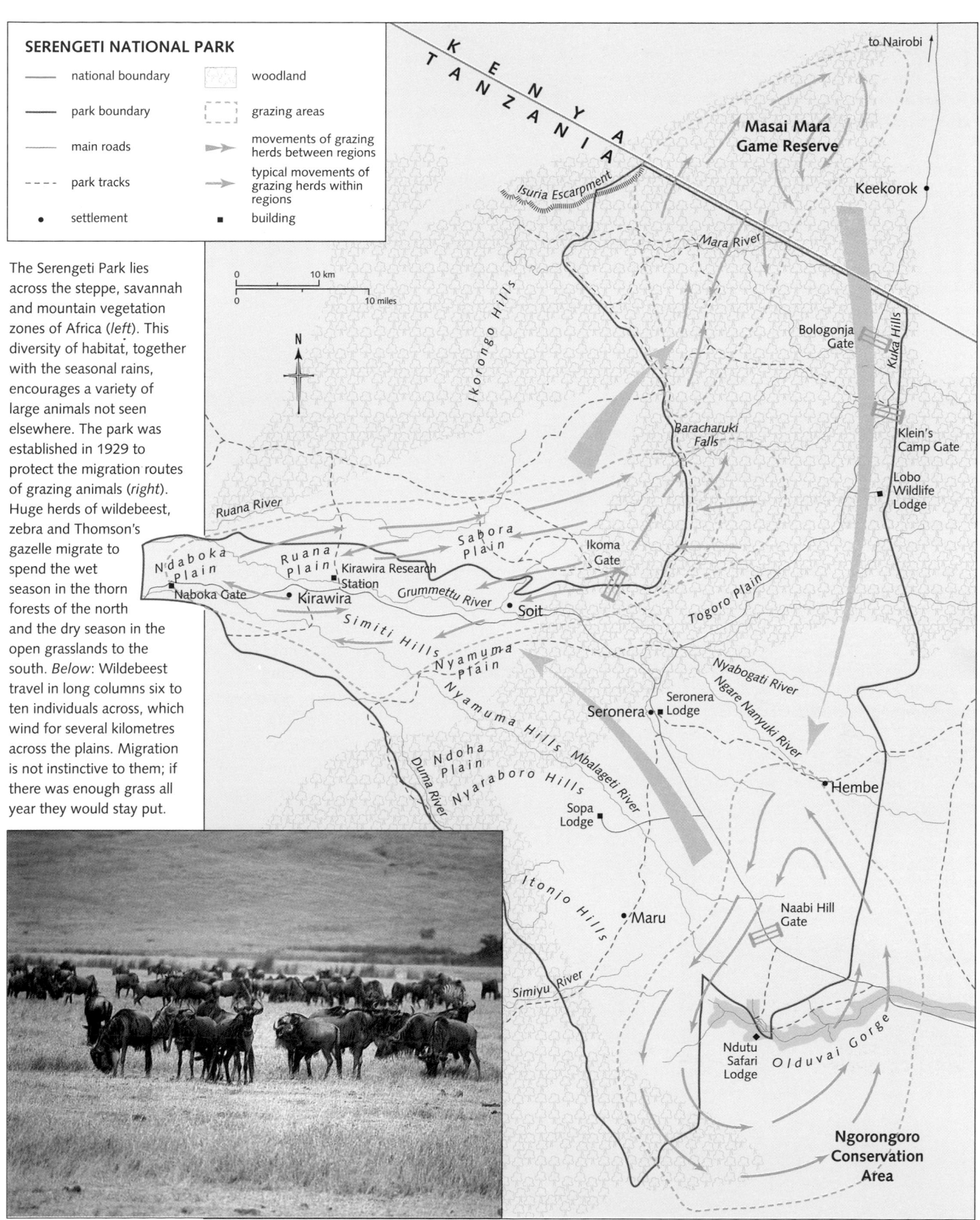

SERENGETI NATIONAL PARK

- —— national boundary
- —— park boundary
- —— main roads
- - - - park tracks
- • settlement
- woodland
- grazing areas
- → movements of grazing herds between regions
- → typical movements of grazing herds within regions
- ■ building

The Serengeti Park lies across the steppe, savannah and mountain vegetation zones of Africa (*left*). This diversity of habitat, together with the seasonal rains, encourages a variety of large animals not seen elsewhere. The park was established in 1929 to protect the migration routes of grazing animals (*right*). Huge herds of wildebeest, zebra and Thomson's gazelle migrate to spend the wet season in the thorn forests of the north and the dry season in the open grasslands to the south. *Below*: Wildebeest travel in long columns six to ten individuals across, which wind for several kilometres across the plains. Migration is not instinctive to them; if there was enough grass all year they would stay put.

Smallpox
The eradication of a disease organism

Diseases like smallpox are organisms in the same sense that plants or animals are, and like other organisms they adapt by the persistence of mutations. Biologists now understand that most disease organisms adapt in order to live in equilibrium with their hosts. However virulent the virus or bacillus, it will not survive if it kills off all of the host organisms.

Diseases therefore tend to go through phases. Over time the host develops an immunity to one strain of the virus and is able to resist the disease. But then another strain arises that is more destructive. Before the host is completely destroyed, it once again develops sufficient immunity to build up resistance. The most successful factor in humans' building up resistance to infectious diseases seems to have been improved nutrition, followed up by immunization techniques. Resistance to smallpox is not radically improved by better nutrition, but the disease is relatively easily prevented by immunization.

Smallpox was the first disease to be effectively immunized against. This was done through the injection of cowpox, which was pioneered in 1796 by the English physician Edward Jenner. In 1801, Jenner predicted that "the annihilation of smallpox – the most dreadful scourge of the human race – will be the final result of this practice." Throughout the next century, immunization programmes were introduced to most industrialized countries. In 1967, the World Health Organization (WHO) put into action a plan to eradicate the smallpox virus from every part of the world. The successful outcome of this campaign is widely regarded as one of the greatest triumphs of 20th-century medicine.

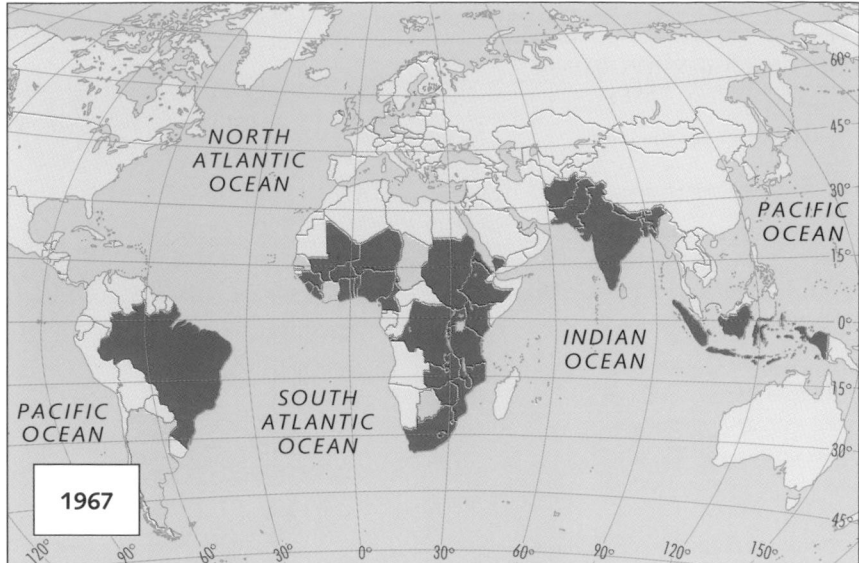

1967

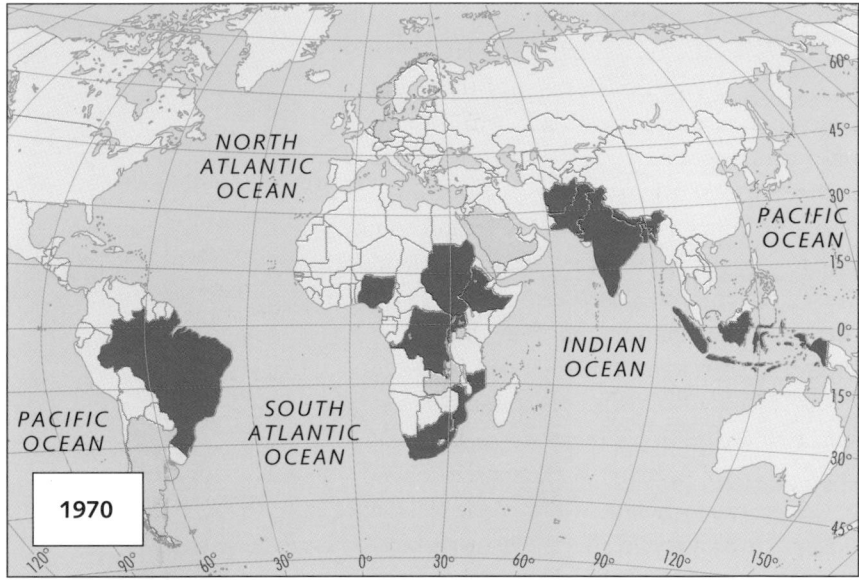

1970

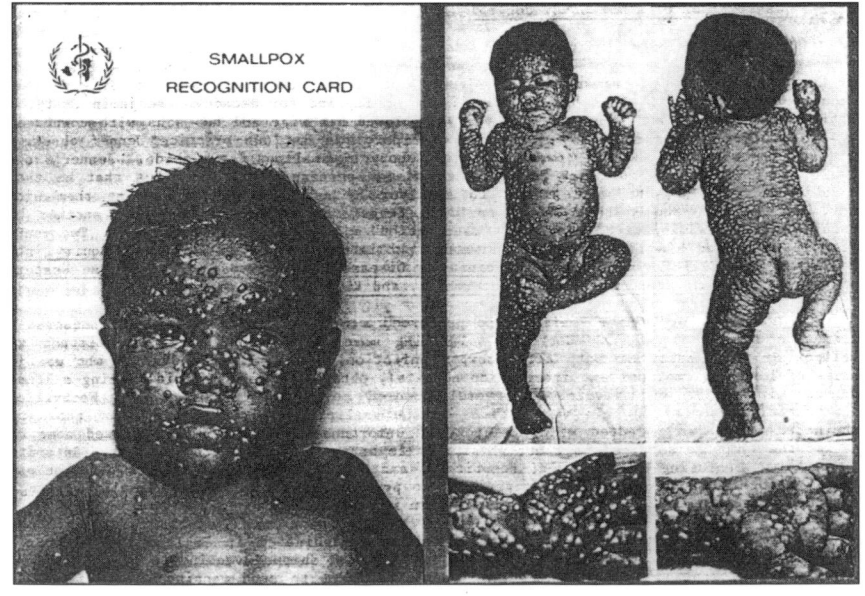

SMALLPOX
RECOGNITION CARD

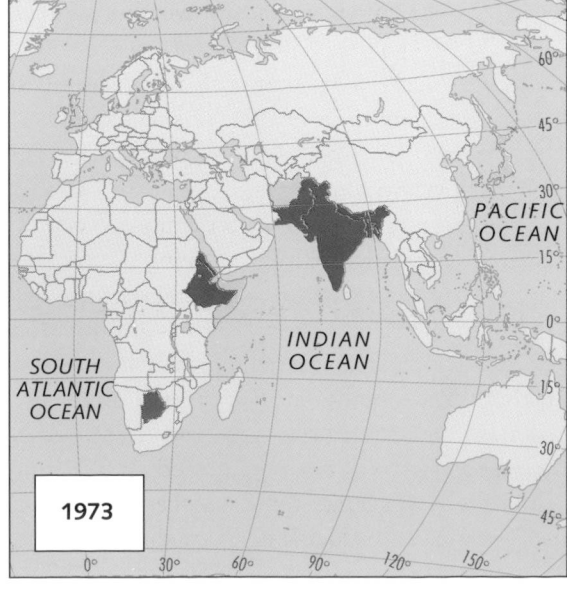

1973

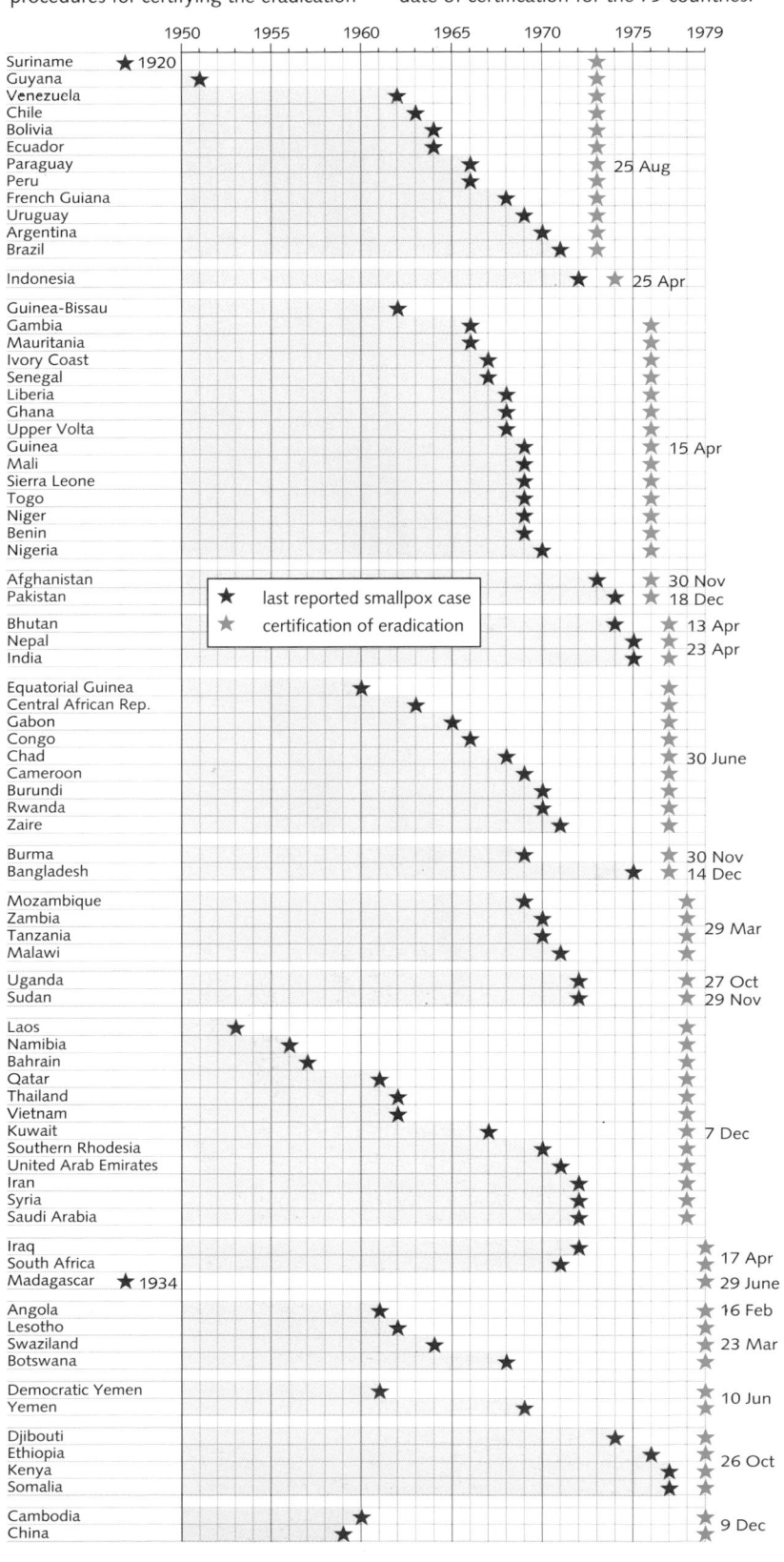

COUNTRIES REPORTING SMALLPOX CASES

- smallpox endemic
- imported cases

SOUTH ATLANTIC OCEAN

INDIAN OCEAN

1976

The shrinking smallpox map

In 1967, at the outset of the WHO campaign, smallpox was endemic in most of the countries of South America, Africa and southern Asia (*top left*). Cases also appeared in Europe, largely as a result of being imported from these regions by people migrating from ex-colonies to European countries in search of work. By 1970 the disease had disappeared from most of West Africa (*left*), and by 1973 it had been eradicated from the Americas, from all of Africa except Botswana and Ethiopia, and from Indonesia (*bottom left*). By 1976, the only country with endemic smallpox was Ethiopia (*above*). The last country to report a case of smallpox was Somalia in 1977.

The extraordinary rapidity of the elimination of smallpox was a result of a readily available vaccine and of a vast and well-coordinated global operation. As part of the campaign, health workers showed Smallpox Recognition Cards (*far left*) in the towns and villages of affected countries all over the world to help in the early identification of smallpox cases. This allowed the immunization of people who had been in contact with the disease.

In most cases vaccinations are used to control diseases, by strengthening the immune system's response to them, rather than to eliminate them. In fact the smallpox virus is not extinct – it lives on in laboratories, where it is stored for research purposes. Smallpox was a great success story, but disease organisms are as variable as any other group. Many diseases have never been eradicated. Instead the conditions in which they thrive have been reduced in the light of better biological understanding.

Documenting the death of a disease

In countries without medical reporting systems, WHO had to set up its own procedures for certifying the eradication of smallpox. The table below shows the month of the last reported case and the date of certification for the 79 countries.

Genes and Language
Reconstructing human relationships

Recent human evolution has been reconstructed in detail from evidence provided by fossils and artefacts (*see pages 30–31, 116–117*), but there has been much debate over the antiquity of modern human groups. The single origin hypothesis is that *Homo sapiens* arose in Africa 250,000 years ago and spread first to the Middle East, then to Europe, Asia and Australasia, and finally to the Americas. The multiple origins hypothesis is that an older worldwide human species, *Homo erectus*, gave rise to modern groups in isolated populations in Africa, Asia and Europe up to 1 million years ago, and that each group has evolved independently since that time.

Genetics provides a powerful test. The genetic material, or genome, of every organism is unique, except in the case of identical twins. How much of this variation in human genomes is between groups and how much is between individuals? Studies have revealed that the proportion of genome variation between groups is 10 per cent – much less than normal individual variation. This suggests that the groups have a recent origin. The average genomes were then compared among the groups and a tree of relationships drawn up. Groups with the most similar genomes had the most recent common ancestors and those with very different genomes had split longest ago. Recent studies point to a central position for Africans and strongly indicate that *Homo sapiens* originated in Africa 250,000 years ago.

Languages evolve, and it is possible to draw up an evolutionary language tree as a check on genetic and fossil evidence. European languages are closest to Iranian and some southwest Asian languages. Japanese pairs with Korean, and the two pair with

Blood groups
Prior to the development of sophisticated genetic techniques, blood groups were used to trace the relationships among the different human populations. These distributions (*left*) are similar to those shown on the map opposite (*bottom*).

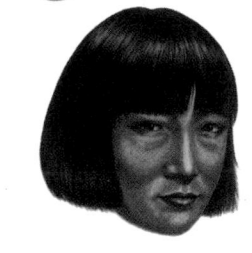
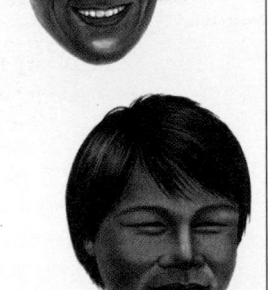

FREQUENCY OF BLOOD GROUP O	
	90%
	80%
	70%
	60%

COMPARATIVE GENE FREQUENCY IN PRESENT-DAY POPULATIONS

number of gene substitutions

0.0
0.1
0.2
0.3
0.4
0.5
0.6
0.7
0.8
0.9
1.0
1.1
1.2
1.3

Gurkha
Veddah
Turk
English
Korean
Eskimo
Lapp
Maori
North Amerindian
New Guinean
Australian
Ethiopian
South Amerindian
Bantu
Ghanaian

Genetic tree (*left*)
This tree showing the relationships among modern human groups is based on comparisons of their genetic material. The tree is calibrated against the number of gene substitutions, and the populations that show the least number of genetic substitutions are the most closely related. Initial studies have suggested that Africans and Europeans form one group, and that Asiatics, Native Americans and Australasians form another. The genome tree can be calibrated against a time-scale by a comparison of human genomes with those of chimps and more distantly related species. The

calibration suggests that the initial split between the groups happened 150,000 years ago.

This genetic evidence can be used with a combination of paleontological, linguistic and archaeological evidence to reconstruct the movements of the human groups throughout the past 10,000 years.

Tibetan, then with Mongol and Eskimo languages. These then pair with Native American languages, proving the link between Asiatic groups and the peopling of the Americas. The tree shows a basal split between Africans and other groups, confirming the African origin of all modern groups.

At the end of the last ice age, Caucasians inhabited Europe, the Middle East to northern India, and North Africa. By AD 1000 they had expanded north to the Arctic Circle, east to occupy India, and south over North Africa. In Africa, Negroes moved south and Pygmies and Bushmen declined. By 10,000 years ago, early Mongoloids occupied China, eastern Siberia and much of the Americas. By AD 1000 they had been displaced by later Mongoloids, who came to inhabit much of Asia and northern North America. In the past 1,000 years much of the Americas and Australia has been colonized by white Europeans and Negroes. Pygmies and Bushmen have lost further ground to the Negroes, and white Europeans have taken over large parts of western Russia and a strip of Russia through to the Pacific.

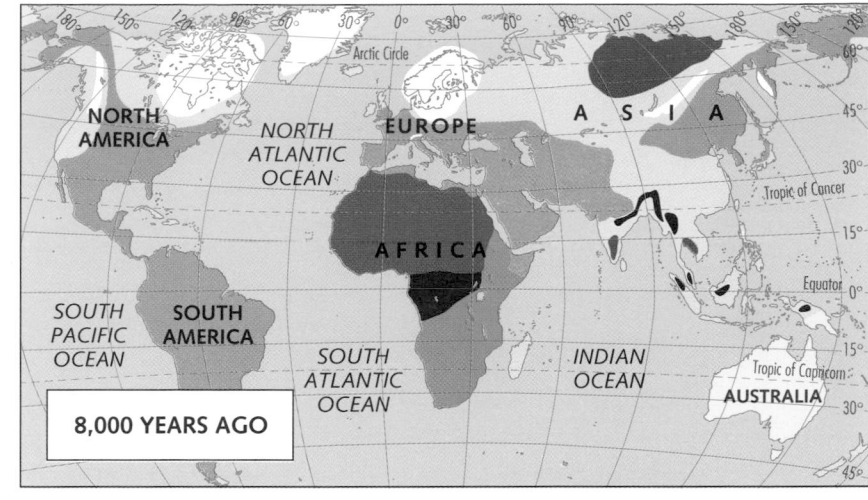

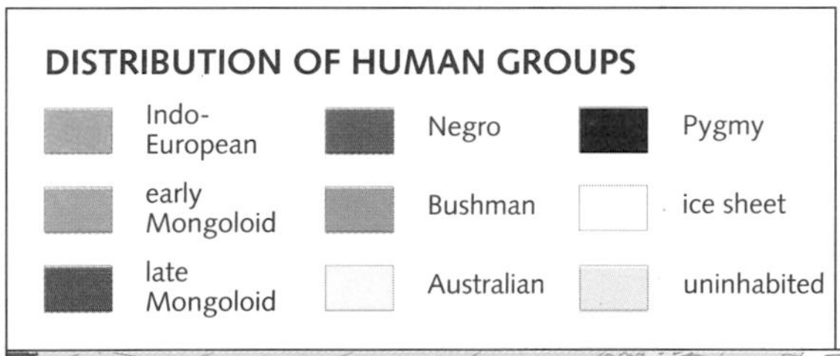

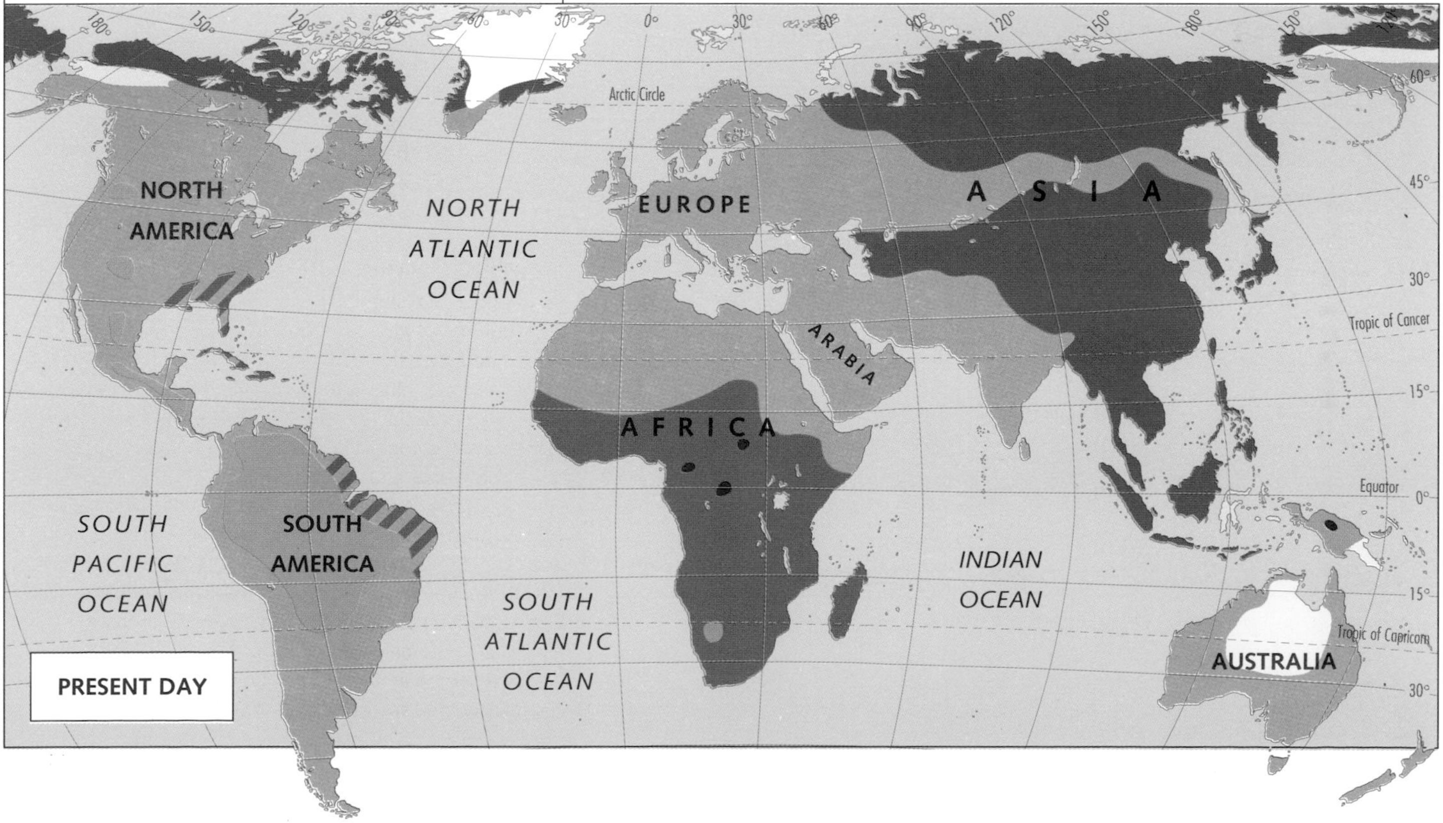

8,000 YEARS AGO

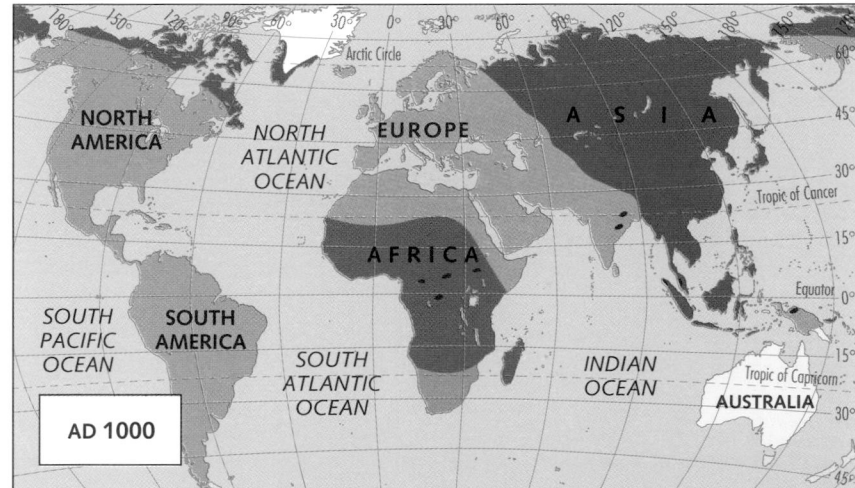

AD 1000

DISTRIBUTION OF HUMAN GROUPS

- Indo-European
- early Mongoloid
- late Mongoloid
- Negro
- Bushman
- Australian
- Pygmy
- ice sheet
- uninhabited

PRESENT DAY

Timeline
Tectonic history and the development of life

Tectonic History	Development of Life
Pre-Archean 4,600 million years ago Formation of the Earth, Moon and planets. Magnetic field formed at some time in this period.	

PRECAMBRIAN	

4,000 million years ago	

Tectonic History	Development of Life
Archean 3,900 million years ago First permanent crust formed. 3,500 million years ago Atmosphere and seawater formed. 3,000 million years ago Greenstone belts – strips of microcontinent – appear.	3,300 million years ago Oldest sedimentary rocks. First stromatolites. Atmosphere with some carbon dioxide. 3,100 million years ago More developed algae and bacteria. 2,900 million years ago Massive stromatolites formed by photosynthesizing blue-green algae.

2,500 million years ago	
Proterozoic 2,500 million years ago Buildup of free oxygen in atmosphere. 2,300 million years ago First large-scale glaciation. 2,000–1,000 million years ago Rapid growth of continents by accretion of microcontinents. Possible formation of a supercontinent. Southern continents combine into Gondwanaland. 700–600 million years ago Major glaciation, affecting every continent.	2,200 million years ago Stromatolites common. Atmosphere contains free oxygen. 1,800 million years ago Diversification of species of prokaryote algae (cellular forms with no nucleus). 1,400 million years ago Bacteria formed into colonies – first step towards multicellular organisms. Atmosphere rich in oxygen. 1,200 million years ago Evolution of eukaryote cells. These cells have a nucleus containing DNA, and the capacity for sexual reproduction. 800 million years ago Evidence of sexual reproduction in eukaryote cells. Filamentous and tubular algae. Appearance of fungi. 600 million years ago Appearance of diverse species of soft-bodied, multicellular organisms (Ediacaran fauna).

550 million years ago	

Tectonic History	Development of Life
PALEOZOIC	

550 million years ago	
Cambrian 550 million years ago Laurentia and Baltica positioned in tropics. Gondwanaland stretches from 50°N to the South Pole. Volcanic episodes in the Caledonian region.	550 million years ago Worldwide emergence of marine invertebrate groups with shells and skeletons. Trilobites, brachiopods, archeocyathids, echinoderms and molluscs all common. Stromatolites decline in abundance.

500 million years ago	
Ordovician 500 million years ago Baltica drifts closer to Laurentia – separated by the first Iapetus Ocean. 450 million years ago Taconic Orogeny in north-east Laurentia, caused by collision on offshore island arc.	480 million years ago First definite vertebrates – jawless freshwater fish. Freshwater plants assumed to be present. 450 million years ago Possible emergence of first land plants.

440 million years ago	
Silurian 425 million years ago Caledonian Orogeny begins, as Baltica and Avalonia collide with Greenland and Laurentia. Baltica and Laurentia drift near to the African part of Gondwanaland. They are separated by an early version of the Tethys Ocean.	440 million years ago Abundance of jawless fish. First fish with jaws – freshwater acanthodians. Giant sea-scorpions (eurypterids) emerge. 420 million years ago First land plants. Vascular plants including lycopsids and psilopsids present, but very rare. First insects and arachnids.

410 million years ago	
Devonian 400 million years ago New phases of the Caledonian disturbances as Gondwanaland rotates clockwise and collides with the eastern margin of Laurentia. The Tethys Sea opens up. 360 million years ago New disturbances along the Gondwanaland/Laurentia boundary, in the final phase of the Caledonian Orogeny. Siberia is the only major block not connected with the Laurentia/Baltica/Gondwanaland land mass.	400 million years ago Age of fish. Jawed and armoured fish become abundant and diversify. Evolution of modern types of fish with bony skeletons and scales. Spore-bearing plants become more common on land – though they are still tied to aquatic habitats. 370 million years ago The first amphibians evolve from fish and reach the land. Emergence of seed ferns, while true ferns cover some lowland areas in dense forest.

350 million years ago	

Tectonic History	Development of Life

350 million years ago	
Mississippian 350 million years ago Laurentia and Gondwanaland remain associated, though separated by ocean as sea levels rise. Widespread limestone formation.	340 million years ago First true reptiles. Emergence of distinct floras associated with different climatic conditions. Glossopteris flora dominates Gondwanaland.

315 million years ago	
Pennsylvanian 310 million years ago Renewed contact between Gondwanaland and Laurentia causes the start of the Appalachian Orogeny. Gondwanaland has continued to turn clockwise. A major glaciation begins to cover large parts of the southern continents in ice. The Hercynian Orogeny results from the collision of northern Gondwanaland and northern Europe.	300 million years ago Evolution of huge lycopsid plants in swamp forests. Amphibians and reptiles diversify in humid tropical conditions, as do insects. Abundance of giant flying insects and cockroaches.

290 million years ago	
Permian 270 million years ago Angaraland (Siberia and Kazakhstan) begins to collide with Baltica, creating the Urals. Last part of supercontinent of Pangea is in place. Pangea stretches from 60°N to the South Pole.	270 million years ago As conditions become drier and hotter, reptiles thrive at the expense of amphibians. Evolution of warm-blooded reptiles (therapsids), the precursors of the mammals. 250 million years ago Mass extinction of marine life. Groups made extinct include trilobites, rugose corals and crinoids. Other marine invertebrates severely affected. Fish are generally unaffected.

250 million years ago	

(vertical right margin) CARBONIFEROUS

Tectonic History	Development of Life		Tectonic History	Development of Life		Tectonic History	Development of Life
MESOZOIC			**TERTIARY**			**QUATERNARY**	

MESOZOIC

250 million years ago

Triassic
250 million years ago
Pangea moves north to straddle the Equator. Many of the continents are now in warm, arid climates. Asian microcontinents begin to move away from Australia and Gondwanaland.

250 million years ago
Ammonites survive the mass extinction at the end of the Paleozoic and thrive in the Mesozoic. Evolution of thecodont reptiles, which become dominant.

220 million years ago
Dinosaurs evolve from thecodont reptiles. First mammals emerge from warm-blooded therapsid reptiles.

205 million years ago

Jurassic
180 million years ago
Africa and South America begin to split from North America, opening up the Central Atlantic.

150 million years ago
Formation of the Rocky Mountains begin.

210–145 million years ago
Dinosaurs become dominant, reaching their largest size. Diversification of flying reptiles (pterosaurs) and aquatic reptiles (plesiosaurs). Birds evolve and spread widely. Continued diversification of insects.

145 million years ago

Cretaceous
120 million years ago
Africa moves further south and begins to split from Europe. India splits from Africa and Antarctica and begins to move north. Australia splits from Antarctica as Gondwanaland starts to break up.

100 million years ago
South America and Africa begin to split apart – the first time they have been separated since the Precambrian period.

85 million years ago
The Central Atlantic stabilizes and links to the South Atlantic, which is still opening. Changes in seafloor spreading in the Atlantic and Pacific oceans push Central America and South America together. South America approaches North America, and a narrow ocean basin is squeezed between them. The Andean region becomes a subduction zone.

145–65 million years ago
Continuing dominance of land by dinosaurs. Mammals remain small. Reptiles diversify – turtles, snakes, lizards are abundant. Emergence of flowering plants (angiosperms). These dominate the land plant kingdom by the end of the Cretaceous.

65 million years ago
Mass extinction of marine and land life forms. The principal casualties are the dinosaurs, the marine reptiles and the ammonites.

65 million years ago

TERTIARY

65 million years ago

Paleocene
Continued uplift in western North America.

65 million years ago
Reptile groups (other than dinosaurs) survive the mass extinction. Mammals and birds also survive and flourish. Emergence of early horse, elephant and bear groups of mammals. Compositae family of plants emerge.

53 million years ago

Eocene
40 million years ago
Uplift of the Rockies and formation of the West Coast mountains completed.

50 million years ago
Grasses emerge and diversify rapidly along with Leguminosae and Compositae plants.

40 million years ago
Grazing animals and monkeys emerge. Mammal groups (whales, dolphins) return to the sea. Foraminifera diversify.

36 million years ago

Oligocene
30 million years ago
Japanese islands split from Asia, opening up the Japan Sea.

25 million years ago
Northern North Atlantic opens between Greenland and northern Europe. Africa moves north to close the Tethys Ocean and collide with Europe. The Alpine Orogeny continues for 15–20 million years.

35 million years ago
The first apes emerge. Large mammals and birds spread over the Earth. Grasses cover large areas of land.

23 million years ago

Miocene
20 million years ago
India begins to collide with Asia in the Himalayan Orogeny.

15 million years ago
Outpourings of basalt lavas in southern Siberia (Baikal Rifts), Central Europe (Rhine Graben), East Africa and Antarctica. Rifts begin in East Africa – first stages in the creation of a new ocean.

10–11 million years ago
Separation of great apes and hominid apes. Radiation of hominid primates culminates in Sivapithecus – an ape showing many characteristics of living apes and humans.

5 million years ago

Pliocene
3 million years ago
Antarctica isolated as South America moves away – the last pieces of Gondwanaland break apart. The Earth's climate cools dramatically.

3–4 million years ago
Emergence of Australopithecus, first hominids.

2 million years ago

QUATERNARY

2 million years ago

Pleistocene
2 million years ago
Nebraskan glaciation (North America) and Donau glaciation (Europe).

700,000–600,000 years ago
Aftonian interglacial period.

600,000 years ago
Kansan glaciation (North America) and Gunz glaciation (Europe).

500,000 years ago
Yarmonth interglacial period.

400,000 years ago
Illinoian glaciation (North America) and Mindel glaciation (Europe).

200,000 years ago
Sangamonian interglacial period.

70,000–10,000 years ago
Wisconsinan glaciation (North America) and Würm glaciation (Europe).

18,000 years ago
Peak of last glaciation.

2–1.75 million years ago
Homo habilis, a possible early form of Homo erectus, emerges in East Africa.

1 million years ago
Homo erectus disperses from Africa as far as China and Java.

500,000 years ago
Homo sapiens appears in Africa and migrates to Europe.

250,000 years ago
Modern humans (Homo sapiens sapiens) emerge in southern Africa.

50,000 years ago
Modern humans reach the Middle East.

35,000 years ago
Modern humans reach Europe as Cro-Magnon man.

30,000–20,000 years ago
Modern humans enter North America via the Bering land bridge and move south.

10,000 years ago

Holocene
Continued rifting in East Africa indicates future seafloor spreading zone.

10,000 years ago
Homo sapiens sapiens reaches every continent except Antarctica, and is the only surviving hominid.

| | | | 550 | 410 | 250 | 65 | 2 |

Glossary

Adaptation
Species change their characteristics to suit their environment. This adaptation is achieved through the persistence of favourable mutations from one generation to the next.

Algae
Generally aquatic plants that include seaweeds and microscopic forms. Distinct from the blue-green algae.

Allopatric speciation
The creation of new species by the physical or geographical separation of populations. Once separated, each population will evolve independently until it is no longer able to breed with other populations.

Alluvial
River-based, or formed by a river.

Ammonite
Group of cephalopods that lived from the Devonian to the Cretaceous period. Ammonites reached their peak of diversity during the Mesozoic era, when their curled shells showed great variation in size and shape. They are used by geologists for dating Mesozoic rocks.

Angaraland
The ancient continent that comprised present-day Siberia and Kazakhstan. This was the site of the massive volcanic eruptions at the end of the Paleozoic era that might have contributed to the mass extinction of life forms at that time.

Angiosperm
Flowering plants, including all deciduous trees and grasses. The angiosperms emerged in the Cretaceous period, around 120 million years ago, and spread rapidly. They now account for 85 per cent of all plant species.

Antarctic convergence
The line where the cold waters of the Southern Ocean meet the warm waters of the Atlantic, Pacific and Indian oceans. Water is driven down by the converging currents, and is replaced by nutrient-rich upwelling water. This makes the Southern Ocean one of the richest marine environments on Earth.

Anthropology
The study of human evolution and human society from the first emergence of the species.

Australasian region
The biogeographical region taking in Australia and its surrounding islands as far as the Wallace Line, which divides the Malay archipelago.

Australopithecine
The earliest hominids, literally 'southern ape'. Australopithecine remains have been found in eastern and southern Africa.

Baleen whales
Group of whales distinguished by horny sheets across their mouths, known as baleen. These are used to filter seawater for krill and plankton.

Benthic
The benthos is the deepest part of the ocean. Benthic describes the conditions, as well as the organisms that live there.

Bergmann's rule
The rule that, within any warm-blooded species, there is a relationship between size and climate. In warmer temperatures individuals are smaller, presumably to aid heat loss. In colder areas they are larger.

Beringia
The area of land that formed across the Bering Strait during the Pleistocene period. This land bridge was hundreds of kilometres wide. At least two waves of humans are known to have crossed from Siberia into North America via Beringia.

Binomial system
The system of naming plants and animals first devised by Carolus Linnaeus. Each species carries two names: the genus name first and the species name second.

Biodiversity
Alternative term for the diversity of life forms. (See **Diversity**)

Biogeographic region
Region of the world that contains life forms not found elsewhere. Boundaries between regions prevent the migration of plants and animals from one region to another.

Biogeography
The study of the distribution of plants and animals. Biogeography has traditionally concerned itself with plant geography, as plants essentially create the basic environments in which animals also live.

Biomass
The total mass of organic matter in

any one place. Biomass is a way of measuring the organic productivity of an environment.

Blue-green algae

These were among the first celled organisms to evolve. They are simple organisms that are more akin to the viruses than to true algae. Photosynthesizing blue-green algae (cyanobacteria) are thought to have produced much of the free oxygen in the Earth's atmosphere.

Boreal

The vegetation zone encompassing the northern and mountainous parts of the northern continents.

Breeding range

The range within which an animal breeds. An important concept in the study of migrating animals, particularly birds, as their habitat range is highly variable according to the seasons.

Cambrian

Period of the geological time-scale from 550 to 500 million years ago. The beginning of the period is marked by the earliest presence of shelled fossils, which are preserved in great quantities in Cambrian rocks.

Carboniferous

Geological period dating from 350 to 290 million years ago. Characterized by limestones and coal measures formed in warm climates. Reptiles first emerged in the Carboniferous.

Cenozoic

The current era in the history of the Earth, dating from 65 million years

ago. Includes the Tertiary and the Quaternary periods.

Cephalopod

Class of marine molluscs that includes ammonites. The shell is often multi-chambered and used for buoyancy. Modern examples include the squid, the octopus and the cuttlefish.

Clade

A group that has a single common ancestor, and includes all the descendants of that ancestor.

Cladistics

The technique for allocating groups of plants and animals into clades on the basis of morphology, in order that their evolutionary relationships may be worked out.

Cladogram

Diagram showing the relationships between groups of organisms, using branches that are based on specific named characteristics.

Class

A division of life forms, below an order. (See **Taxonomy**)

Cline

A gradation in a characteristic – size or colour, for example – that occurs within a species range, across a hybrid range or between species that share the characteristic.

Coelacanth

A fish that was thought to have been extinct since Jurassic times but that was sensationally fished up in the Indian Ocean in 1938. Further specimens have since been found, but

the mysterious 140-million-year gap in the fossil record remains.

Cold-blooded

Those animals that have no internal method of controlling their body temperature but that rely instead on the ambient temperature of their environment. Cold-blooded animals, such as reptiles, are restricted to the warmer parts of the Earth, although they have adopted winter hibernation habits to enable them to occupy temperate zones.

Comparative anatomy

The study of differences in anatomical features. The development of comparative anatomy, by Georges Cuvier in particular, allowed the classification of animals to be fully undertaken and comparisons with fossil remains to be made.

Continental drift

The continuous movement of the continents across the surface of the Earth. The phrase was coined before any mechanism for the movement was known.

Counterpart species

Two species, usually from the same family, that occupy the same ecological niche (such as seed-eating) in different places.

Cretaceous

Geological period that lasted from 145 to 65 million years ago. Named after the great deposits of chalk (*creta* in Latin) that formed at this time across the northern continents. The Cretaceous saw the culmination of the age of the dinosaurs.

Deep time
An expression that is used to signify the enormity of the geological time-scale. Acceptance of the fact that the Earth is millions of years old helped Charles Darwin to formulate his theory of evolution, which depends on a vast time-scale.

Devonian
Period of the Earth's history from 410 to 350 million years ago. Known as the age of fish, because of the great diversification of this group, the Devonian also saw the further evolution of land plants and the emergence of the first amphibians.

Dinosaur
A member of the great group of reptiles that came to dominate the land during the Mesozoic era of the Earth, from about 225 to 65 million years ago.

Disjunct distribution
The occurrence of an organism in two geographically separated places, with no obvious link. These puzzling distributions have helped us to understand how plants and animals move around the Earth.

Dispersal
General term for the movement of an organism or group of organisms away from its place of origin, or simply from one place to another.

Dispersal biogeography
The theory according to which the distributions of plants and animals are best explained by the dispersal of populations away from a historic centre of origin.

Divergence
The result of the independent evolution of separated groups. Each group develops different characteristics and eventually forms a separate species.

Diversity
Technically, the total number of species. Often used more generally to mean the variety of life forms.

DNA
Deoxyribonucleic acid, the complex molecule that is the basis for all life forms on Earth. DNA is the messenger that carries information from one generation to the next, from parent to offspring. Errors in DNA encoding lead to mutations, which are essential for evolution to proceed.

Ecology
The study of the total environment.

Ecological niche
The space in the biological environment into which an organism or group of organisms fits. Organisms diversify rapidly when there are many niches to fill: after a mass extinction, for example, or the colonization of a new piece of land.

Ecosystem
An ecological entity that is, to some extent, self-sustaining. The expression is used to signify a web of relationships that would be disturbed by changes from outside.

Edentate
Literally, 'toothless' mammal. The group includes ground sloths, anteaters and armadillos.

Endemic
A plant or animal that is unique to a certain place.

Endemic centre
Region with an abnormally high number of endemic species or groups. Endemic centres are thought to be the centres of origin from which species disperse, or refuges in which they have survived adverse conditions.

Entomology
Study of insects, the most numerous and diverse group of organisms.

Eocene
Geological epoch lasting from 53 to 36 million years ago. Many mammal groups appeared in the Eocene.

Ethiopian region
The biogeographic region that comprises the whole of the African continent south of the Sahara Desert, as well as Madagascar.

Euphotic zone
The layer of ocean that can be penetrated by sunlight, extending to about 20 metres below the surface. Almost all life in the oceans depends on the microscopic plants that use the sunlight to photosynthesize in the euphotic zone.

Evolution
Changes in the characteristics of life forms, brought about by small mutations that persist to succeeding generations. Evolution has no direction. It works simply by the greater chance of survival gained by favourably mutated individuals.

Exotic species
Species that come from outside a particular environment. They are usually introduced by human agency, not always intentionally.

Extinct/extinction
The total eradication of a species or group. Also taken to mean the eradication of a species from a particular area. Extinction is a natural process that is fundamental to the workings of evolution, although human intervention is accelerating this process.

Family
A division of the classification of life, lower than a class. (See **Taxonomy**)

Fauna
The totality of animal life found in a given place.

Faunal/floral exchange
The meeting of two groups of fauna or flora (by the formation of a land bridge, for example) will lead to an exchange, with some of each population crossing into the newly available territory.

Fecundity
The ability to produce offspring. Some plants and animals produce millions of offspring to increase the chances of some surviving to maturity.

Flora
The totality of plant life found in a given place.

Foraminifera
Minute aquatic organisms with a single or multi-chambered cell. They are important indicators of age in various geological periods.

Fossil
The preserved remains of a plant or animal. Although actual parts of the organism may be preserved, the hard or soft parts are usually replaced by minerals during the formation of the rock. The minerals are sufficiently different from the surrounding rock for the presence of the organism to be preserved. Traces of animals, such as worm burrows and reptile footprints, are also often preserved as fossils.

Gene
The unit that carries information from one generation to the next and that has a specific effect on the offspring. Genes are carried on long chains known as chromosomes, contained in the nuclei of cells.

Gene flow
If two populations of a species are separated, they may evolve into distinct species. If individuals continue to travel between the populations and interbreed, a gene flow is maintained and the two populations will not diverge into separate species.

Genetic diversity
There is variation in the genetic make-up of individuals in any species. This is essential to the future of the species. It is reduced by inbreeding between close family members.

Genome
The total gene content of a cell, which is held in the nucleus of every cell in the organism.

Genus (pl. **genera**)
Unit of classification, above species. All life forms bear their genus and species names.

Geographic speciation
See **Allopatric speciation**.

Geological time-scale
The 4,600-million-year span of the Earth's history is divided up into eons, eras, periods, epochs and stages. In this book only the names of eras and periods have been used. These were originally devised on the basis of fossils found in rock strata. The precise ages of boundaries between periods vary slightly from place to place. For details of the divisions of the geological time-scale see pages 186–187.

Geology
The study of the rocks of the Earth's crust, including their history and the development of life over the history of the Earth.

Glaciation
An episode of cooling, leading to the extension of ice sheets and glaciers over a large portion of the Earth's surface. The glaciations of the Pleistocene period had profound effects on the life of the Earth, which are still in evidence today.

Glossopteris
Type of flora that became widespread across Gondwanaland during the Permian about 280 million years ago. Fossils of *Glossopteris* on widely separated southern continents are evidence for the existence of Gondwanaland and continental drift.

Gondwanaland

The great southern supercontinent, which consisted of South America, Africa, Antarctica, Australia and India, as well as parts of Southeast Asia. Gondwanaland existed from at least late Precambrian times (700 million years ago) to about 150 million years ago. The present distribution of many plants and animals in the southern hemisphere is explained by reference to the breakup of Gondwanaland into its constituent continents.

Gradient of diversity

A gradual change in the density of a species over a distance. The Earth has a gradient of diversity from the poles, where diversity is low, to the tropics, where it is high.

Gradualism

The theory that the evolution of new species takes place on a regular and continuous basis. This notion was an accepted part of evolutionary theory from Darwin until the development of the idea of punctuated equilibrium in the 1970s.

Gymnosperm

Seed-bearing but non-flowering plant such as the conifer. Gymnosperms played a significant part in the development of land plants, as they are more efficient reproducers than spore-bearing plants.

Holarctic region

The biogeographic region encompassing all of North America, Europe, Saharan Africa, Arabia and Asia north of the Himalayas. Usually subdivided into the Nearctic and Palearctic regions.

Holocene

The most recent episode of geological history, lasting from 10,000 years ago to the present day.

Hominid

Group of primates, of which the modern human, *Homo sapiens sapiens,* is the only survivor. Remains of the earliest known hominid were discovered in 1994, and are around 4.5 million years old.

Homology

A physical structure that has evolved once and then adapted to different uses in different groups of descendants. The mammal limb is the classic example, being used as a leg, an arm, a wing, a paddle and a digger by different animals.

Hybrid

Any organism produced by the union of two separate species or subspecies. Usually shows some characteristics of each of its progenitors.

Ice age

A term that is used to denote glaciation or a glacial episode.

Ichthyosaur

A fish-like reptile of the Mesozoic. Similar in shape to modern dolphins, the ichthyosaur died out 65 million years ago.

Immunization

The boosting of the body's immunity to a disease by the introduction of a mild form of that disease or of a related disease. An example is the encouragement of resistance to smallpox by the injection of cowpox.

Inbreeding

Breeding between close family members of a species. Inbreeding reduces genetic diversity and encourages the expression of potentially harmful recessive genes.

Indian and Oriental region

Biogeographic region encompassing the Indian subcontinent, south China and Southeast Asia north of the Wallace Line.

Industrial melanism

Air pollution caused by coalburning and leading to the predominance of dark forms of certain creatures, notably the peppered moth.

Interglacial

Comparatively warm periods that occur between glaciations or ice ages. We are presently in an interglacial period, since the Earth has been getting gradually warmer for about 13,000 years, although it is still cooler than its average temperature throughout its history.

Jurassic

Period of the geological time-scale from 205 to 145 million years ago. Dinosaurs became dominant on land, with other reptiles, the pterosaurs, plesiosaurs and ichthyosaurs, taking to the air and sea. Named after the Jura Mountains in eastern France.

KT event

Mass extinction occurring 65 million years ago, at the boundary between the Cretaceous (K) and Tertiary (T) periods. Dinosaurs and ammonites were among the groups that did not survive into the Tertiary, but this may have been for different reasons. An asteroid impact in the Gulf of Mexico certainly played a part in creating conditions that were unfavourable to some life forms.

Land bridge
Natural bridge formed between two separated land masses or islands by falling sea levels. This was common throughout the Pleistocene glaciations, when seawater was locked up in extensive ice sheets. Land bridges were important routes for migrating plants and animals, including humans.

Laurasia
Ancient supercontinent consisting of present-day North America, northern Europe and Siberia. Laurasia split up when the North Atlantic Ocean began to open 180 million years ago.

Learned behaviour
Behaviour patterns that are passed on to the next generation by example, rather than by genetic means. Learned behaviour is not an evolutionary phenomenon, whereas adaptive behaviour is.

Lungfish
Fish that developed lungs alongside gills. Lungfish are seen as the precursors to amphibians, and therefore represent a key stage in the emergence of life onto land.

Macaronesia
Small biogeographic region with a distinctive suite of plant life. Consists of the Canary Islands, Madeira, the Azores, the Cape Verde Islands and a slice of the northwest African coast.

Marsupial
A group of mammals that give birth to underdeveloped young, which are then nurtured in a pouch in the mother's body. The interpretation of the particular distribution of marsupials in Australia and South America is an important element in historical biogeography.

Mass extinction
The general term for a period in which extinctions take place at a faster than normal rate. The causes of mass extinctions, and the reasons why certain groups survive while other groups die out, are a key part of evolutionary biology.

Meiofauna
Minute fauna occupying the gaps between sand grains, the discovery and study of which has led to large-scale revisions in our estimates of the number of species on Earth.

Mesozoic
Era of geological time from about 250 to 65 million years ago. Mesozoic means 'middle life', and the era is often called the age of the reptiles, as dinosaurs were the dominant land animal for much of the time. The Mesozoic comprises the Triassic, Jurassic and Cretaceous periods.

Migration
The movement of any plant or animal. Used to describe the seasonal cycle of animal movements as they seek to maintain their food supply.

Mimicry
Certain organisms have evolved to look similar to those from unrelated groups. This is particularly striking in the case of certain tropical butterfly species. Batesian mimicry entails species mimicking poisonous species to fool likely predators. In Mullerian mimicry, two or more poisonous species mimic each other to their mutual advantage.

Miocene
Geological epoch lasting from 23 to 5 million years ago and a crucial time in the evolution of apes.

Molecular phylogeny
Technique for discovering evolutionary relationships by studying the DNA of organisms. This modern method is the most reliable way of confirming classifications based on observable characteristics.

Mollusc
Large group of invertebrates including bivalves, gastropods and cephalods – some of the most important fossil and living marine animals.

Monophyletic group
A group of organisms that has a common ancestor. The monophyletic group is the equivalent of a clade.

Monotreme
Egg-laying animals (duckbilled platypus and echidna) that are found only in Australia. They were once more widespread.

Montane
Literally, 'of mountains'. Montane plants and animals often become isolated in high country, a process that helps us to unravel the life history of a region.

Morph

Literally, a 'form'. A morph is a subspecies that is identifiable by its observable physical characteristics.

Morphology

The shape, structure and form of organisms. Plants and animals can be classified according to their morphology or according to other kinds of characteristics.

Mutability

Before Darwin, it was generally believed that species were 'immutable': that is, no species could change and go on to form a new species. Mutability, or the ability of species to change, is a key tenet of evolutionary theory.

Mutation

An error in copying genetic information from one generation to the next, resulting in an unexpected change in a characteristic. If this change helps the mutated individuals to survive to maturity in greater numbers, it will be passed on to future generations in greater quantities and become part of the genetic make-up of the species. Mutations occur on a regular basis, and are essential to the development of new species.

Native species

Species that are present in a location prior to a significant event, usually the introduction of exotic species by human beings.

Natural selection

Those individuals, species or groups that survive to maturity in the greatest numbers pass on their characteristics to successive generations in greater numbers. This natural selection of favourable characteristics was a key part of Darwin's formulation of an evolutionary theory.

Neanderthal

The early humans that lived in Europe up to 30,000 years ago. Named after the Neander Valley in Germany, where their remains were first identified, they are assumed to have been replaced by modern humans.

Nearctic region

Biogeographic region encompassing Greenland and North America as far south as the Rio Grande.

Neotropical region

Biogeographic region that comprises South and Central America, including the islands of the Caribbean.

Oligocene

Geological epoch from 36 to 23 million years ago. The Oligocene saw the emergence of the first apes, and the spread of large birds and mammals, particularly as grasslands spread across large parts of the Earth.

Order

Unit of classification above a family and below a class. (See **Taxonomy**)

Ordovician

Period of the geological time-scale from 500 to 440 million years ago. The earliest known definite vertebrates emerged during this period. Freshwater plants, and possibly the first land plants, are known from Ordovician rocks.

Ornithischian

One of the two great subgroups of dinosaurs. Ornithischians had superficially bird-shaped hips, but they are not the precursors of birds.

Outbreeding

The union of two parents who are not closely related. Outbreeding demands a large population, and is a way of maintaining genetic diversity.

Palearctic region

Biogeographic region comprising Europe, Saharan Africa, Arabia and Asia north of the Himalayas and the Yangtze River.

Paleocene

Epoch of the Earth's history from 65 to 53 million years ago. This is the earliest epoch of the Tertiary period, coming immediately after the mass extinction that killed off the dinosaurs. The Paleocene saw the emergence of new groups of mammals, which radiated rapidly in the absence of the dinosaurs.

Paleontology

The branch of geology specializing in the study of ancient life forms.

Paleozoic

Geological era stretching from the start of the Cambrian 550 million years ago to the end of the Permian period 250 million years ago. Paleozoic means 'ancient life'.

Pangea

The supercontinent formed at the end of the Paleozoic era 250 million years ago. Pangea consisted of all the major continental blocks, but began to break up almost as soon as it had formed. Its literal meaning is 'all of the Earth'.

Paramo and puna
Types of vegetation formed above the tree line in the high Andes, consisting of tussocks of grass.

Pelagic
The middle depths of seas and oceans, between the euphotic and benthic zones.

Peripheral isolate
Part of a population or a species that becomes geographically isolated from the main stock. If isolation is maintained, the divergence of characteristics between the two populations will lead to the development of a new species.

Permian
Last period of the Paleozoic era, from 290 to 250 million years ago. The end of the Permian saw the extinction of around 90 per cent of marine species.

Phenotype
An organism's characteristics, as produced by the interaction of its genes and the environment. Can be a species or a subspecies.

Photosynthesis
Chemical process by which plants manufacture food using sunlight. Photosynthesizing plants absorb carbon dioxide and produce oxygen.

Phylogeny
An evolutionary tree, showing the descent and the evolutionary relationships of a species or group.

Phylum
The largest unit in the classification of the plant and animal kingdoms.

Placental
Mammals that give birth to fully developed young, as compared with marsupials. The distribution of the two groups is an indication of continental movements over the last 70 million years.

Plankton
Free-floating microscopic organisms that are a rich food source for marine animals. Phytoplankton (planktonic plants) produce food by photosynthesis and are the basis of the marine food chain.

Plate tectonics
The overall theory that is used to explain the mechanisms by which the geological processes of the Earth are driven. Plate tectonics describes how the Earth's crust is divided into rigid plates, how these plates are moved around on the surface by currents in the underlying mantle, and how new plates are continually formed and old ones destroyed. The principle of continental movement had long been argued for, but it lacked a plausible driving mechanism until the theory of plate tectonics was formulated.

Pleistocene
Epoch lasting from 2 million to 10,000 years ago and notable for the cooling of the Earth that led to a series of ice ages.

Plesiosaur
Marine reptile from the Mesozoic era, often with a long neck. A fish-eater.

Pliocene
The geological epoch lasting from 5 to 2 million years ago, part of the Tertiary period.

Precambrian
Era covering geological history from the oldest rocks (about 4,000 million years ago) to the start of the Cambrian period (550 million years ago). The Precambrian was beyond the reach of early geologists, who relied on fossil evidence, of which there is very little in the Precambrian, for the comparative dating of rocks.

Primate
A member of the group of mammals that includes monkeys, apes and human beings.

Proboscidean
Member of the mammoth, mastodon or elephant family.

Protozoan
Early unicellular life form.

Pterosaur
A close relative of the dinosaurs that evolved the ability to fly.

P-Tr event
A mass extinction that occurred 250 million years ago, at the boundary of the Permian and Triassic periods. This great decimation of marine life, in which over 90 per cent of species may have been made extinct, marks the end of the Paleozoic and the beginning of the Mesozoic era.

Puna
See **Paramos and puna**

Punctuated equilibrium

The theory that species remain stable and in equilibrium for extended periods, but that these periods are then punctuated by short bursts of intense speciation. This theory was proposed in the 1970s as an argument against the prevailing theory of gradualism.

Quaternary

Geological period covering the last 2 million years of the Earth's history and including the Pleistocene and Holocene epochs.

Radiation

Radiation occurs when the genesis of new groups and species greatly exceeds the rate of extinction, leading to increased diversity.

Recessive

Recessive genes are genes that bring about changed characteristics only when they occur in pairs. Both parents therefore need to be carriers, and even then only one offspring in three will have the double gene.

Reef

Large structures that are formed by the skeletons of corals and other marine organisms.

Refuge

A location in which certain conditions persist despite changes in the overall environment. This phenomenon allows species to remain in those areas, while becoming extinct elsewhere. If overall conditions revert to their former state, as in the Amazon rainforests, the species will spread outwards again.

Relict

Species that have persisted in certain places while being wiped out elsewhere. This can lead to separate or disjunct distributions.

Replacement species

A species may find its range restricted by changes in vegetation, but individuals can colonize neighbouring territories by successfully exploiting accidental mutations. They then develop into a new but closely related species. Where close relatives occupy adjacent ranges in this way, they are known as replacement species.

Ring species

Variations within a species are common. They can become more pronounced as distance from the original population increases. If the species distribution is circular (as is the case of the gulls of the Arctic rim), then by the time the increasing variation has come back around to the original group, the two populations are unable to breed with each other and must therefore be classified as separate species.

Saurischian

One of the two dinosaur subgroups. Saurischians had a more reptile-like hip bone than the ornithischians. They include the sauropods, the largest of all dinosaurs.

Savannah

A vegetation type consisting of grassland with scattered tree clumps. The few species of plant that are found in savannah regions provide nutrition for a large variety of mammalian herbivores.

Seed fern

An early land plant that shows significant development since the time of its spore-producing forebears.

Silurian

Geological period from 440 to 410 million years ago. The first fish with jaws and the first definite vascular land plants emerged at this time, although they remained scarce.

Speciation

The formation of a new species from an existing population.

Species

A group of organisms defined by their ability to breed with one another and their inability to breed naturally with members of other species.

Steppe

Open stretches of grassland found in temperate climates.

Stereotypy

The tendency of extant life forms to follow a limited number of body plans. Stereotypy is used by some biologists as an argument for greater variation on Earth in the past, with fewer species showing greater variability in morphology.

Strain

Variant within a species with recognizable characteristics. Principally used in relation to artificial breeding or diseases.

Stratigraphy

The study of the spatial and chronological relationships of the rock strata of the Earth's crust.

Subspecies
Group within a species that has recognizable defining characteristics but that retains the ability to breed with other members of the species outside this group.

Taxonomy
The arrangement of organisms into a system of classification. The units of classification in both animal and plant kingdoms are, in ascending order: species, genus, family, order, class and phylum. In order to accommodate diverse phyla, terms such as subclass and superphylum are used.

Tectonic
Large-scale forces or movements in the Earth's crust.

Tethys Ocean
Ancient seaway that separated Gondwanaland from Laurasia for most of the Mesozoic and Tertiary eras. The Tethys shrank to a sea, and then disappeared entirely. Its marine fauna is found in fossil beds across Europe and Asia.

Territory
An area that is occupied exclusively by an individual, by a mating pair or by a group within a species.

Tertiary
The geological period from 65 to 2 million years ago, known as the age of mammals because of their extraordinary radiation after the extinction of the dinosaurs.

Tetrapod
Four-legged animal. The first tetrapods were a transitional form between fish and amphibians and represent an important stage in the movement of life onto land.

Thermoregulation
The ability of an organism to control its own body temperature. The development of thermoregulation in mammals and birds has enabled them to inhabit virtually every environment on Earth.

Tree line
Altitude, or latitude, above which trees will not grow. Areas above the tree line are colonized by particular groups of flora and fauna.

Tree of life
General term for the chart of all life forms, showing inferred relationships between groups as branches from a central trunk. These usually show only living groups, presumably due to difficulties of presentation, and therefore omit the enormous diversity of extinct organisms.

Triassic
Period of the geological time-scale from 250 to 205 million years ago. First period of the Mesozoic era. Reptiles begin to dominate. The late Triassic sees the emergence of the first dinosaurs, pterosaurs, turtles, lizards and the like. In the sea, the ammonites survived the mass extinction at the start of the Triassic and began to radiate.

Tundra
A type of vegetation that appears over a large area at a higher latitude than the tree line in the Nearctic and Palearctic biogeographic regions. Tundra consists of lichens, sedges, grasses and small bushes.

Ungulates
Hoofed grazing mammals that are now classified into four separate groups but probably all arose at the same time in response to the evolution of grasslands.

Variation
Those changes within a species that do not lead to the establishment of a new species.

Vascular plant
Plant with a system for circulating fluids internally. Land plants developed this system for the distribution of nutrients, a task that is done by seawater in marine plants.

Vicariance biogeography
The theory according to which the separate distributions of a single organism are best explained by a split in existing widespread populations caused by the development of physical barriers. This then leads to separate evolutionary development.

Index

Items set in **bold** are pictures, illustrations or maps.

GENERAL

PLACES

A

B

Select Bibliography

Baker, R., *Migration, Paths Through Space and Time* (Holmes & Meier, London, 1982)

Beehler, Bruce McP., *Birds of New Guinea* (Princeton University Press, Princeton, 1986)

Benton, M.J., *Vertebrate Palaeontology* (Chapman & Hall, London, second edition, 1997; first edition, 1990)

Benton, M.J. and Harper, D.A.T., *Basic Palaeontology* (Addison-Wesley Longman, London and New York, 1997)

Bramwell, D. (ed.), *Plants and Islands* (Academic Press, London, 1979)

Brenchley, P. (ed.), *Fossils and Climate* (John Wiley, Chichester, 1984)

Briggs, D.E.G. and Crowther, P.R. (eds.), *Palaeobiology: a Synthesis* (Blackwell Scientific, Oxford, 1990)

Briggs, J.C., *Biogeography and Plate Tectonics* (Elsevier, Amsterdam, 1987)

Burton, John A., *Atlas of Endangered Species* (David and Charles, Newton Abbott, 1991)

Coleman, Neville, *Encyclopedia of Marine Animals* (Blandford, London, 1991)

Collins, N.M., Sayer, J., and Whitmore, T.C., *Conservation Atlas of Tropical Rainforests, Asia and the Pacific* (Macmillan, London, 1991)

Collinson, A.S., *Introduction to World Vegetation,* 2nd edition (Unwin Hyman, London, 1988)

Cox, C.B. and Moore, P.D., *Biogeography, an Ecological and Evolutionary Approach*, 5th edition (Blackwell Scientific, Oxford, 1993)

Darwin, Charles, *Journal of Researches into the Geology and Natural History of the Various Countries Visited by HMS Beagle* (London, 1839)

Darwin, Charles, *On The Origin of Species* (John Murray, London, 1859; Penguin Classics, London, 1985)

Daubenmire, R., *Plant Geography with Special Reference to North America* (Academic Press, New York, 1978)

De Beers, Gavin, *Atlas of Evolution* (Thomas Nelson, London, 1964)

Desmond, A.J. and Moore, J., *Darwin* (Michael Joseph, London, 1991)

Domnovan, S.K., *Mass Extinctions--Processes and Evidence* (Belhaven, London, and Wiley, New York, 1989)

Dorst, Jean and Dandelot, Pierre, *A Field Guide to the Larger Mammals of Africa* (Collins, London, 1970)

Dott, R.H. and Batten, R.L., *Evolution of the Earth*, 4th edition (McGraw-Hill, New York, 1988)

Druett, Joan, *Exotic Intruders, the Introduction of Plants and Animals Into New Zealand* (Heinemann, Auckland, 1983)

Ellenberg, H., *Vegetation Mitteleuropas mit den Alpen in ökologisher Sicht* (2nd edition, Ulmer, Stuttgart, 1978)

Endler, J.A., *Geographic Variation, Speciation, and Clines* (Princeton University Press, Princeton, 1977)

Forey, P.L. and others (eds.), *Cladistics, a Practical Course in Systematics* (Clarendon Press, Oxford, 1992)

Frankel, O.H. and Soulé, M.E., *Conservation and Evolution* (Cambridge University Press, Cambridge, 1981)

Futuyman, D.J., *Evolutionary Biology* (Sinauer, Sunderland, Massachusetts, 1986)

Gamble, Clive, *Timewalkers, Prehistory of Global Civilization* (Alan Sutton, Stroud, 1993)

Good, R., *The Geography of the Flowering Plants* (Longman, London, 1974)

Gorman M.L., *Island Ecology* (Chapman & Hall, London, 1979)

Gould, J.L. and Gould, C.G., *Sexual Selection* (Scientific American Library, New York, 1989)

Gould, S.J., *Wonderful Life, The Burgess Shale and the Nature of History* (Hutchinson, London, 1989)

Gould, S.J., *Eight Little Piggies, Reflections in Natural History* (Jonathan Cape, London, 1993)

Gould, S.J. (ed.), *The Book of Life* (Ebury, London and New York, 1993)

Grant, P.R., *Ecology and Evolution of Darwin's Finches* (Princeton University Press, Princeton, 1986)

Gray, William, *Coral Reefs and Islands* (David & Charles, Newton Abbott, 1993)

Greenwood, P.H., *Cichlid Fishes of Lake Victoria, East Africa: the Biology and Evolution of a Species Flock* (British Museum, Natural History, London, 1974)

Gressit, J., *Biogeography and Ecology of New Guinea* (Junk, The Hague, 1982)

Groves, R.H. and Di Castri, F., *Biogeography of Mediterranean Invasions* (Cambridge University Press, Cambridge, 1991)

Grzimek, Bernard, *Encyclopedia of Mammals*, 5 volumes, English edition (McGraw-Hill, New York, 1990)

Harrison, Colin, *An Atlas of the Birds of the Western Palearctic* (Collins, London, 1982)

Heaney, L.R. and Patterson B.D., *Island Biogeography of Mammals* (Academic Press, London, 1986)

Hengeveld, R., *The Dynamics of Biological Invasions* (Chapman & Hall, London, 1989)

Humphries, C.J. and Parenti, L.R., *Cladistic Biogeography* (Clarendon Press, Oxford, 1986)

Keast, A., *Ecological Biogeography of Australia*, 3 volumes (Junk, The Hague, 1981)

Keast, A., *Mammals, Evolution and the Southern Continents* (State University of New York, Stony Brook, New York, 1972)

Kettlewell, H.B.D., *The Evolution of Melanism* (Oxford University Press, Oxford, 1973)

Lack, David, *The Life of the Robin* (London, 1943)

Lack, D., *Darwin's Finches* (Cambridge University Press, Cambridge, 1947 and 1983)

Lack, D., *Ecological Adaptations for Breeding in Birds* (Methuen, London, 1968)

Lack, D., *Ecological Isolation in Birds* (Harvard University Press, Cambridge, Massachusetts, 1971)

Lever, Christopher, *Naturalized Animals of the British Isles* (Hutchinson, London, 1977)

Manchester Museum, *The Geological Column*, 7th edition (Manchester Museum, Manchester, 1992)

MacArthur, R.H., and Wilson, E.O., *Theory of Island Biogeography* (Princeton University Press, Princeton, 1967)

MacArthur, R.H., *Geographical Ecology, Patterns in the Distribution of Species* (Princeton University Press, Princeton, 1972)

Maynard Smith, J., *Evolutionary Genetics* (Oxford University Press, Oxford, 1989)

Mayr, E., *Populations, Species, and Evolution* (Harvard University Press, Cambridge, Massachusetts, 1970)

Mayr, E., *The Growth of Biological Thought* (Harvard University Press, Cambridge, Massachusetts, 1982)

Mead, Chris, *Bird Migration* (Country Life Books, Feltham, 1983)

Mielke, H.W., *Patterns of Life, Biogeography of a Changing World* (Unwin Hyman, London, 1989)

Nelson, Gareth and Rosen, Donn E., *Vicariance Biogeography, A Critique* (Columbia University Press, New York, 1981)

Nitecki, Matthew (ed.), *Extinctions* (Chicago University Press, Chicago, 1984)

Osborne, Roger and Tarling, Donald, *The Historical Atlas of the Earth* (Viking Penguin, London, 1995; Henry Holt, New York, 1996)

Raup, D.M., *Extinction: Bad Genes or Bad Luck?* (Norton & Co., New York, 1991)

Ridley, M., *Evolution* (Blackwell Scientific, Oxford, 1993)

Ruse, M., *The Darwinian Revolution; Science Red in Tooth and Claw* (University of Chicago Press, Chicago, 1979)

Serjeant, G.R., *Sickle Cell Disease*, 2nd edition. (Oxford University Press, Oxford, 1992)

Shorrocks, Bryan, *The Genesis of Diversity* (Hodder & Stoughton, London, 1978)

Sims, R.W., *Evolution, Time and Space, the Emergence of the Biosphere* (Academic Press, London, 1983)

Skelton, Peter (ed.), *Evolution, A Biological and Paleontological Approach* (Addison-Wesley/Open University, Wokingham, 1993)

Stanley, S.M., *Earth and Life through Time* (Freeman, San Francisco, 1986)

Stevens, Graeme R., *New Zealand Adrift, the Theory of Continental Drift in a New Zealand Setting* (Reed, Wellington, 1980)

Sutcliffe, Anthony J., *On the Track of Ice Age Mammals* (British Museum, Natural History, London, 1985)

Tarling, D.H., *Plate Tectonics and Biological Evolution* (Carolina Biological Supply Company, Burlington, 1992)

Tivy, Joy, *Biogeography, A Study of Plants in the Ecosphere*, 3rd edition. (Longman, London, 1993)

Tudge, C., *Global Ecology* (Natural History Museum, London, 1991)

Warwick, T., 'The distribution of the muskrat in the British Isles', *Journal of Animal Ecology*, 3, pp.250–67.

Wegener, Alfred, *The Origin of Continents and Oceans*, English translation of 4th edition (Dover, New York, 1966)

Werger, M.J.A. (ed.), *Biogeography and Ecology of Southern Africa*, 2 volumes (Junk, The Hague, 1978)

Whitmore, T.C., *Biogeographical Evolution of the Malay Archipelago* (Clarendon Press, Oxford, 1987)

Whitmore, T.C., and Prance, Ghillean T. (eds.), *Biogeography and Quaternary History in Tropical America* (Clarendon Press, Oxford, 1987)

Williamson, M.H., *Quantitative Aspects of the Ecology of Biological Invasions* (Royal Society, London, 1986)

Wilson, E.O., *The Diversity of Life* (Penguin, London, 1994)

World Health Organization, *Global Eradication of Smallpox* (WHO, Geneva, 1980)